# The Patentability of Synthetic Biology Inventions

Ilaria de Lisa

# The Patentability of Synthetic Biology Inventions

## New Technology, Same Patentability Issues?

Springer

Ilaria de Lisa
Faculty of Law
Leibniz University Hannover
Hannover, Germany

ISBN 978-3-030-51205-7 ISBN 978-3-030-51206-4 (eBook)
https://doi.org/10.1007/978-3-030-51206-4



This Springer imprint is published by the registered company Springer Nature Switzerland AG.
The registered company address is: Gewerbestrasse 11, 6330 Cham, Switzerland

*To my grandfather Ugo*

# Acknowledgements

I would like to thank Professor Christian Heinze (University of Hannover) for his encouragement to pursue this research topic and guidance during the drafting of the PhD thesis on which this book is based upon.

Likewise, I would like to express my gratitude to the Max Planck Institute for Innovation and Competition in Munich, as much of the research for this book was undertaken there. I am also grateful to my employer, Gleiss Lutz, for its support during the drafting of this book.

I also wish to thank all the people who offered advice and comments on my work, including Professor Matteo Winkler, Dr. Konrad Duden, Dr. Ralph Krafczyk, Dr. Andrew Pitts and Dott.ssa Isabella Damiani.

I also appreciate the editorial guidance offered by the Springer team.

Lastly, I would like to thank my entire family for their support and encouragement; in particular my husband, Enrico, and our children, Tancredi and Ludovica. I am also deeply thankful to my parents, Anna and Luigi. A special acknowledgement goes to my grandfather Ugo, to whom this book is dedicated, for truly leading by example both academically and in life.

# Contents

# About the Author

**Ilaria de Lisa** completed her PhD in patent law at the University of Hannover. She holds an LLM in patent and European law from the University of Cambridge and a law degree from the University of Milan. She has been a Visiting Scholar at the University of California, Berkeley.

Her research focuses on patent and competition law.

Ilaria de Lisa works as an attorney at law in Munich.

# Abbreviations

| | |
|---|---|
| Biotech Directive | EU Directive 98/44 on the legal protection of biotechnological inventions |
| BPAI | US Board of Patent Appeals and Interferences |
| CAFC | US Court of Appeals for the Federal Circuit |
| CCPA | US Court of Customs and Patent Appeals |
| cDNA | Complementary DNA |
| CJEU | Court of Justice of the European Union |
| Commission | European Commission |
| Directive | EU Directive 98/44 on the legal protection of biotechnological inventions |
| ECHR | European Convention on Human Rights |
| ECJ | European Court of Justice |
| EGE | European Group on Ethics in Science and New Technologies |
| EPC | European Patent Convention, as last amended |
| EPO Guidelines | Guidelines for examination in the European Patent Office |
| Epo | Erythropoietin |
| EPO | European Patent Office |
| FP | EU Framework Programme |
| gDNA | Genomic DNA |
| GMO | Genetically Modified Organism |
| HESC | Human Embryonic Stem Cell |
| HGP-write/GP-write | Human Genome Project-write |
| iGEM competition | International Genetically Engineered Machine competition |
| Implementing Regulations | Implementing Regulations to the Convention on the Grant of European Patents, as last amended |
| IPC | International Patent Classification |
| IPRs | Intellectual property rights |

| | |
|---|---|
| JPO | Japanese Patent Office |
| Recital | Recitals to EU Directive 98/44 on the legal protection of biotechnological inventions |
| Rules | Implementing Regulations to the Convention on the Grant of European Patents, as last amended |
| SynBio | Synthetic biology |
| TBA | Technical Board of Appeal of the EPO |
| TEU | Treaty on the European Union |
| TFEU | Treaty on the Functioning of the European Union |
| TRIPS | Agreement on Trade-Related Aspects of Intellectual Property Rights |
| UPC | Unified Patent Court |
| USPTO | United States Patent and Trademark Office |
| WTO | World Trade Organization |

# Chapter 1
# Introduction

"*What I cannot create, I do not understand*". These words were written on a blackboard at Caltech in 1988 by *Richard Feynman*, renowned physicists and Nobel laureate.[1] Little did he know that two decades later this sentence would become the motto of a promising new biotechnology and be inscribed in the first living organism to have been fully constructed synthetically.[2]

Synthetic biology, that is the name of this new and promising technology, aims to create biological systems that do not exist in nature and that are tailored upon human needs and design.[3] In its quest, it employs engineering principles to increase understanding and control over biological systems. This objective is ambitious, given that a transition from describing biological systems to making them clashes against the complexity of nature, which so far humans were not able to master.

Nevertheless, the potential of this approach is immense. This is why synthetic biology has been hailed as *"the miracle solution that will boost growth whilst preserving the environment"* and as the defining technology of the century.[4] With its increasing capabilities and large spectrum of applications, it promises to provide cheaper medicaments, to develop biofuels to reduce dependence on fossil-derived

---

[1]Eichenlaub (2015).

[2]On this, see Sect. 2.2.2.2. This sentence was used as a watermark in the synthetic bacterium constructed by the J. Craig Venter Institute in 2010, which represents one of the most relevant and contentious works in the field of synthetic biology. To distinguish this synthetic bacterium from the naturally occurring one, watermarks sequences were added to it. Such watermarks are DNA sequences that have been added to the synthetic DNA. At first, the team encoded an incorrect version of Feynman's statement. This mistake was later rectified (Ewalt 2011). The relevance of this motto for synthetic biology is evidenced in the number of publications that cite those words as the basis of what synthetic biologists are trying to achieve. For example, SCHER, SCENIHR, SCCS Scientific Committees (2015, p. 30) and Ribarits et al. (2014, p. 92).

[3]The term "synthetic biology" is often abbreviated as "SynBio".

[4]Rousseaux (2015) and Johnson (2013, p. 5). Commentators have been critical of this idea (Rousseaux 2015).



I. de Lisa, *The Patentability of Synthetic Biology Inventions*,
https://doi.org/10.1007/978-3-030-51206-4_1

ones and to detect and remediate contaminated sites. The fascination with this discipline is thus understandable, as the world's population continues to grow, alongside concerns over the strain that this could put on global resources, quality of life and sustainability. High hopes for its development were expressed also by investors, who considered synthetic biology one of the key technologies of the twenty-first century.

It is therefore unsurprising that synthetic biology caught the interest of policy makers and regulators.[5] Numerous studies have been conducted on the topic, mostly focusing on the repercussions that future synthetic biology inventions will have on the safety, ethical and regulatory framework; intellectual property rights (IPRs) were considered as well, albeit predominantly from a policy perspective.[6] This included discussions on which protection regimes would be most suitable in this field and their possible downsides. In contrast, reports and literature on the patentability of synthetic biology inventions are scarce. This could be attributed to a number of reasons. First, questions on synthetic biology have often been dismissed by arguing that this discipline represents nothing more than an incremental change compared to prior technologies. As a result, it would not require a specific patent analysis, as it is unlikely to raise any new or different issues. Yet, in the absence of a thorough examination, this assumption cannot be accepted.

The second reason approaches the patentability of synthetic biology from the opposite end of the spectrum. By focusing on futuristic SynBio products, it assumes that these inventions will raise limited patentability issues, since they would not present some of the characteristics that rendered biotechnological patents so controversial. For instance, human-designed biological systems having no equivalent in nature would likely avoid much of the criticism drawn by prior biotech patents. However, this approach may be reductive, as it focuses on scientific developments that are decades away, while disregarding current conditions and concerns.

Third, it is often assumed that the innovative character of synthetic biology as a field would automatically translate to the patentability of its inventions, thus reducing the need for a detailed patent analysis. Lastly, the absence of case law and the novelty of this field can also explain the lack of patent studies on it.[7]

---

[5]The general public has so far showed limited interest in this discipline.

[6]McLennan (2018), Holman (2015), Conference of the parties to the Convention on Biological Diversity (2014), OECD (2014), Torrance and Kahl (2013), *de Miguel Beriain and Romeo Casabona (*2014*)*, Holman (2011), *Deutsche Akademie der Naturforscher Leopoldina (*2010*)*, Commissione federale d'etica per la biotecnologia nel settore non umano (CENU) (2010), Then and Hamberger (2010), Schmidt et al. (2010), OECD and The Royal Society (2010), European Commission's Directorate-General for Health & Consumers (2010), Torrance (2010), Balmer and Martin (2008) and Rai and Boyle (2007).

[7]Authors noted that there have been a few patent infringement cases in this sector. Specifically, in 2007, Codon Devices, Duke University and MIT sued Blue Heron Biotechnology for infringement of a DNA synthesis patent. The case was later settled. Equally, in the same year, Febit Biotech sued Codon Devices for infringement relating to a patent for producing synthetic nucleic acids. Lastly, in 2012, the company DNA2.0 sued Genome Compiler Corporation for infringement of a patent for genetic sequence design software (McLennan 2017, p. 64).

This analysis tries to fill that void by examining the application of patent criteria to SynBio inventions pursuant to European law. In particular, the enquiry will centre upon the subject matter and morality requirements set in Articles 52 and 53 EPC and in the Biotechnology Directive.[8]

The patentable subject matter requirement has generally been underplayed both before the EPO and Courts. This trend might continue for synthetic biology, as commentators often assume that this new technology will not raise subject matter concerns. Still, this conclusion is not straightforward, as two factors may influence the patentability assessment in Europe. First, the Biotechnology Directive may not be applicable to all types of SynBio inventions, with both substantial and procedural law repercussions. Second, recent developments in other jurisdictions (e.g. USA) show an increased attention towards this requirement as well as a heightened patentability threshold.

As for the morality requirement, the necessity to analyse its application to synthetic biology inventions is dictated both by the general relevance of this proviso for biotechnologies and by the morality questions specifically posed by this field. Indeed, a number of publications highlighted the morally problematic nature of SynBio inventions. These objections are often connected to the belief that synthetic biologists are usurping the role of God or imposing a reductive view of life and nature. Still, the significance of the morality clause for synthetic biology is not clear-cut, considering the specific requirements set by the EPC and the Biotechnology Directive for the application of this exception. Therefore, this issue needs to be assessed in more detail.

To conclude, a brief overview of the issues that may emerge in Europe under the novelty and inventiveness criteria will be presented in order to determine whether those issues are really unlikely to be raised in synthetic biology, as argued by some commentators.

The analysis will be introduced by an overview of the technical characteristics of synthetic biology and of its most relevant policy issues. The legal assessment will be initiated by outlining the norms applicable to this discipline and by providing an overview of the current patent panorama. Specific patent applications presented before the EPO and the USPTO will be analysed in order to provide a detailed assessment of the patent issues arising in this field. Indeed, patent offices are so far the only authorities that have decided on synthetic biology and, therefore, the indications provided by them are particularly valuable. Furthermore, the patent assessment was needed to integrate an otherwise wholly theoretical examination. Indeed, since no SynBio patents were yet litigated before Boards and Courts, this evaluation would otherwise be limited to normative and academic sources.[9]

---

[8] *Convention on the grant of European patents of 5 October 1973, as revised on 29 November 2000 (European Patent Convention - EPC)* (1973) and *Directive 98/44/EC of the European Parliament and of the Council of 6 July 1998 on the legal protection of biotechnological inventions (Biotechnology Directive or Biotech Directive or Directive)* (1998).

[9] This situation is not likely to change soon, since years will pass before those inventions are litigated. The time shift between the emergence of a technology, the first wave of patents on it and

The assessment will encompass current synthetic biology inventions as well as some futuristic lines of research. Attention has purposively been placed on both, given that concentrating on the patentability of futuristic inventions would be premature, as synthetic biology is still far from realising them.[10] However, should synthetic biology reach those ambitious goals and follow the same innovative path traced by synthetic chemistry almost two centuries ago, it would open extraordinary horizons and revolutionise our understanding of nature.[11]

## References

Balmer A, Martin P (2008) Synthetic biology social and ethical challenges. Biotechnology and Biological Sciences Research Council (BBSRC)

Commissione federale d'etica per la biotecnologia nel settore non umano (CENU) (2010) Biologia sintetica - Riflessioni etiche. Commissione federale d'etica per la biotechnologia nel settore non umano - Svizzera

Conference of the parties to the Convention on Biological Diversity (2014) Potential positive and negative impacts of components, organisms and products resulting from synthetic biology techniques on the conservation and sustainable use of biodiversity, and associated social, economic and cultural considerations (No. UNEP/CBD/COP/12/INF/11). Pyeongchang

Convention on the grant of European patents of 5 October 1973, as revised on 29 November 2000 (European Patent Convention - EPC), 1973

de Miguel Beriain I, Romeo Casabona CM (eds) (2014) Synbio and human health: a challenge to the current IP framework? Springer, Dordrecht

Deutsche Akademie der Naturforscher Leopoldina (ed) (2010) Realising European potential in synthetic biology: scientific opportunities and good governance, EASAC policy report. Halle (Saale)

Directive 98/44/EC of the European Parliament and of the Council of 6 July 1998 on the legal protection of biotechnological inventions (Biotechnology Directive or Biotech Directive or Directive), 1998. OJ L 213, 30.7.1998, pp 13–21

---

the first patent controversies is quite dramatic. For example, at the beginning of the millennium, the patent system was focusing primarily on technologies from the 1980s (Eisenberg 2002, p. 3). Similarly, the technology assessed in the *Myriad* case was also rather old, as it has been patented and litigated for over three decades (Obranovich 2016, p. 11). On this, see Sect. 4.1.6.5. In the meantime, decisions over their patentability will be taken by patent offices, thus conferring them law and policy-making powers (Schneider 2014, pp. 155–156).

[10]For example, while the idea that synthetic biology will turn biological systems into engineered machines is appealing, the road to achieve this kind of sophistication is still very long. Such focus on futuristic inventions detracts from a realistic analysis of the patent issues raised by synthetic biology. Futuristic cases tend to present extreme scenarios that polarise ideas and facilitate the taking of a stand. By contrast, current synthetic biology developments might find themselves in a grey area. Furthermore, an appraisal of those developments cannot be avoided, as the patent system will need to deal with those "intermediate" inventions before this discipline transitions to the design and production of completely new biological systems. Additionally, patent issues could emerge from lines of research introduced for the first time by synthetic biology.

[11]On this, see Sect. 2.1.2.3.

Eichenlaub M (2015) What did Richard Feynman mean when he said, "What I cannot create, I do not understand"? [WWW Document]. www.quora.com. URL https://www.quora.com/What-did-Richard-Feynman-mean-when-he-said-What-I-cannot-create-I-do-not-understand. Accessed 27 Feb 2020

Eisenberg RS (2002) How can you patent genes? Am J Bioeth 2:3–11

European Commission's Directorate-General for Health & Consumers (2010) Synthetic biology from science to governance. European Commission's Directorate-General for Health & Consumers, Brussels

Ewalt DM (2011) Craig Venter's genetic typo [WWW Document]. www.forbes.com. URL https://www.forbes.com/sites/davidewalt/2011/03/14/craig-venters-genetic-typo/#2526982f7f07. Accessed 27 Feb 2020

Holman CM (2011) Copyright for engineered DNA: an idea whose time has come? W Va Law Rev 113:699–738

Holman CM (2015) Developments in synthetic biology are altering the IP imperatives of biotechnology. Vanderbilt J Entertain Technol Law 17:385–462

Johnson RA (2013) Synthetic biology: 10 policy reasons it matters to U.S. foreign policy. State Department Briefing

McLennan A (2017) Patent law and the emerging science of synthetic biology: an examination of principle and practice. Biotechnol Law Rep 36:59–116

McLennan A (2018) Regulation of synthetic biology: biobricks, biopunks and bioentrepreneurs. Edward Elgar, Cheltenham

Obranovich T (2016) Biotechnology and patentability: navigating unchartered waters. Manag Intellect Prop 256:8–11

OECD (2014) Emerging policy issues in synthetic biology. OECD, Paris

OECD, The Royal Society (2010) Symposium on opportunities and challenges in the emerging field of synthetic biology: synthesis report. OECD - The Royal Society

Rai A, Boyle J (2007) Synthetic biology: caught between property rights, the public domain, and the commons. PLOS Biol 5:389–393

Ribarits A, Stepanek W, Wögerbauer M, Peterseil V, Kuffner M, Topitschnig C, Brüller W, Hochegger R, Gansberger M, Widhalm I, Leonhardt C (2014) Synthetic biology. Bundesministerium für Gesundheit (Österreich)

Rousseaux A (2015) Synthetic biology: life reconstructed by engineers and multinationals [WWW Document]. www.multinationales.org. URL http://multinationales.org/Synthetic-Biology-Life. Accessed 27 Feb 2020

SCHER, SCENIHR, SCCS Scientific Committees (2015) Final opinion on synthetic biology III: risks to the environment and biodiversity related to synthetic biology and research priorities in the field of synthetic biology. Scientific Committees of the European Commission

Schmidt M, Kelle A, Ganguli-Mitra A, de Vriend H (2010) Synthetic biology - the technoscience and its societal consequences. Springer, Dordrecht

Schneider I (2014) Exclusions and exceptions to patent eligibility revisited: examining the political functions of the "discovery" and "ordre public" clauses in the European Patent Convention and the arenas of negotiation. In: de Miguel Beriain I, Casabona CMR (eds) Synbio and human health. Springer, Dordrecht, pp 145–173

Then C, Hamberger S (2010) Synthetische Biologie - Teil 1: Synthetische Biologie und künstliches Leben - Eine kritische Analyse. Testbiotech e.V. - Institut für unabhängige Folgenabschätzung in der Biotechnologie

Torrance AW (2010) Synthesizing law for synthetic biology. Minn J Law Sci Technol 11:629–665

Torrance AW, Kahl LJ (2013) Bringing standards to life: synthetic biology standards and intellectual property. St Clara High Technol Law J 30:199–230

# Chapter 2
# Synthetic Biology

## 2.1 Synthetic Biology

### *2.1.1 Definition of Synthetic Biology*

#### 2.1.1.1 Origin of the Term

Synthetic biology is an emerging scientific field that is increasingly featured in the public debate and in the media. This growing exposure is caused by the great benefits that this field promises to deliver in the health, energy, and food sectors, just to name a few, as well as to the concerns it raises from a scientific, ethical, safety and regulatory point of view.

While synthetic biology has only recently reached the spotlight, the origin of both the term and the concept has roots in the past.

In 1904, Dutch botanist *Hugo de Vries* declared that "*evolution has to become an experimental science, which must first be controlled and studied, then conducted and finally shaped to the use of man*".[1] With those words, de Vries highlighted what would become the pillar of synthetic biology: the transition from a descriptive and analytical approach to the study of life to one of synthesis and creation.

Another step in the direction of the principles of synthetic biology was set out in 1912, this time by German-American physiologist *Jacques Loeb*. While he argued that "*we are not yet able to give an answer to the question as to how life originated on the earth... nothing indicates, however, at present that the artificial production of living matter is beyond the possibilities of science*".[2] Therefore, in his opinion, "*we must either succeed in producing living matter artificially, or we must find the*

[1] Campos (2009, pp. 6–7).
[2] Keller (2002, p. 18).



I. de Lisa, *The Patentability of Synthetic Biology Inventions*,
https://doi.org/10.1007/978-3-030-51206-4_2

*reason why this is impossible*".[3] The ability to control life at need and at will is one of the cornerstones of Loeb's position and, alongside his belief that biology could be construed as an engineering science rather than a natural one, it illustrates the similarities between his approach and that of modern synthetic biology.[4]

The term "synthetic biology" was also first coined in 1912. In that year, French biologist and chemist Stéphane Leduc wrote a book titled *La biologie synthétique*.[5] In the section on *Methods in biology*, he postulated that biology is no different from other scientific fields. Therefore, biology will go through a descriptive, analytic, and synthetic phase.

The term "synthetic biology" in his current connotation reappeared in 1974 at the birth of the era of recombinant DNA. In that year, geneticist *Waclaw Szybalski* addressed the next steps of biology by stating:

> Up to now we are working on the descriptive phase of molecular biology... But the real challenge will start when we enter the synthetic biology phase of research in our field. We will then devise new control elements and add these new modules to the existing genomes or build up wholly new genomes. This would be a field with an unlimited expansion potential and hardly any limitations to building 'new better control circuits' and... finally other 'synthetic' organisms.[6]

The idea of an approach to biology based on wilful creation and engineering principles has permeated the twentieth century.[7] Yet, the possibility to realise this goal became available only at the turn of the millennium. At that time, the completion of the human genome project and the decreased cost of DNA sequencing and synthesis offered the possibility to concretely realise the objectives of synthetic biology.

The first examples of applied synthetic biology surfaced in 2000, when two publications on a toggle switch and on a biological clock "*illustrated the feasibility and predictability of engineering sophisticated functions into biological systems using standard components*".[8]

Only four years later, in 2004, the first conference on synthetic biology called Synthetic Biology 1.0 (First International Conference on Synthetic Biology) was organised in Cambridge, USA. The fact that the organisers of the conference chose to call it "synthetic biology" was "*anything but inevitable or foreordained*" considering that:

---

[3]Keller (2002, p. 18).

[4]Campos (2009, pp. 10–12).

[5]Leduc (1912).

[6]Committee on Science, Technology, and Law et al. (2013, p. 9).

[7]For a more extensive overview of the history of synthetic biology (Campos 2009).

[8]SCHER, SCENIHR, SCCS Scientific Committees (2015a, p. 12), Gardner et al. (2000) and Elowitz and Leibler (2000).

> Although the new field of 'synthetic biology' clearly shared significant aims and goals with the earlier 'synthetic biology' approaches... the new coinage seems to have come through no direct historical or verbal link to the earlier efforts to engineer biology.[9]

It took a while before the expression "synthetic biology" established itself as the name for this new technology. For instance, in 2006, the term "intentional biology" was employed and the expression "synthetic genomics" was still used in a number of publications between 2010 and 2013, albeit in much smaller numbers compared to "synthetic biology".[10]

#### 2.1.1.2 Range of Definitions

Despite having established itself as the term of choice for this new technology, synthetic biology is still devoid of a specific definition. Over time, more than 30 definitions have been formulated and no consensus was ever reached over the exact content of this expression and its boundaries.[11] While this has not hindered the scientific development of the subject, it still caused uncertainties in the field. For example, one expert used the phrase "*I know it when I see it*"[12] when asked to offer a definition of synthetic biology, while another scientist pointed out that, "*if you ask five people to define synthetic biology, you will get six answers*".[13]

Although defining synthetic biology is problematic, it is a necessary step; otherwise, how could laws and regulations be effectively drafted if it is unclear what synthetic biology is and which activities fall within its scope[14]? Similarly, in the absence of an agreed definition, it would be possible for companies to set their own definition and thus to choose whether or not to include their products in the field of synthetic biology, with possible repercussions on regulations and public perception.[15]

[9]Campos (2009, p. 18). Scientists Drew Endy and Robert Carlson first considered the name "synthetic biology" in 2001 upon suggestion of scientist Carlos Bustamante, who had highlighted the analogies between this emerging field and the route taken by synthetic chemistry almost two centuries before (Carlson 2006).

[10]Some studies have considered the terms "synthetic biology" and "synthetic genomics" as interchangeable (Ribarits et al. 2014, pp. 8–82). Conversely, other reports considered synthetic genomics as a sub-field of synthetic biology (European Group on Ethics in Science and New Technologies to the European Commission 2009, p. 14).

[11]The interdisciplinarity of synthetic biology might also be responsible for this lack of consensus.

[12]This expression was coined in 1964 by Justice *Potter Stewart* of the US Supreme Court to describe his threshold test for obscenity in *Jacobellis v. Ohio* (U.S. Supreme Court 1964, p. 197).

[13]Ferry (2015) and "What's in a name?" (2009, p. 1073).

[14]SYBHEL (2014, p. 16).

[15]This was the case for a cleaning products producer. In 2014, Ecover released a laundry detergent containing oil that had been manufactured by algae, whose genetic code had been modified via synthetic biology (Strom 2014). The use of an ingredient indirectly produced via synthetic biology—that is, the oil itself was not genetically modified, but had been produced via a synthetic biology algae—generated much press. This led the producing company to present its understanding

What clearly emerges from an analysis of the definitions currently used for synthetic biology is the ambiguous and uncertain use of this term not only amongst the general public, but also and especially in the scientific community.

Definitions of synthetic biology have spanned from broad formulations to lists of principles and parameters. Some researchers have opted for inclusion and exclusion criteria or suggested defining synthetic biology via its main research fields.[16] The creation of a sliding scale model that would place works on a scale between classical biotechnologies and extreme synthetic biology was also considered.[17] Equally, methodologies were used as defining elements.[18]

---

of synthetic biology, by stating that it is: "*The process of creating DNA from scratch or inserting human-made DNA into an organism*" (Thomas 2014). On the basis of this definition, the product would indeed fall outside the realm of synthetic biology (Thomas 2014). Still, this result could change if one were to use a different definition. Similarly, Solazyme, the US biotechnology company that supplied Ecover with such oil, used to describe itself as a synthetic biology company, although it has taken off such reference from its website since then. Nowadays, the company prefers to refer to itself as a twenty-first century oil company (Ferry 2015; Strom 2014; Thomas 2014). Equally, the Enogen corn produced by Syngenta is considered synthetic biology by the ETC Group, but not by the company itself. Along the same lines, a modified corn developed by the US company Agrivida was considered synthetic biology by some commentators, but not by the company itself, which instead referred to it just as engineering (Conference of the parties to the Convention on Biological Diversity 2014a, p. 21). Another trend is the frequent use of the expression "synthetic biology" by start-ups and the reluctance to use it by companies that have been operative in classic genetic engineering (Conference of the parties to the Convention on Biological Diversity 2014a, p. 18).

[16]SCHER, SCENIHR, SCCS Scientific Committees (2014, pp. 11–16). Delimiting the scope of synthetic biology via its research areas has also been a common approach. The most frequent examples were bottom-up and top-down approaches, the design of minimal genomes, the fabrication of biological toolkits, and xenobiology research (Porcar and Peretó 2012, pp. 81–82; European Group on Ethics in Science and New Technologies to the European Commission 2009, p. 14; Schwille and Sundmacher n.d., pp. 2–8). A survey carried out amongst representatives of funding agencies from six European countries showed that only four areas were unanimously considered to belong to synthetic biology. Those are: biocircuits, both with and without standard biological parts, expanded genetic alphabet and DNA with a different backbone. Surprisingly, the quest for a minimal genome and the creation of artificial life were not universally considered as belonging to the realm of synthetic biology (Pei et al. 2012, p. 5). The variety of approaches is such that it would be problematic to find a common denominator between them and it may also be difficult, as well as highly controversial, to establish which research endeavours should be included. Definitions could also vary based on the desired outcomes or aims for which they are formulated such as, for example, funding and risk assessment (European Commission's Directorate-General for Health & Consumers 2010, p. 28; European Group on Ethics in Science and New Technologies to the European Commission 2009, p. 13). Finding a definition is also complicated by the consideration that synthetic biology may just be an umbrella term (Kronberger 2012, p. 130; O'Malley et al. 2008, p. 57; de Lorenzo and Danchin 2008, p. 822).

[17]SCHER, SCENIHR, SCCS Scientific Committees (2014, p. 26).

[18]Gaisser et al. (2008, p. 4). Most of the surveyed definitions did not explicitly mention quantitative or qualitative limits for the existence of synthetic biology. An exception was the working definition formulated by FERA (Food and Environmental Research Agency), which provided a two-prong definition for products and applications in the food and feed sector. The quantitative requirement concerned: "*Substantially large synthetic parts of genetic material caused to function in a*

An analysis of the definitions formulated so far illustrates the criteria that were most often considered as falling within the realm of synthetic biology. Those are: the application of engineering principles, the existence of a rational design, the use of standards, the construction of novel systems or novel functions, the intentionality and control over the creation. In addition to those, elements such as the size or complexity of the intervention in the biological system were also taken into consideration. On the other hand, ad hoc approaches, the absence of standards and of engineering features in the development of biological products were considered exclusion criteria given their incompatibility with the principles on which synthetic biology is based upon.[19] No exclusion criteria were formulated on the basis of taxonomical categories (i.e. the different groups in which living beings are classified, e.g. bacteria and plants).[20]

The principles behind synthetic biology have also been used as a reference point in trying to define this field. Synthetic biology is seen as move away from previous discovery-based biosciences, which relied on a descriptive approach, and as a step towards a hypothesis and synthesis methodology. Equally, the development of commercial outputs is fundamental in synthetic biology. This is confirmed by the fact that the majority of the research in this sector is geared towards commercial applications.[21] Hence, synthetic biology aims to transform biology into a production technology. Lastly, the aims of synthetic biology were also considered in delimiting this discipline. For example, the EU TESSY project (Towards a European Strategy for Synthetic Biology) specifically mentioned that the purpose of synthetic biology is "*to engineer and study biological systems that do not exist as such in nature*".[22]

---

*biological system*", while the qualitative part focused: "*On the engineering aspect of novel, synthetic (i.e. artificial) organisms, and in particular on 'how much of the produced organism was designed as new*'" (British Food & Environment Research Agency 2014, p. 29). The Scientific Committees appointed by the European Commission considered this possibility, but decided against it on the following basis: "*The SC considered applying parameters such as the degree of novelty of function or construction, as well as the number, size or complexity of synthetic or modified genetic elements. In each case, major SynBio activities occur at both extremes of the possible gradient of parameter values (e.g., very little vs. very high degree of functional novelty). In the remaining cases, the relevant aspects were so abstract that they are impossible to operationalise. For example, it is not meaningful to discriminate 'how much rational design or engineering concepts had gone into designing a specific organism', as the outcome of the activity will be identical, independent of the psychological process or conceptualisation by the 'engineer'*" (SCHER, SCENIHR, SCCS Scientific Committees 2014, p. 26).

[19]Porcar and Peretó (2012, p. 82).

[20]Kuzma and Tanji (2010, p. 96).

[21]Conference of the parties to the Convention on Biological Diversity (2014a, p. 9). Synthetic biology is certainly not the first science trying to exploit nature for commercial purposes. Still, its placing at the very centre of its investigative efforts and from the very beginning the development of new useful products and functions sets it apart from more descriptive sciences.

[22]Gaisser et al. (2008, p. 4). TESSY defined synthetic biology as a discipline: "*To engineer and study biological systems that do not exist as such in nature, and use this approach for (i) achieving better understanding of life processes, (ii) generating and assembling functional modular components, (iii) developing novel applications or processes*" (Gaisser et al. 2008, p. 4). This definition

Those approaches sparked analogies between the world of synthetic biology and that of software and computers. From this perspective, DNA could be seen as a quaternary code—as opposed to the binary code of computer programming—thus suggesting that the work of synthetic biologists could somehow mirror that of software designers. Such perspective would fit with the notion that biological life can be seen and handled as information.[23]

The aspects outlined above have been taken into consideration by scholars and authorities when formulating a definition of synthetic biology.[24] One frequently cited definition states that synthetic biology is "*(i) the design and construction of new biological parts, devices and systems, and (ii) the re-design of existing, natural biological systems for useful purposes*".[25]

Similar to the above two-pronged definition is the one formulated by ERASynBio, which describes synthetic biology as "*the engineering of biology: the deliberate (re)design and construction of novel biological and biologically based parts, devices and systems to perform new functions for useful purposes, that draws on principles elucidated from biology and engineering*".[26] The reference to engineering principles is also present in the definition formulated by the Royal Academy of Engineering, which is used also by the OECD. In their opinion, "*synthetic biology aims to design and engineer biologically based parts, novel devices and systems as well as redesigning existing, natural biological systems*".[27] The purposes of synthetic biology were also cited in the definition formulated by the Scientific Committees appointed by the European Commission, which held that "*synthetic biology is the application of science, technology and engineering to facilitate and accelerate the design, manufacture and/or modification of genetic*

was criticised for being formulated in terms so generic that it would be impossible to distinguish synthetic biology from genetic engineering. Also, its broad formulation would include transgenic organisms within the realm of synthetic biology (Commissione federale d'etica per la biotecnologia nel settore non umano (CENU) 2010, p. 6).

[23]van den Belt (2009a, p. 259). This analogy explains the use of software metaphors in synthetic biology.

[24]For additional definitions, SCHER, SCENIHR, SCCS Scientific Committees (2014, pp. 55–60), Conference of the parties to the Convention on Biological Diversity (2014a, pp. 9–11), British Food & Environment Research Agency (2014, p. 29), Ribarits et al. (2014, p. 9), OECD (2014, pp. 17–20), Kelley (2014, pp. 19–20), Commissione federale d'etica per la biotecnologia nel settore non umano (CENU) (2010, pp. 5–7), OECD and The Royal Society (2010, p. 8), Then and Hamberger (2010, p. 9), Cserer and Seiringer (2009, p. 33), "What's in a name?" (2009), European Group on Ethics in Science and New Technologies to the European Commission (2009, pp. 13–14), Balmer and Martin (2008, pp. 6–7), Health Council of the Netherlands (2008, p. 17), Panke (2008, p. 9) and European Commission, Directorate-General for Research (2005, p. 10).

[25]Conference of the parties to the Convention on Biological Diversity (2014a, p. 9), Ribarits et al. (2014, p. 5) and Schmidt and Giersch (2012, p. 286). Commentators noted that the vast majority of the definitions contain references to "biological systems" and avoid using to the term "life" (Commissione federale d'etica per la biotecnologia nel settore non umano (CENU) 2010, p. 6).

[26]ERASynBio (2014, p. 6).

[27]OECD (2014, p. 18) and Royal Academy of Engineering (Great Britain) (2009, p. 6).

*materials in living organisms*".[28] Yet another approach to synthetic biology could be found in the description provided by the ETC Group, an international civil society organisation, which dubbed this field "*genetic engineering on steroids*".[29] Likewise, the British Food and Environmental Agency used the formulation "*extreme GM*" as its "*working understanding of synthetic biology*".[30] Lastly, a detailed definition was formulated by the US Presidential Commission for the Study of Bioethical Issues. According to it:

> Synthetic biology is the name given to an emerging field of research that combines elements of biology, engineering, genetics, chemistry, and computer science. The diverse but related endeavours that fall under its umbrella rely on chemically synthesized DNA, along with standardized and automatable processes, to create new biochemical systems or organisms with novel or enhanced characteristics. Whereas standard biology treats the structure and chemistry of living things as natural phenomena to be understood and explained, synthetic biology treats biochemical processes, molecules, and structures as raw materials and tools to be used in novel and potentially useful ways, often quite independent of their natural roles. It joins the knowledge and techniques of biology with the practical principles and techniques of engineering. 'Bottom-up' synthetic biologists, those in the very earliest stages of research, seek to create novel biochemical systems and organisms from scratch, using nothing but chemical reagents. 'Top-down' synthetic biologists, who have been working for several decades, treat existing organisms, genes, enzymes, and other biological materials as parts or tools to be reconfigured for purposes chosen by the investigator.[31]

#### 2.1.1.3 Radically New Technology or Extension of Earlier Practices

Ever since the inception of synthetic biology, questions have been raised on whether this discipline constitutes a revolutionary technology that will cause a paradigm shift in life sciences or whether it is merely an incremental change that represents the most recent highpoint in the development of biotechnologies. The approach taken on this matter has repercussions on the regulatory panorama of synthetic biology, since viewing this discipline as a revolutionary technology may lead to demands for additional specific norms on policy, safety, legal and ethical issues.[32]

Scholarly opinions on this point are divergent. Those who believe that synthetic biology is an incremental technology have pointed out that rebranding traditional genetic and metabolic engineering as synthetic biology served funding purposes,

---

[28]SCHER, SCENIHR, SCCS Scientific Committees (2014, p. 5).

[29]ETC Group (2007, p. 1).

[30]British Food & Environment Research Agency (2014, p. 52).

[31]Presidential Commission for the Study of Bioethical Issues (2010, p. 36).

[32]Viewing synthetic biology as an incremental technology has repercussions on the evaluation of the ethical questions posed by this field. In this case, it could be argued that these ethical issues will be for the most part similar to the ones already posed by other biotechnologies and that, consequently, the framework established until now would offer a valid point of reference (Parens et al. 2009, p. 11). From a safety perspective, it was mentioned that synthetic biology shares similarities with recombinant DNA technologies. For this reason, it was maintained that: "*Placing a new name on an old technology does not create a new hazard*" (OECD 2014, p. 123).

since one could take advantage of the hype surrounding this new technology to attract investment or serve as a marketing banner.[33]

Other authors have expressed more uncertain views on whether synthetic biology is indeed a truly innovative field.[34] To add to the uncertainty, most current and near-term applications of synthetic biology are based on techniques connected to traditional genetic engineering.[35]

Notwithstanding this, the novel character of synthetic biology has also been defended, as a great degree of innovation in its "*systematic, application-driven engineering perspective to biology*" was detected.[36]

Interestingly, the divergent views expressed on the innovative nature of synthetic biology do not appear to be dependent upon the affiliation of the individuals or groups writing about it, but rather to be impacted by the topics and goals for which such formulations are being used. Indeed, it was claimed that:

> There is no simple correlation between the nature of social groups (such as regulatory institutions, scientific community, NGOs, companies etc.) and their pronouncements about synthetic biology. In fact, views on whether synthetic biology should follow previous experience or be treated differently often vary even within the same report, depending on which topic (such as financial, ethical or environmental implications) is under discussion.[37]

#### 2.1.1.4 Negative Connotation

Synthetic biology has been associated with negative connotations due to its name, its research endeavours and its connection to biotechnologies that have attracted criticism in the past. As synthetic biology products are now starting to reach the market, the consequences of these negative impressions are being acknowledged and are becoming more relevant.

Already at the dawn of synthetic biology, at the Synthetic Biology 2.0 conference in 2006, the negative connotations of the world "synthetic" were discussed, as this

---

[33]"What's in a name?" (2009, p. 1071). *Ashcroft* and *Dawson* pointed out the: "*Tendency in the field to take a new sexy science, coin a neologism, start a journal and a society, and bid for Centre funding*" (SYBHEL 2014, p. 17). Studies show that both journalists and the public viewed synthetic biology as a traditional gene technology (Gschmeidler and Seiringer 2012, p. 163). Others have gone even further and argued that, by observing the impact of synthetic biology on the distinction between artefacts and living organisms and other ontological questions, this new field may only be a: "*Late and unexceptional offshoot*" development in comparison to prehistoric agriculture (Greco 2013; Preston 2013, p. 649).

[34]Douglas and Stemerding (2014, pp. 1–2), Kronberger (2012, p. 130), Then and Hamberger (2010, p. 8), Deutsche Akademie der Naturforscher Leopoldina (2010, p. 5) and Balmer and Martin (2008, pp. 29–30).

[35]Conference of the parties to the Convention on Biological Diversity (2014a, p. 10).

[36]European Commission, Directorate-General for Research (2005, p. 10).

[37]Zhang et al. (2011, p. 7).

word could conjure up images of "*monstrous life forms let loose by maniacal scientists*".[38]

Similarly, it was feared that synthetic biology may follow the path of Genetically Modified Organisms (GMOs) when it comes to public acceptance, especially in Europe.[39] However, the situation may take a different course this time, considering that addressing ethical, safety, social, and legal issues has been central for synthetic biology ever since its inception.[40]

#### 2.1.1.5 Engineering Principles

Despite the lack of consensus over the definition of synthetic biology, one distinctive aspect of this discipline has unequivocally emerged: the application of engineering principles to biology.[41] This approach is based on the idea that living systems are intrinsically complex, as they developed through evolutionary pressure.[42] In particular, in the case of synthetic biology, reducing the complexity of biological systems is supposed to help controlling them, rendering their behaviour more predictable and designing them in a rational and systematic manner. The means to achieve these goals have been located in the core principles of engineering, which could be applied at all biological levels (e.g. molecules, cells, organisms).[43]

This approach has led to analogies with the world of electronics, even though the characteristics of biological systems set synthetic biology apart from any other engineering-related discipline. Indeed, biological systems can self-replicate and repair, undergo mutation and evolution and are thus substantially different from other engineering objects.[44] Such dissimilarities, coupled with the complexity of biological systems, led scholars to debate whether synthetic biology could ever successfully apply engineering principles to biology. While currently the consensus on this point is negative, some scholars have argued that it is merely a question of time before biology is transformed into an engineering discipline with the help of synthetic biology.[45]

---

[38]Cserer and Seiringer (2009, p. 31) and Balmer and Martin (2008, p. 6). Others noted that: "*The term 'synthetic biology' is now tainted with negative connotations and should be avoided in public. Consumers prefer natural to synthetic, and 'syn' brings up negative connotations of 'sin.' The term 'genetically engineered' should also be avoided because it's gotten too much public backlash. Some alternative terms suggested at the meeting were 'fermentation derived' and 'nature identical'*" (Perls 2014).

[39]Kronberger (2012, p. 132).

[40]Torgersen and Hampel (2012, p. 139).

[41]The actual use of engineering principles in current synthetic biology products is still quite limited (Davies 2019).

[42]Calvert (2010, pp. 98–99).

[43]Kronberger (2012, p. 130).

[44]"What's in a name?" (2009, p. 1073) and Schwille and Sundmacher (n.d., p. 2).

[45]Jefferson et al. (2014a, p. 45).

A number of engineering principles have been considered relevant in this field. The most important one is the concept of standardisation. As hinted by the word itself, this principle demands the "*definitive description and characterization of parts*" and extends also to the standardisation of measurement protocols and documentation systems.[46] Standardisation in biology is exemplified by the concept of BioBricks™, which are standard biological parts that can be assembled together in order to build new biological systems or devices.[47] This approach facilitates exchanges amongst research groups as well.

Another important principle is decoupling, which relates to the "*process of breaking down the construction of complicated entities into manageable semi-independent tasks*".[48] On the other hand, abstraction refers to the possibility to separate "*the overall system into meaningful subsystems, which in turn might be once more separated into meaningful subsystems, and so on*".[49] Similarly, the decoupling of design and production aims at separating these two tasks and assigning them to specialised individuals, in the same way a car is not assembled by the people who have designed it.[50] To enable decoupling, modularisation is needed. This concept, whose application is still hypothetical, predicts a "*functional unit that is capable of maintaining its intrinsic properties irrespective of what it is connected to*".[51] The concept of orthogonality describes instead systems "*where modifying one component does not result in side effects to other components in the system*".[52] This principle would be used to increase the predictability of the behaviour and of the interactions of the modified system within the cell and the environment.[53]

### 2.1.2 Difference from Other Technologies

Understanding synthetic biology is understanding the connection between this field and other biotechnologies. This task is not trivial, considering the existing "*overlap in terms of vocabulary, actors and concerns*" and in terms of research endeavours.[54]

---

[46]Panke (2008, p. 8) and O'Malley et al. (2008, p. 57).

[47]On this, see Sect. 2.2.1.3.

[48]O'Malley et al. (2008, pp. 57–58).

[49]Panke (2008, p. 7).

[50]Panke (2008, pp. 8–9).

[51]Calvert (2010, p. 98). This concept is fundamental for BioBricks™. On this, see Sect. 2.2.1.3.

[52]Conference of the parties to the Convention on Biological Diversity (2014a, p. 17). Refactoring refers instead to a software concept, which would be used in synthetic biology to: "*Remove all uncharacterised functional elements and molecular interactions, which might lead to unpredictable system behaviour*" (SCHER, SCENIHR, SCCS Scientific Committees 2014, p. 12).

[53]Conference of the parties to the Convention on Biological Diversity (2014a, p. 17).

[54]Kronberger (2012, p. 132).

Consequently, a clear separation between them is not always possible and it is likely that this uncertainty will persist in the future.

The principles that characterise synthetic biology are critical to the assessment of whether an activity falls within this area or not. Considering that both synthetic biology and related technologies operate in similar application and research areas, what sets synthetic biology apart from other biotechnologies are really its tools and approaches.[55] Overall, it can be said that both synthetic biology and traditional biotechnologies aim at solving technological problems via the manipulation of biological material. However, prior biotechnologies have approached this task empirically and with an ad hoc methodology, whereas synthetic biology is implementing engineering principles in order to rationally manipulate biological systems.[56]

#### 2.1.2.1 Genetic Engineering

The methods used and the complexity of the modifications carried out by synthetic biology are often considered the core of what sets this field apart from traditional genetic engineering.[57] Genetic engineering has been associated with the reading and analysis of DNA, whereas synthetic biology is connected to DNA writing and synthesis. Moreover, synthetic biologists intend to plan and manufacture new biological systems on the basis of engineering principles, whereas genetic engineers proceed through a trial and error method and have mostly focused on the adaptation and slight modification of already existing biological systems.[58]

When it comes to the products of synthetic biology and genetic engineering, the discussion becomes more controversial, as it steps into the realm of GMOs. Synthetic biology products have been compared to GMOs for the high hopes and expectations they generate and for some possible overlaps on regulatory and policy issues. Nevertheless, from a biological perspective, differences between the products of synthetic biology and genetic engineering are poised to become more marked, as GMO products usually contain a single modification that is related to a new feature (e.g. pest resistance), whereas synthetic biology aims for more extensive ones.[59] For instance, manipulations on the scale of whole genomes—either in terms of number of base pairs or loci—have been considered a main trait of synthetic biology, which is not present in traditional genetic engineering.[60] Nevertheless, at this point in time, when the potentials of synthetic biology are only starting to be realised, examples from the food and feed sector show that there is still a large area of overlap between

[55]"What's in a name?" (2009, p. 1073).

[56]European Commission, Directorate-General for Research (2005, p. 11).

[57]Conference of the parties to the Convention on Biological Diversity (2014a, p. 10).

[58]Open Science (2012, p. 7) and Then and Hamberger (2010, p. 9).

[59]Rousseaux (2015).

[60]Ribarits et al. (2014, p. 111).

SynBio products and GMOs.[61] Additionally, the tools used by synthetic biology build on the ones used by genetic engineering.[62] Yet, it is to be expected that with time the gap between the two will widen, as synthetic biology achieves more complex modifications and further implements engineering principles.[63]

#### 2.1.2.2 Metabolic Engineering

The field of metabolic engineering is so closely related to synthetic biology that it is unclear whether it represents a discipline within it or rather a separate one.[64] According to *San Yup Lee*:

> Originally, synthetic biology sought to redesign and rebuild biological parts and systems without specific biotechnological objectives, whereas metabolic engineering aimed at purposeful modification of metabolic and other cellular networks to achieve desired goals, such as overproduction of bioproducts. Recently, it has become more difficult to distinguish the two disciplines as each is employing the other's approaches... and both are moving towards integration with systems biology.[65]

Others have differentiated the two by stating that synthetic biology, contrarily to metabolic engineering, designs biological systems to make them perform tasks that are different from what they would normally do.[66]

#### 2.1.2.3 Synthetic Chemistry

Synthetic chemistry has been considered a precursor of synthetic biology in more ways than one.[67]

Before the advent of synthetic chemistry, organic chemistry was a discovery-based science focused on understating the properties of natural compounds by

---

[61] British Food & Environment Research Agency (2014, p. 49).

[62] Recombinant DNA, Polymerised Chain Reaction and automated DNA sequencing are the tools of choice of genetic engineering. The first two are concerned with writing DNA, while the third with reading it out. Synthetic biology adds to these tools by using three additional approaches: automated construction of DNA, standards and abstraction (Commissione federale d'etica per la biotecnologia nel settore non umano (CENU) 2010, p. 7).

[63] Technically, it is hard to draw the line between this and more traditional recombinant DNA technologies. Indeed: "*A piece of DNA can be synthesized, identical in sequence to an existing gene, and inserted into an organism. Such an organism could also be constructed using traditional rDNA techniques, and even the scientists who produced the organisms may be unable to tell which was produced with which technology. However, as synthetic DNA constructs become more and more complex... it becomes nearly impossible to accomplish the same engineering feats through traditional rDNA technology*" (Carter et al. 2014, p. 9).

[64] ERASynBio (2014, p. 7) and O'Malley et al. (2008, p. 62).

[65] "What's in a name?" (2009, p. 1072).

[66] "What's in a name?" (2009, p. 1071).

[67] van den Belt (2009a, pp. 258–259).

examining compositions and reactions. The field proceeded along those lines, since it was commonly believed to be impossible to synthesise organic molecules. The situation drastically changed in 1828, when German chemist *Friedrich Wöhler* first synthesised the organic chemical Urea from an inorganic one.[68] This experiment profoundly affected the world of chemistry, as it showed that even organic compounds could be built from scratch starting from simpler pieces. This led to discussions on the boundaries between natural and artificial, which are common also today in the field of synthetic biology.[69]

With time, synthetic chemistry moved from the synthetic reproduction of already existing substances to the creation of new ones. This trajectory is similar to the one synthetic biology is trying to explore nowadays, that is, the passage from reproducing natural systems to creating novel ones.[70] According to *Boldt* and *Müller*:

> Synthetic chemistry has shown the way: from systematic analysis of chemical processes to synthesis of novel products. Synthetic biology does the same, but in the realm of the living.[71]

#### 2.1.2.4 Systems Biology

Systems biology is a descriptive scientific discipline that aims to provide a "*quantitative and predictive understanding of biological systems*" by gathering information on the interactions of the various parts that compose a biological system.[72] This field emerged at the turn of the millennium and focuses on systems rather than specific genes, just like synthetic biology.[73]

Systems biology lays the analytical and conceptual framework necessary for synthetic biology and provides the understanding of biological systems needed by the latter to develop new products and applications. Synthetic biology could thus be seen as the "*design counterpart of systems biology*".[74] Synthetic biology can be further distinguished from systems biology based on three aspects. First, modelling in systems biology is an expression of basic research, while in synthetic biology it is focused on the design of products. Second, large data gathering and integration is pivotal in systems biology, but not in synthetic biology. Lastly, synthetic biology tries to reduce complexity, whereas systems biology embraces it.[75] Still, the two disciplines are strongly correlated and are bound to heavily influence each other in

---

[68]Yeh and Lim (2007, p. 521).

[69]van den Belt (2009a, p. 259) and Carlson (2006).

[70]European Commission, Directorate-General for Research (2005, p. 11).

[71]"What's in a name?" (2009, p. 1071).

[72]Health Council of the Netherlands (2008).

[73]O'Malley et al. (2008, p. 62).

[74]European Commission, Directorate-General for Research (2005, p. 11).

[75]O'Malley et al. (2008, pp. 62–63).

their development.[76] It was even argued that "*synthetic and systems biology are the ultimate synergetic partners for ushering in an era of rapid and probably systematic biological discovery*" and that synthetic biology is "*the other side of the coin of systems biology*".[77]

#### 2.1.2.5 Other Related Disciplines

Further disciplines overlap or contribute to synthetic biology. For instance, genomics is a discipline focused on the study of the genome (i.e. the genetic material of an organism).[78] It analyses its interactions with the environment as well as diseases that could be connected to it.[79] Genomics revolves around reading and understating the genome, whereas synthetic biology concentrates on the writing of the genome, which has proven to be a much more intricate task.[80] Still, synthetic biology is benefiting from the knowledge gathered through genomics.

Lastly, synthetic biology must be differentiated from molecular biology. The latter adopts a descriptive approach to study biology at the molecular level, for example by analysing the structure and function of DNA, RNA, and proteins,[81] while synthetic biology focuses on design and synthesis. Still, the existence and development of synthetic biology would not be possible in the absence of these discovery sciences.[82]

## 2.2 Techniques, Research Areas and Applications

### *2.2.1 Techniques Used in Synthetic Biology*

The realisation of the goals of synthetic biology depends on the progress of its enabling techniques. In recent years, programs for the design and optimisation of biological systems have improved and have become more user-friendly. Likewise, the production of those designs via DNA synthesis has improved in quality and speed, while becoming cheaper. Those techniques, coupled with the use of

---

[76] Health Council of the Netherlands (2008, p. 19) and European Commission, Directorate-General for Research (2005, p. 11).

[77] "What's in a name?" (2009, p. 1073) and O'Malley et al. (2008, p. 62).

[78] Genomics is not to be confused with genetics, which is the study of genes and their role in inheritance (National Human Genome Research Institute 2014).

[79] National Human Genome Research Institute (2014).

[80] OECD (2014, p. 20).

[81] Mandal (2014).

[82] O'Malley et al. (2008, p. 62).

standardised parts, facilitate the modelling and realisation of products of synthetic biology and will thus be examined in more detail in the following sections.

Such DNA design and production techniques are based on the premise that DNA is both an information storing medium and a sequence of chemicals. From a chemistry perspective, DNA is made up of chemical bases that are attached to a sugar molecule and a phosphate molecule. Hence, the synthesis of DNA consists of a chemical process to reproduce and assemble these different substances. On the other hand, these same chemical bases store hereditary material and information, much like an alphabet or a software code. As such, this code could be sequenced, amended and even reformulated. While emphasis could be placed on either one of these two aspects, the informational and chemical properties of DNA are inextricably intertwined.[83]

#### 2.2.1.1 Designing Techniques

Design techniques used in synthetic biology enable the modelling of new biological circuits, which could be designed by scientists to achieve the best possible functionality. Such designed circuitry could be markedly different from anything found in nature. Although this is the ultimate goal, at the moment the ability to design new complex and functional biological systems is still limited, as natural circuits still represent the starting point of the work of synthetic biologists.[84]

The development of these tools is supported by bioinformatics, which offers the possibility to model circuits before using them in living biological systems.[85] As in other engineering disciplines, Computer-Aided Design software (CAD) have been introduced. Such design programs, which are becoming increasingly sophisticated, support the gene assembly process and predict the behaviour and performance of the new system via simulations. They can also operate in conjunction with DNA databases and registries of standardised parts.[86] For the moment, these bio-CAD programs have not reached the precision of their engineering counterparts given the complexity of biological systems.[87] However, it will be fundamental for the development of synthetic biology to enhance these design techniques, as the existing gap between our limited ability to design and our evolved DNA synthesis skills is the main hurdle to fulfil the promises of this field.[88]

[83]On this, see Sect. 2.2.5.

[84]Carter et al. (2014, p. 9).

[85]British Food & Environment Research Agency (2014, p. 86).

[86]Nowogrodzki (2018) and SCHER, SCENIHR, SCCS Scientific Committees (2014, p. 13).

[87]Yarris (2011).

[88]Rice (2010).

#### 2.2.1.2 Production Techniques

> The gene-building field is doing for biology what Johannes Gutenberg did for printing – turning what was once a laborious and uneven artisanal effort into affordable and accurate mass production.[89]

This quote shows the paradigm shift that occurred in the techniques for the production of DNA over the past decades: from methods that were expensive and labour-intensive to cheap and reliable techniques that permit the mass production of genes. Such techniques revolve around two concepts: DNA sequencing and DNA synthesis. The former is focused on determining the order of the nucleic acid sequence, while the latter is concerned with the actual artificial production of the DNA sequence. This differs from traditional recombinant DNA techniques. Prior methods allowed the rearrangement of genes and genomes, but did not offer the possibility to write specific DNA sequences, which is instead offered by DNA synthesis. A metaphor used to describe these two techniques compared synthetic DNA technology to a typewriter and recombinant DNA technology to copy/cut/paste/delete functions.[90]

The cost and speed of DNA sequencing have been improving over time.[91] Equally, when it comes to DNA synthesis, contemporary methods enable the synthesis of both completely new sequences as well as of sequences that are a duplicate of natural existing ones.[92] The costs associated with this technique have drastically decreased in the past years.[93] At the same time, the velocity of writing DNA strands increased exponentially.[94] During this period, not only did the velocity and costs of production improve, but also the length of the sequences that could be manufactured.[95]

Even the process for making DNA has undergone revolutionary changes in the last years. Nowadays, scientists have the possibility to type and then submit a DNA sequence via computer to a gene manufacturing company. After production, such genes are then shipped back to the scientists within a few days.[96] According to American scientist and entrepreneur *Craig Venter*, a similar process could be devised to permit biological teleportation. In his opinion, it should be possible to make a "*digital copy of an organism's DNA in one place and sending the file to a*

[89]Krieger (2015a).

[90]Nista (2015).

[91]OECD (2014, p. 57).

[92]SCHER, SCENIHR, SCCS Scientific Committees (2015a, p. 37).

[93]Krieger (2015a) and Zhang (2015).

[94]ETC Group (2007, p. 6).

[95]If in 2004 the longest strand amounted to 32,000 base pairs, in 2008 this number had already reached 580,000 base pairs. This size corresponds to that of viruses and minimal bacteria, which explains how by that time several viruses (e.g. the polio virus, the virus of the Spanish flu) could already be synthesised (Then and Hamberger 2010, p. 11).

[96]Krieger (2015a).

*device somewhere else that can then recreate the original life-form*".[97] Interestingly, these creations via DNA synthesis might be identical to the ones found in nature and could not be distinguished from them.[98]

Progress is not limited to the production of DNA, but extends also to its editing. In recent years, a wide array of genome-editing tools to alter virtually every existing DNA sequence in a reliable and fast manner have emerged. Such tools can be applied to higher organisms (e.g. plants and animals) and allow the introduction of several modifications, even at the same time.[99]

Notwithstanding the above improvements, the synthesis of new genomes remains a complex task due to biological complexity and errors in the design and production of biological systems. Mistakes and inaccuracies can emerge at every stage of the preparation and even low error rates (e.g. 1 in 10,000 base pairs) can be highly problematic.[100] This explains why testing and debugging newly-made genomes constitutes one of the most challenging and time-consuming tasks in their production, even accounting for 95% of the time spent on projects.[101] Hence, the assembly of a genome *ex novo* is more complex than merely editing an already existing one.[102]

#### 2.2.1.3 Standardisation

Standardisation of biological parts is one of the main goals of synthetic biology. It is based on the assumption that biological components can be assembled to create more complex systems and that these parts could become standardised enough to be exchanged between different devices and laboratories.[103] Such characteristics have sparked analogies with Lego. Still, the possibility to standardise biological parts has not been univocally accepted.[104]

---

[97]Corbyn (2013).

[98]SCHER, SCENIHR, SCCS Scientific Committees (2015a, p. 37) and Then and Hamberger (2010, p. 11).

[99]SCHER, SCENIHR, SCCS Scientific Committees (2015a, p. 37). The most revolutionary and promising methods include Zinc-Finger Nucleases (ZFNs), Transcription Activator-Like Effector-Based Nucleases (TALENs) and CRISPR-Cas (Gaj et al. 2013, p. 397). The technology behind CRISPR-Cas has been at the centre of a heated patent battle between scientists and universities claiming priority in the invention (Cynober 2019).

[100]Ribarits et al. (2014, p. 11) and OECD (2014, p. 58).

[101]OECD (2014, p. 62). While the availability of cheap DNA synthesis is bound to help scientists and companies experimenting with different gene combinations, it still does not remove the complexity of synthetic biology. Indeed, it has been pointed out that: "*Making DNA is not the hardest part of synthetic biology. Synthetizing a gene is chemistry; getting a gene to work in a cell is biology, and that comes with all of biology's messiness*" (Zhang 2015).

[102]Ribarits et al. (2014, p. 11).

[103]Helman et al. (2007, p. 6).

[104]Commentators noted that: "*The idea… is either dismissed as a research question because it's irrelevant or dismissed as a research question because it's impossible*" (Frow and Calvert 2013, p. 44), while others pointed out that: "*There is no such thing… because even a standard component*

The best-known example of standard biological parts was initiated by American scientists Drew Endy and Tom Knight and is known under the name BioBricks™. Such standard biological parts perform a specific task (e.g. turn genes on and off, transmit signals between cells) and are compatible with other BioBricks™, which offers the possibility to combine them to produce longer circuits.[105] The term BioBricks™ refers not only to the parts themselves, but also to the assembly standard used to make sure that parts are compatible with each other. BioBricks™, which are available in an open source format, are collected in physical form at the Registry Repository, while information on them is stored in the Registry of Standard Biological Parts.[106]

In spite of this, researchers rarely use parts contained in repositories. Indeed, most of the sequences mentioned in publications do not operate in the same vectors and do not have standardised cloning sites. This is the reason why most standards exist only within companies or research groups.[107] Nevertheless, the setting of precise and agreed standards is fundamental for the development of synthetic biology.[108] For this reason, standardisation efforts were undertaken on both sides of the Atlantic by national institutions.[109]

### 2.2.2 Current Research Areas

Engineering living systems is proving more complex than traditional engineering. The complexity of the living, coupled with our limited understanding of the inner workings of nature, currently impede the full implementation of engineering principles in biology. This is exemplified by the unpredictability of the interactions between cells, their parts and the environment and by the fact that biological systems seem to be devoid of any logical formula. Indeed, studies show that ample sections of the genome may be deleted without damaging the functioning of an organism, while minor mutations could prove fatal for it.[110]

In order to devise biological systems that are reliable, predictable and perform desired functions, synthetic biologists developed two approaches: the bottom-up and the top-down. These processes, while complementary and partially overlapping,

---

*works differently depending on the environment. The expectation that you can type in a (DNA) sequence and can predict what a (genetic) circuit will do is far from reality and always will be*" (Endy 2013, p. 19).

[105]ETC Group (2007, p. 16) and iGEM (n.d.-a).

[106]Ribarits et al. (2014, pp. 39–40). Related to the BioBricks™, is the BioBricks Foundation, which aims to ensure that the engineering of biological systems is performed ethically and openly (BioBricks Foundation n.d.-a).

[107]SCHER, SCENIHR, SCCS Scientific Committees (2014, p. 12).

[108]Jain (2015).

[109]Parliamentary Office of Science and Technology (POST) (2015, p. 2) and Basulto (2015a).

[110]OECD (2014, pp. 62–63).

stem from opposing starting points. The top-down approach tackles biological complexity by stripping an existing organism of genetic elements that are not essential to its functioning and is performed in vivo. Conversely, the bottom-up approach, which is performed in vitro, intends to build living systems from scratch using as components non-living building blocks or standard parts. This innovative and challenging approach, based on the engineering concept of modularity, postulates that biological systems can be built by assembling independent functional modules, whose characteristics will determine the features of the final biological product.[111]

#### 2.2.2.1 Minimal Genome

The quest for a minimal genome represents the best example of the top-down approach of synthetic biology and, as hinted by its name, it is focused on determining the smallest number of genes needed to support basic life functions.

Based on the assumption that even the genome of the smallest organism contains redundant sections, synthetic biologists are trying to strip those genomes to a minimum in order to "*minimise the metabolic burden on the cell*" so that "*the remaining cellular energy can be directed toward the manufacture of a desired industrial product*".[112] This should also make the cell easier to control. The purpose of this approach is to redesign genomes to render them more efficient or to make them perform new functions.[113] On the down side, the removal of genes usually affects the robustness of the organism and its ability to survive in the wild.

The starting point of this approach is understanding which genes support basic metabolism and cell replication. In the past, theoretical studies postulated that as few as 206 genes would be needed for the minimal genome, while other scientists believed that a viable cell could exist with 256 genes.[114] Others, such as *Kitney*, argued instead that "*there is in fact no universal minimal cell, only a minimal cell for a given set of conditions*", while *Koonin* argued that the minimal genome might be something to which synthetic biologists get closer and closer to without ever reaching it.[115]

Empirical studies showed that these predictions on the size of the minimal genome were too optimistic. Of the 482 genes that compose the organism with the

---

[111]Presidential Commission for the Study of Bioethical Issues (2010, pp. 43–46) and European Group on Ethics in Science and New Technologies to the European Commission (2009, pp. 19–20).

[112]SCHER, SCENIHR, SCCS Scientific Committees (2015a, p. 29).

[113]Tucker and Zilinskas (2006, p. 27).

[114]Ribarits et al. (2014, pp. 20–21) and Mushegian and Koonin (1996, p. 10268). There are also examples of naturally reduced genomes. A number of free-living prokaryotes in different environments have all about 1400 genes. This circumstance was seen as an indication that nature itself was able to reduce the genome of those free-living prokaryotes to a minimum number of genes (Porcar et al. 2011, p. 3).

[115]Coghlan (2016) and Twilley (2016).

shorter known genome, *Mycoplasma genitalium*, only 100 could be removed.[116] Their deletion had a positive effect on the growth rates of the organism, possibly due to the reduced amount of energy allocated to non-essential processes.[117] The study of the minimal genome of this bacterium was first undertaken by Craig Venter during his Minimal Genome Project. In 2008, Venter and his team removed the non-essential genes and then assembled a new version of the bacterium, this time using chemically synthesised DNA.[118] Subsequently, Craig Venter applied for a patent on the 381 genes that were believed to constitute the minimal bacterial genome.[119] This sparked controversy, as some commentators viewed those claims as contrary to public morality, safety and as too wide.[120]

In 2016, Venter and his team made further progress in the quest for the minimal genome. Taking as a starting point the *Mycoplasma laboratorium* bacterium they designed in 2010 (so called JCVI syn1.0) and its 901 genes, they identified the essential, nonessential and quasi-essential genes that made it up.[121] They achieved this by deleting one by one sections of the genome to see which repercussions it would have. At the end, a 473-gene genome was achieved and was named JCVI syn3.0. Interestingly, 149 of these 473 genes do not have a known function.[122] This lack of understanding was addressed by *Koonin* by stating that "*anyone who claims that she or he understands how a cell works is either ignorant or ridiculously arrogant*".[123] At the moment, scientists are attempting to install the genome of JCVI syn3.0 into a synthetic liposome containing the machinery needed to convert DNA into protein. The goal is to see whether a cell with both synthetic "hardware" and "software" could survive.[124]

Venter and his team filed patents covering both JCVI syn3.0 and the process to make it.[125] Some commentators doubted the practical relevance of such research, as

---

[116] For comparison, humans have approximately 23,000 genes (Specter 2009).

[117] OECD (2014, p. 63).

[118] Specter (2009).

[119] The ETC Group nicknamed this invented bacteria *Synthia*, in order to provide the public with an easier and catchier name (The Economist 2007).

[120] The Economist (2007). The ETC group was also very critical of the work of Venter and his team and demanded the withdrawal of the related patents or, otherwise, their rejection on public morality and safety grounds (Balmer and Martin 2008, p. 16).

[121] Quasi-essential genes are needed for growth, but are not absolutely required to keep the organism alive. Some of these genes were left in the genome of JCVI syn3.0 in order to guarantee a workable growth rate of the bacterium.

[122] Twilley (2016), Coghlan (2016), Lowe (2016), NewScientist.com (2016) and Dvorsky (2016).

[123] Twilley (2016).

[124] Powell (2018).

[125] Dvorsky (2016).

other organisms (e.g. yeast) have been successfully used and studied by man, although their genome is not minimal.[126]

Attempts to achieve a minimal genome have also been applied to *E. coli*, which is one of the most used organisms in biology. This led to the deletion of parts of the genome, which had a positive effect on the properties of the bacterium.[127] This was achieved despite the fact that more than 20% of its genes do not have a proper functional characterisation, although *E. coli* is one of the best-known and studied organisms.[128]

The quest for the minimal genome serves also another purpose. Considering that organisms evolved due to evolutionary pressure rather than to be fit for industrial processes, it was considered ideal to develop a *chassis* organism. The word chassis, which was originally used in mechanical engineering, describes a simplified cellular structure in which modular biological parts can be inserted.[129] This biological production platform would operate efficiently and would not contain any features that would interfere with the purposes for which it is designed.[130] The hope is that in the future those chassis could be combined with BioBricks™ to provide platforms in which the latter could be embedded to create new organisms or functions.[131]

#### 2.2.2.2 Synthetic Life

One of the main promises of synthetic biology is the creation of artificial life. This entails the development of synthetic organisms, which are bound to have a revolutionary effect not only in scientific terms, but also from an ethical perspective.

Bottom-up approaches have been devised with the aim to build functional genomes using synthesised DNA. The difference in this line of research compared to traditional genetic engineering lays in the manufacturing of cells using genome sequences that are computer-designed.[132] Recent experiments have shown that "*genomes can be designed in the computer, chemically made in the laboratory*

[126]Twilley (2016). Syn3.0 is tailored to grow on a specific diet and in a petri dish, which would make it more complex to use in comparison to other organisms that are easier to maintain (e.g. yeast) (Twilley 2016).

[127]SCHER, SCENIHR, SCCS Scientific Committees (2015a, p. 30), Conference of the parties to the Convention on Biological Diversity (2014a, p. 15) and Ribarits et al. (2014, p. 23).

[128]Panke (2008, p. 6).

[129]Conference of the parties to the Convention on Biological Diversity (2014a, p. 15) and Frow and Calvert (2013, p. 48).

[130]Schwille and Sundmacher (n.d., pp. 3–4).

[131]Calvert (2010, p. 102).

[132]Conference of the parties to the Convention on Biological Diversity (2014a, p. 15).

*and transplanted into a recipient cell to produce a new self-replicating cell controlled only by the synthetic genome*".[133]

The steps undertaken in this field are remarkable. In 2002, researchers first synthesised the poliovirus, producing the first synthetic organism.[134] Soon after, in 2003, Craig Venter announced the creation of a synthetic virus from scratch in just 14 days and in 2007 his team transplanted the genome of one organism into another.[135] This last research showed that a foreign genome could take control of another living cell.[136] At the time, Craig Venter considered this step as pivotal towards the creation of synthetic life. Other scientists were instead more critical and believed that, while his studies might be interesting, they still concerned organisms that were unsuitable for industrial purposes.[137] In 2008, Craig Venter and his team reached another relevant milestone in the road towards creating the first synthetic organism by synthesising from scratch the genome of a bacterium.[138] This last achievement was saluted as a new approach to reach the minimum genome, this time from the bottom-up.[139]

In 2010, Craig Venter and his team announced a new and controversial milestone towards the creation of artificial life. By announcing the creation of the world's first synthetic cell, he and his team declared that they had produced the first self-replicating synthetic genome that was inserted into a cell of a different species. This achievement, called JCVI syn1.0, marked the first time that a synthetically produced genetic material completely replaced the natural existing one in a bacterial cell.[140] This cell was also the first one controlled entirely by human-made genome, which means that all the features and characteristics of the cell were determined by scientists.[141] In describing their work, *Venter* and his team highlighted the main points of their research:

> We report the design, synthesis, and assembly of the... Mycoplasma mycoides JCVI-syn1.0 genome starting from digitized genome sequence information and its transplantation into a M. capricolum recipient cell to create new M. mycoides cells that are controlled only by the synthetic chromosome. The only DNA in the cells is the designed synthetic DNA sequence.[142]

---

[133]Then and Hamberger (2010, pp. 6–7). In spite of this, some commentators noted that: "*There's no such thing as an artificial gene. DNA is DNA. What matters to a gene is sequence, not how you made it*" (Garthwaite 2014).

[134]European Group on Ethics in Science and New Technologies to the European Commission (2009, p. 18).

[135]Balmer and Martin (2008, p. 15) and Health Council of the Netherlands (2008, pp. 34–35).

[136]Wade (2007).

[137]Wade (2007).

[138]Singer (2008).

[139]Panke (2008, pp. 48–50) and Singer (2008).

[140]Palca et al. (2010) and Presidential Commission for the Study of Bioethical Issues (2010, pp. 20–42).

[141]McLennan and Rimmer (2012).

[142]Gibson et al. (2010, p. 52).

In lay terms, their work could be described as follows. Scientists used very accurate sequences of the 1 million-base-pair long genome of a bacterium called *Mycoplasma mycoides* to produce a synthetic version of its genome.[143] This synthetic genome was then inserted into a cell (*Mycoplasma capricolum*) that had previously been emptied of its own genome. After the insertion, this new cell was booted up and showed the ability to grow and reproduce. The new cell behaved just like a *Mycoplasma mycoides* one and it no longer contained sequences that previously existed in the *Mycoplasma capricolum* cell, which meant that the cell was run by its wholly artificial genome.[144] This project costed approximately $40 million.[145]

The main struggles connected to this work related to the complexity of producing an error-free genome, to jump-start life and to boot up a genome in a different cell.[146] Those characteristics led commentators to compare the synthetic DNA to software that is being booted up in a new hardware.[147] On this, *Venter* stated that "*bacterial cells are software-driven… machines. If you change the software, you build a new machine*".[148] To differentiate the artificial genome from the natural one, scientists inserted watermark sequences in it and disabled the part of the genome that rendered it infectious to humans.[149] This was needed because Venter and his team reproduced exactly the genome of an already existing organism and it would otherwise be impossible to distinguish the two. Such similarity was necessary, given that our limited understanding of biological complexity precludes the construction of completely new genomes. According to the scientists, this work:

> Stands in sharp contrast to various other approaches to genome engineering that modify natural genomes by introducing multiple insertions, substitutions, or deletions. This work provides a proof of principle for producing cells based on computer-designed genome sequences. DNA sequencing of a cellular genome allows storage of the genetic instructions for life as a digital file. The synthetic genome described here has only limited modifications from the naturally occurring M. mycoides genome. However, the approach we have developed should be applicable to the synthesis and transplantation of more novel genomes as genome design progresses.[150]

While *Venter* announced this result as the "*first…species… whose parent is a computer*" and believed that this was both a philosophical and technical

[143]For comparison, the genome of fruit flies is approximately 165 million base-pair long, while the human genome contains more than 3 billion base pairs (Presidential Commission for the Study of Bioethical Issues 2010, p. 39).

[144]Gibson et al. (2010, pp. 52–56) and Dillow (2010).

[145]Nista (2015).

[146]Krieger (2013a) and Gibson et al. (2010, p. 55).

[147]Nista (2015) and Rousseaux (2015).

[148]Rice (2010).

[149]Ballantyne (n.d.).

[150]Gibson et al. (2010, p. 55).

landmark,[151] others were more sceptical.[152] *Venter* responded to claims that he did not create life from scratch, by stating that:

> A baker who starts with flour, sugar, and eggs gets credit for his creation, not accusations that he should have begun with atoms and molecules. Besides... once the original cell reproduced, the question should be considered settled... in what we call the synthetic cell there is not a single molecule of the original form.[153]

Aside from the reactions of the scientific community on whether this work was truly ground-breaking, also the general public weighted heavily in the debate, especially on whether synthetic life was actually created. Commentators expressed doubts on how to define life in the age of synthetic biology, especially considering that this field blurs the line between what is natural and what is synthetic.[154] Vitalism-inspired philosophies and "playing God" arguments featured prominently in media headlines.[155]

Venter and his team are not the only ones working on synthetic life projects. At the moment, a new research endeavour promises to bring forward our knowledge of biological systems, while laying the foundation for the future design of genomes for specific purposes. This project, known as Synthetic Yeast Genome Sc2.0, aims to synthesise the entire genome of a yeast (around 6000 genes) with "*a built-in diversity generator that will enable researchers to discover how yeast, as a model organism, deals with genetic change and how genomes might be improved to create more robust organisms*".[156] This project involves several universities worldwide and is attempting to synthesise for the first time the genome of a eukaryotic cell.[157]

This line of research holds a very high potential. Differently from the bacterium used by Venter's team, yeast represents a more complex organism that has often been used as a model for genetic manipulation due to its resistance. Yeast is also well-known and widely employed by men for industrial purposes (e.g. production of beer and bread) and shares several chromosomes with humans.[158]

By 2014, the first artificial yeast chromosome had already been prepared. It was based on the smallest of the 16 yeast chromosomes and presented some differences compared to the natural version. Unnecessary sequences were removed and landing sites to create on-demand mutations were added. Such insertions should be helpful in

---

[151] Wade (2010).

[152] Wade (2010), Palca et al. (2010) and Presidential Commission for the Study of Bioethical Issues (2010, pp. 1–2).

[153] Corbyn (2013).

[154] Commissione federale d'etica per la biotecnologia nel settore non umano (CENU) (2010, p. 5) and Balmer and Martin (2008, p. 4).

[155] Corbyn (2013) and van den Belt (2009a, p. 264).

[156] GenScript (n.d.).

[157] Eukaryotic cells contain membrane-bound organelles, such as the nucleus, and are more complex. Humans belong to this category. Conversely, prokaryotes do not have a nucleus and have a much simpler structure.

[158] Villa (2011) and Helman et al. (2007, pp. 16–17).

determining which mutations can be borne by it and to discover useful variations.[159] Apart from the development of foundational tools, the project is likely to lead to the development of new chemicals and to the streamlining of old production methods.[160] Furthermore, research on the most industrially relevant microbe is likely to have major commercial repercussions. This is confirmed by the fact that some companies have already expressed their interest in this line of research.[161] Nonetheless, the results of this project will be shared under an open access format.[162]

In 2016, researchers considered the possibility to synthesise the human genome from scratch. This project, known as Human Genome Project-write (HGP-write or GP-write), would have enormous potential in the health sector (e.g. growing transplantable human organs, new and improved gene therapies) and could be financed via a large public-private initiative similar to the one used for the HGP.[163] The project, managed and directed by a consortium, is currently ongoing.[164] The promoters of the projects have claimed that the "*intellectual property developed in GP-write will encourage broad access and use through the use of patent pooling and common licensing agreements*".[165]

#### 2.2.2.3 Xenobiology

Xenobiology is the "*design, engineering and production of biological systems with non-canonical biochemistries and/or alternative genetic codes*".[166] It investigates unusual life-forms based on a biochemistry which is not available in nature.[167]

The main goals of this line of research are understanding the origin of life, how and why DNA and organisms are shaped the way they are and whether life could have taken other forms. Researchers see in xenobiology a possibility to create new and industrially interesting compounds and an advanced form of biocontainment.[168] Furthermore, this technology could be used for medical purposes in the fight against

[159]Young Rojahn (2014).

[160]Young Rojahn (2014).

[161]Gateway to Research - Research Councils UK (n.d.).

[162]Gateway to Research - Research Councils UK (n.d.).

[163]Callaway (2016, p. 163) and Boeke et al. (2016, pp. 5–12). The Human Genome Project (HGP) was: "*An international research effort to sequence and map all of the genes - together known as the genome - of members of our species, Homo sapiens. Completed in April 2003, the HGP gave us the ability, for the first time, to read nature's complete genetic blueprint for building a human being*" (National Human Genome Research Institute n.d.). Apparently, the initial name HGP-write was changed to GP-write, as it could spark moral and ethical issues (Shivam 2019).

[164]GP Write (2018).

[165]GP Write (n.d.-a). Equally, the promoters of the project clarified that GP-write does not involve ova or embryos since it is focused on cell culture from many organisms (GP Write n.d.-b).

[166]SCHER, SCENIHR, SCCS Scientific Committees (2015a, p. 33).

[167]Conference of the parties to the Convention on Biological Diversity (2014a, p. 16).

[168]SCHER, SCENIHR, SCCS Scientific Committees (2015a, p. 34).

diseases such as Ebola and HIV.[169] However, commercial applications in this field are still years away.[170] Another advantage of this line of research is connected to the concept of orthogonality, as those unnatural materials can exist next to natural ones (e.g. XNA and DNA) without interacting or interfering with them.[171]

An interesting research area in this field is the expansion of the genetic alphabet. As known, DNA (deoxyribonucleic acid) and RNA (ribonucleic acid) are at the basis of the hereditary and the evolutionary process of all creatures on earth and represent the code of life. DNA is composed of four bases A (adenine), G (guanine), C (cytosine), T (thymine), which pair together as A-T and C-G to form a double helix structure. RNA follows the same structure, except for the fact that it contains a U base (uracil) instead of the T one.[172] Short sections of DNA create genes, which in turn constitute chromosomes. This new research field increases the number of base pairs, by introducing new ones that are not found in nature. For each new base pair, the genetic alphabet increases twofold.[173]

In 2014, two bases called X and Y were created. A single pair of them was inserted in an *E. coli* bacterium where they were able to reproduce, albeit at a slower rate than usual.[174] The reproduction continued for almost a week and at the end of it the bacterium substituted these "alien" bases with natural ones.[175] This represents the first time that unnatural synthetic DNA built in a lab was inserted in a cell, where it was accepted and duplicated instead of being destroyed.[176] By 2019, scientists had developed four new DNA letters.[177]

Commenting on this line of research, scientists stated that a wholly unnatural-DNA-based cell could be possible, while the creation of a wholly synthetic organism may prove very problematic due to the degree of integration of natural DNA into cellular systems and in the environment.[178] Still, the potential of this field is remarkable considering that, as *Romesberg* said:

> If you read a book that was written with four letters, you're not going to be able to tell many interesting stories... If you're given more letters, you can invent new words, you can find new ways to use those words and you can probably tell more interesting stories.[179]

---

[169]Connor (2014).

[170]Conference of the parties to the Convention on Biological Diversity (2014a, p. 17).

[171]On this, see Sect. 2.1.1.5.

[172]RNA is generally considered to be single stranded. However, in some cases, RNA can form double strands and, in some cases, even double helices.

[173]SCHER, SCENIHR, SCCS Scientific Committees (2015a, p. 34).

[174]Pollack (2014). However, strictly speaking, it would be the DNA that contained the bases that reproduced.

[175]Callaway (2014).

[176]Rita et al. (2014). Even before this, in 1989, modified forms of Cs and Gs were created and this expanded bases system, called AEGIS, was commercially licensed to a company that produced short DNA molecules for genetic testing (Callaway 2014; ETC Group 2007, p. 21).

[177]Dengler (2019).

[178]Callaway (2014).

[179]Callaway (2014).

The advantages of this extended code have also been pointed out by those who believe that the earth will soon undergo a sixth wave of extinction and who believe that xenobiology could recreate biodiversity across the globe.[180]

Other research groups focus instead on changing the backbone of DNA. This entails moving away from deoxyribose or ribose acid and use instead another chemical structure (e.g. TNA if threose is used as a backbone substance). Such non-DNA or RNA structures are generally called xeno nucleic acids (XNA).[181] To our knowledge, natural evolution on earth has never generated any XNAs.[182] Interestingly, the use of a different backbone renders XNA invisible to natural biological systems. This effectively impedes the exchange of genetic information between them and the environment. This feature constitutes a very powerful biosafety tool, a so called "genetic firewall".[183] Still, given its novelty, commentators have pointed out that both XNAs and new base pairs would need to be tested to make sure that they do not pose risks for humans or for the environment.[184]

The possibility of creating non-canonical amino acids is also being explored in order to go beyond the 20 found in most species, which in turn would have repercussions on the number and type of proteins available. This field is also likely to be affected by research on the expansion of DNA bases, as the introduction of two additional bases could raise the number of amino acids from 20 to 172.[185]

Industry-wise, a number of companies tapped into this market. For example, Maxygen developed a synthetic DNA sequence that encodes novel proteins, whose metabolising properties are far stronger than those of naturally existing ones.[186] This approach was taken even further by the company Amunix, which devised synthetic proteins not based on naturally existing ones.[187]

### 2.2.3 *Futuristic Research Areas*

Having been pinned as the solution to the problems of mankind, synthetic biology is not only focused on projects that have direct present relevance, but is also engaging in research endeavours that are still theoretical and that could be realised only

[180] Krumins (2015) and Rousseaux (2015). However, the same authors expressed concerns over the idea of a diversity triggered by scientists in a lab and the possible negative effects of such unnatural organisms on native ones.

[181] SCHER, SCENIHR, SCCS Scientific Committees (2015a, p. 34).

[182] Schmidt (2010, p. 328).

[183] SCHER, SCENIHR, SCCS Scientific Committees (2014, p. 15), Deutsche Akademie der Naturforscher Leopoldina (2010, p. 12) and Schmidt (2010, pp. 325–327).

[184] SCHER, SCENIHR, SCCS Scientific Committees (2015a, p. 35).

[185] SCHER, SCENIHR, SCCS Scientific Committees (2015a, p. 35) and Krumins (2015).

[186] Holman (2015, pp. 429–430).

[187] Holman (2015, p. 430).

decades from now. Amongst those various research areas, the two that seem to have caught the attention of the public and scientific community the most are the possibility to bring back to life extinct organisms and the possibility of building completely autonomous and artificial cells using non-living raw materials.

#### 2.2.3.1 De-extinction

Concerns over a possible sixth great extinction wave affecting a vast number of species have sparked suggestions to use synthetic biology to restore biological diversity.[188] Such attempts include not only currently endangered species, but consider also bringing back to life animals that have been extinct for thousands of years. This quest is facilitated by discoveries indicating that fossils might be able to preserve cells much longer than expected.[189] For example, soft tissues were discovered in the bones of a *Tyrannosaurus rex* as well as in the relatively more recent mammoth. While this is no guarantee to find intact DNA in those fossils, it still opens some futuristic scenarios.[190]

In 2005, researchers were able to sequence part of the genome of a mammoth. On a similar note, one year later, a research group announced that it had partially sequenced the genome of a Neanderthal man.[191] While these are interesting developments, they do not guarantee that scientists will be able to organise and place the genome into chromosomes and to recreate the apparatus needed for life. Still, an American research group has tentatively discovered how to place mammoth DNA into the egg of an elephant and was investigating whether the extinct animal could be birthed by an elephant mother.[192]

But it was the possibility of bringing back to life our ancestor, the Neanderthal, that caused the biggest uproar in the public. In an interview in 2013, *George Church* mentioned that technical improvements might enable the creation of a Neanderthal clone. In particular, in an interview, he explained that:

> Church: The first thing you have to do is to sequence the Neanderthal genome, and that has actually been done. The next step would be to chop this genome up into, say, 10,000 chunks and then synthesize these. Finally, you would introduce these chunks into a human stem cell... We developed the semi-automated procedure required to do that in my lab. Finally, we assemble all the chunks in a human stem cell, which would enable you to finally create a Neanderthal clone.
>
> Spiegel: And the surrogates would be human, right? In your book you write that an 'extremely adventurous female human' could serve as the surrogate mother.
>
> Church: Yes.[193]

[188] Takahashi (2016).
[189] Tennant (2015).
[190] Wilson (2008).
[191] Anthony (2013).
[192] Specter (2009).
[193] Bethge and Grolle (2013).

The public outcry that followed this statement led to a retraction from George Church, who clarified that he was not working on cloning Neanderthals, but was merely discussing the possibilities opened by future synthetic biology.[194]

Technology-wise, three methodologies could be employed to reach those results: cloning, genetic engineering and artificial selection. The first would come closer to truly recreating the extinct species, whereas the second would provide a hybrid between the genome of the extinct species and that of one of its closest living relatives. The third option would instead use living species to recreate extinct ones.

The methodology behind the cloning method resembles the one used in the field of synthetic life. A cell is deprived of its DNA and, in its place, the nuclear DNA of another organism is added. This would then be developed into an embryo. The resulting organism would have the nuclear genotype of the donor and the mitochondrial genome of the egg donor. While this method offers the closest realisation of an extinct organism, it still does not offer a perfect copy of it, since the egg would affect the expression of the genes. For this technique to work, it is also necessary to find a close egg donor for the creation and gestation of the organism. Furthermore, the complete and viable nuclear DNA of the extinct species is needed. Therefore, this method is applicable to recently extinct organisms. In 2009, a *Pyrenean ibex* was recreated via this method. This represents the first viable birth of an extinct animal.[195]

By contrast, genetic engineering could make use of a partial gene sequence to integrate and change the genome of a similar species. This process, which has not yet successfully been performed, would generate hybrids that have some of the genome of the extinct species, but are not identical to it. Lastly, artificial selection would use selective breeding techniques to cross living organisms from other species that have some characteristics of the extinct ones. In light of this, the final organism would still belong to the donor species rather than the extinct one.[196]

None of the above techniques would thus lead to the recreation of an exact copy of the extinct organism. In particular, the last two techniques could be considered as either something completely new or as a modified version of a living species.[197]

Lastly, keeping in mind the risks connected to a mass extinction, the Frozen Ark Consortium started collecting DNA and viable cells from endangered species and to preserve them. Although these samples could not be used to re-create extinct animals under current technology, the founders of this project imagine that such technologies could become available in the future, thus enabling the cloning of extinct species.[198]

---

[194] Spiegel Online (2013).

[195] Carlin et al. (2014, pp. 8–9). Still, it has to be noted that the animal survived only a few minutes and that this species had died out in 2000.

[196] Carlin et al. (2014, p. 11).

[197] Carlin et al. (2014, p. 16).

[198] ETC Group (2007, p. 38).

#### 2.2.3.2 Protocells

Research endeavours connected to protocells seek to "*construct new simple forms of living systems, using chemical and physical processes and employing as raw ingredients only materials that were never alive*".[199] In other words, protocells are the precursor to building wholly artificial cells by using non-living materials. This approach, which is strictly bottom-up, aims to reduce biological complexity at the cellular level while creating new biological systems.[200]

Such attempt to reduce biological complexity is inspired by the same idea behind the research on the minimal genome. However, in this case, efforts are concentrated at cellular level, instead of genome level. Equally, in comparison to the work on minimal genomes, this area of research appears more complex, given that protocells cannot take as example or be based upon already existing biological systems. On the contrary, they require "*a detailed master plan, or at least a convincing hypothesis about the minimal set of functional modules to 'jump-start' life – a true intellectual act of synthesis*".[201]

So far, protocells that interact with living cells were devised, but no truly autonomous and artificial cell could yet be developed.[202] Equally, protocells are not yet capable of replication.[203] However, scientists are currently actively working on constructing cell-like growing and dividing systems.[204]

Protocells are expected to act as chemical containers and it is auspicated that in the future they could be employed as chassis in which to insert synthetic DNA to form new living organisms.[205] To reach this goal, protocells must first become functionalised enough to support reproduction, self-maintenance and evolution.[206] Furthermore, since protocells originate from non-living materials, much of the research in this field is connected to chemistry and physics rather than to biology.[207]

Considering that protocells are built via materials that were never alive, they blur the line between living and non-living matter and raise both ethical and safety questions. This explains why commentators have defined protocells as "*artificial*

---

[199] SCHER, SCENIHR, SCCS Scientific Committees (2014, p. 15).

[200] SCHER, SCENIHR, SCCS Scientific Committees (2014, p. 13).

[201] Schwille (2015, p. 688).

[202] SCHER, SCENIHR, SCCS Scientific Committees (2015a, pp. 30–31).

[203] SCHER, SCENIHR, SCCS Scientific Committees (2015a, p. 32). Recently, further research has been undertaken on this topic (Joyce and Szostak 2018).

[204] Powell (2018). Scientists are also devising cell-free systems (CFS) for synthetic biology applications. CFS, which can be thought of as programmable liquids, seem to remove many of the complexities of cell-based systems (Tinafar et al. 2019).

[205] SCHER, SCENIHR, SCCS Scientific Committees (2014, p. 15). In particular, chemistry is relevant since protocells could currently be covered by the regulatory framework pertaining to chemicals (SCHER, SCENIHR, SCCS Scientific Committees 2015a, p. 32).

[206] Conference of the parties to the Convention on Biological Diversity (2014a, p. 102).

[207] Deutsche Akademie der Naturforscher Leopoldina (2010, p. 14).

*cells that have some properties of living systems but are not yet fully alive*".[208] However, for the moment, those concerns have only a theoretical relevance. Indeed, even though protocells could be seen as precursors of truly synthetic cells, this field is still in its infancy. Therefore, it is unlikely to yield relevant commercial applications in the near future.[209]

### 2.2.4 Applications

The hype connected to synthetic biology stems from its promise to profoundly affect an array of disciplines and industries and, ultimately, our lives. Being a platform technology that could cut across a variety of industrial sectors, synthetic biology has been pinned to address climate change and environmental protection, energy concerns, food and land shortage as well as health conditions.

As the field is still in its infancy, most of these expectations have not yet materialised. Still, failures to deliver some highly expected products have increased the pressure on synthetic biologists. Indeed, commentators expressed their concerns on the matter by stating that "*the field has had its hype. Now it needs to deliver*".[210]

#### 2.2.4.1 Health

The development of medicinal treatments for a large number of diseases is one of the strongholds of synthetic biology. By consulting the Synthetic Biology Project database, it emerges that more than 20 products are currently being developed as medicaments. The release of the majority of these medicines is only "*on the horizon*", as only a single product is close to or on the market already.[211] At the same time, other drugs are undergoing trials.[212]

The best-known synthetic biology application in the medical field is the production of a semi-synthetic version of the anti-malaria drug Artemisinin. After resistance against previous treatments spread, a new cure containing Artemisinin in combination with other drugs was developed. However, this treatment has also its drawbacks, given that *Artemisia annua* is not easy to cultivate and that the price and availability of Artemisinin are subject to heavy fluctuations.[213] For this reason, a research group

[208]Conference of the parties to the Convention on Biological Diversity (2014a, p. 16).

[209]SCHER, SCENIHR, SCCS Scientific Committees (2014, p. 15).

[210]European Commission's Directorate-General for Health & Consumers (2010, p. 8).

[211]The information is available at the website www.synbioproject.org in the section on applications of synthetic biology.

[212]Kelley (2014, p. 40).

[213]The plants are mostly cultivated by small farmers in Asia and in Africa (ETC Group 2014a, pp. 1–4; European Commission's Directorate-General for Health & Consumers 2010, p. 10).

supported by the Gates Foundation engineered a bacterium to produce artemisinic acid (the precursor of Artemisinin) microbially and in a streamlined manner.[214] The product was then manufactured on a larger scale in collaboration with a pharmaceutical company and is now available on the market.[215]

While Artemisinin is the best-known example of the use of synthetic biology in the health sector, considerable research is also being carried out in other directions. Xenobiology is being used to develop new diagnostic tests and researchers are working on SynBio-derived cancer treatments. Smart drugs and vaccines are also being investigated, as they could deliver targeted treatments and produce a specific immune response when needed.[216] Equally, phage are being engineered to provide an alternative to antibiotics and to treat certain bugs, which is especially relevant considering the growing resistance to antibiotics.[217] Furthermore, faster and more accurate production of vaccines is also being studied.[218]

Progress has also been made in the field of microbial factories that could manufacture molecules needed for pharmaceutical purposes. If used as little production machines, they could lead to the quicker manufacturing of the required product (e.g. painkillers) in a contained environment that is not vulnerable to external conditions, which would reduce the risk of environmental contamination.[219] The situation would be different in case the synthetic biology organism would be the drug itself, which is engineered to survive for some time in the body of the patient, as this would increase the uncertainties connected to exposure and release.[220] Lastly, researchers have cultured an animal body part (in this case, a rat foreleg) and generated a bio-limb. This could open great possibilities for transplants, as cells could be grafted from the body of the patient.[221]

In conclusion, the role of synthetic biology in the health sector could be summarised as follows:

> While the benefits of synthetic biology to health care may prove monumental, significant hurdles remain. With the exception of semi-synthetic Artemisinin and potential, near-term improvements in vaccine design, most of the anticipated health benefits of synthetic biology remain in the preliminary research stage. We are unlikely to see commercial applications

---

[214]Health Council of the Netherlands (2008, p. 34).

[215]Simone (2014). Even though the availability of another source of Artemisinin has been saluted as a great improvement by many, critics have highlighted the negative impact that semi-synthetic Artemisinin could have on the livelihood of the farmers that produce the natural version. They also dissented with the idea that the synthetic version would provide a cheaper option (ETC Group 2014a, p. 1; Marris 2013). Those statements need to be read in conjunction with reports arguing that the promises made by synthetic biology on this malaria treatment have backfired (Marris 2013).

[216]Maynard (2016) and Simone (2014). Scientists are also using synthetic biology to tackle Coronavirus (Cumbers 2020).

[217]The Economist (2015).

[218]Conference of the parties to the Convention on Biological Diversity (2014a, p. 44).

[219]Krieger (2015b) and Henry (2014).

[220]Paradise and Fitzpatrick (2012, p. 61).

[221]Hanel (2015).

from much of the biomedically oriented synthetic biology research for many years, although the pace of discovery is unpredictable.[222]

#### 2.2.4.2 Food and Agriculture

The population of the world is set to reach 9.5 billion by 2050. 40% of the world's ice-free land is used for agriculture, which also accounts for about 70% of the water consumption on the globe. The use of fertilisers is responsible for the pollution of water wells.[223] Additionally, between 1964 and 2007, the global production of cereals shrank by 10% due to droughts and extreme weather conditions.[224]

To feed a growing population notwithstanding increasingly unstable weather conditions and limited land availability, researches have been considering synthetic biology as a solution to the problems posed by food and agriculture. Reports have identified five areas in which synthetic biology is likely to make an impact on the food industry. Those are health products and processing aids, preservatives, flavours, biosensors (e.g. artificial noses), and food waste processing.[225]

Meat has been in the focus of synthetic biology companies. The environmental effects of meat production as well as the growing request for meat-alternatives have sparked a great interest in the production of meat via synthetic biology techniques. This interest has then led to the founding of around 1000 cultured meat companies around the globe, despite the problematic issue of public acceptance.[226] Milk has also been eyed as a product whose manufacturing chain could be streamlined by using synthetic biology.[227]

Another area that is evolving under the influx of synthetic biology is food flavouring. Recently, many synthetic biology companies shifted their focus from fuels to fine chemicals for food and fragrances, since those products take less time and money to develop and command higher prices in the market ($1 per kg of biofuel compared to the $10–10,000 per kg of fine chemicals).[228] On the other hand, these products may face consumer rejection, especially if the label "natural" usually attached to them is perceived as deceitful by the public.[229]

---

[222]Presidential Commission for the Study of Bioethical Issues (2010, p. 67).

[223]Garthwaite (2014).

[224]Ulivieri (2016).

[225]British Food & Environment Research Agency (2014, p. 45) and OECD and The Royal Society (2010, p. 19).

[226]Burningham (2016). Reports showed that only one fifth of Americans would try lab-grown meat and these types of products have already been dubbed by the press as "Frankenmeat" or "test tube burgers" (Kerr 2016; Harman 2014).

[227]A team of DIY biologists is trying to replicate cow milk via yeasts. Such milk, which would be indistinguishable from the bovine molecule, could be used as milk or to produce vegan cheese (Kerr 2016). The regulatory aspects of such lab-grown meat are also challenging (Mayhall 2019).

[228]Check Hayden (2014, p. 598).

[229]Check Hayden (2014, p. 598).

The first synthetic biology flavour to hit the market was vanillin. In this case, synthetic biology was used to engineer an organism to make it produce vanillin. This means that the vanillin itself was not a synthetic biology product, but rather a product of synthetic biology. For this reason, it would not be subjected to the GMO labelling and could be considered a natural flavour.[230] This situation generated ample controversy.[231] Other SynBio products that are being developed are saffron (the most expensive spice in the world), cocoa butter and synthetic biology alternatives to palm oil.[232]

Agriculture is another field that is likely to be impacted by synthetic biology. Supported by heavy government funding, synthetic biology should be able to enhance crop yield and resistance to external factors (e.g. drought, pest), while reducing the negative environmental impact of agriculture.[233] Similarly, a new generation of targeted pesticides that do not linger in the environment after their use is being developed and so are crops that do not need nitrogen fertilisers altogether.[234] Even though it is unclear whether any synthetic biology products for agriculture have already been commercialised, some suggest that their implementation should be carefully considered, as it could have a negative impact on biodiversity or it could lead to unintended consequences due to the release of engineered organisms into the wild.[235] Lastly, ornamental plants that glow in the dark or grass that is greener and requires less mowing and herbicides have also been manufactured.[236]

#### 2.2.4.3 Energy and Fuel

The production of biofuels has been at the centre of the synthetic biology discourse since the very beginning. The development and manufacturing of biofuels embodied the great hopes and hype of synthetic biology and represented some of its greatest

[230]Watson (2015a), van der Hoeven (2015), McEachran (2015) and Check Hayden (2014, p. 598).

[231]The environmental organisation Friends of the Earth was critical of this product and contacted companies, asking them whether they were planning to use this ingredient (Colwell 2015; Barclay 2014). The producing company countered these claims by adding that they: "*Synthetize genes as a part of that process, but we're not 'printing fake DNA'. All genes are just sequences of data and these sequences change all the time in nature anyway*" (Watson 2015b).

[232]Johnson (2014). In both cases, critics of the projects fear that these products might negatively affect the livelihood of the local populations who manufacture the natural version of these two materials and add further pressure on the manufacturing of sugar feedstock, which is necessary for synthetic biology's fermentation processes (ETC Group 2014b, p. 1; ETC Group 2014c, p. 1).

[233]Kelley (2014, pp. 43–44).

[234]ETC Group and Heinrich Böll Stiftung (2015, p. 11) and Conference of the parties to the Convention on Biological Diversity (2014a, p. 26).

[235]Conference of the parties to the Convention on Biological Diversity (2014a, p. 26).

[236]Synthetic Biology Project (n.d.-a, p. 2).

disappointments.[237] For years, generous funding and investors have been attracted by the possibility of substituting petroleum-based oils with new generations of fuels based on renewable and greener energy sources. Biofuels could be extracted from biomass. Several types of biofuels are currently available. However, current biofuels have considerable drawbacks, such as high production costs and inefficiencies.[238]

Synthetic biology alternatives are thus being devised to overcome those hurdles. For example, the production of butanol via the fermentation of sugars and starches or via cellulose is being assessed, especially considering that this biofuel could be inserted directly into engines powered via traditional gasoline. In this case, synthetic biology is used to enhance the natural butanol production capacity of some enzymes.[239]

Similarly, much hope resides in the use of algae as biofuels. Algae would be an ideal source of fuel as they are low-input, biodegradable, can mitigate greenhouse emissions and can grow both on land and water, even in areas that could not otherwise be subjected to agriculture. Furthermore, algae have a very high yield and are believed to be compatible with already existing technologies.[240] Still, despite heavy investments and intense research in the field, algae biofuels will probably be amongst the last to reach the market due to high technical hurdles.[241] Similar developments are also envisioned for biohydrogen fuels.[242]

By examining the funding and companies behind these projects, it emerges that governmental agencies, major corporations (e.g. ExxonMobil, DuPont) as well as companies specialised synthetic biology (e.g. Amyris, Solazyme) have been involved in the quest to develop biofuels.[243] However, these investments have not paid back mostly because of the competition with other fuels and the inability to reach the anticipated targets for industrial-scale production.[244] The struggle to render biofuels cost-effective (especially if oil prices are low) and to produce at a large scale have hindered the development of many SynBio companies in this sector.[245]

Yet, a number of products reached the market already. The US navy bought biofuels in 2010, while 300 buses in Brazil are fuelled via biodiesel produced through synthetic biology techniques via yeast.[246] Equally, another start-up engineered microorganisms to produce ethanol. To achieve this goal, the genome of these organisms was stripped to a minimum in order to increase their productivity

[237]Kelley (2014, pp. 44–46).

[238]Presidential Commission for the Study of Bioethical Issues (2010, pp. 56–57).

[239]Presidential Commission for the Study of Bioethical Issues (2010, pp. 59–60).

[240]Presidential Commission for the Study of Bioethical Issues (2010, p. 60).

[241]OECD (2014, p. 36).

[242]Presidential Commission for the Study of Bioethical Issues (2010, pp. 61–62).

[243]European Group on Ethics in Science and New Technologies to the European Commission (2009, p. 20) and Balmer and Martin (2008, pp. 11–12).

[244]Kelley (2014, pp. 44–46).

[245]LaMonica (2014) and Savage (2007).

[246]Conference of the parties to the Convention on Biological Diversity (2014a, p. 18).

and make sure that their metabolism was focused on ethanol production rather than growth.[247]

#### 2.2.4.4 Environment

Synthetic biology is displaying its potential in environmental applications. In particular, synthetic biology is likely to make a contribution in the detection of contaminants and pathogens as well as in onsite degradation.

For example, a team of scientists has developed a biosensor that detects arsenic in water by engineering bacteria to produce acid, which can be revealed via a simple pH test.[248] Equally, researchers devised an organism to degrade pesticides.[249] In addition, resources are being devoted to research on bioremediation, which could be used to restore an environment by using organisms capable of degrading poisonous substances. So far, the use of bioremediation has been limited, possibly due to its perception as a less reliable method that requires extensive site assessment.[250] Nevertheless, this technology is particularly promising. Lastly, bacteria engineered via synthetic biology have also been considered in the race to stop desertification, since they could boost the growth of plant roots.[251]

#### 2.2.4.5 Data Processing and Storing

DNA is the code of life and via its four bases—A, C, G and T—encodes the information necessary for organisms to function. Believing that the features that render DNA such a great code could be used for engineering purposes, researchers have programmed cells and DNA to be used for data storage and computational purposes.[252] The idea of using DNA for data storing was sparked by the ever-increasing amount of data that is being generated and that needs to be stored. It is estimated that by 2020 about 35 Zettabytes of data—1 Zettabyte equals $10^{21}$ bytes—will have been produced. Although other storage devices are available, their size,

---

[247]Bullis (2012).

[248]European Group on Ethics in Science and New Technologies to the European Commission (2009, p. 22).

[249]European Group on Ethics in Science and New Technologies to the European Commission (2009, p. 22).

[250]European Group on Ethics in Science and New Technologies to the European Commission (2009, p. 22). Even though these applications seem promising, concerns about their use in the environment exist. The release into nature of organisms that could be difficult to control, could propagate beyond intention, and could affect ecosystems and species in unknown ways have been identified as downsides (Presidential Commission for the Study of Bioethical Issues 2010, pp. 70–71).

[251]Stinson (2015).

[252]Scudellari (2015, p. 15771).

maintenance, cost and risk of decay have led researchers to investigate other options.[253]

Biostorage represents an ideal candidate due to the characteristics of DNA. DNA is a high-density storage medium that is particularly stable over time and can be read and reproduced without becoming obsolete. Furthermore, it represents a static offline system, which protects it from remote internet access. These features render DNA storing ideal for maintaining information that needs to be available long-term, does not need to be accessed frequently and where security is pivotal. Indeed, DNA could be seen as an encryption system that is invisible to the naked eye.[254] This storage method uses the four DNA bases as a code, similar to the binary code used in software. This allows the storing of any sort of information and data.[255] The concentration of DNA storage devices is impressive. A human body could store up to 150 Zettabytes, 70 million copies of a book could be fitted into one drop of liquid and 7 litres could contain all the data in the world.[256] The data are generally inserted in the non-coding areas of a gene and therefore they do not affect the workings of the biological system or its functions.[257]

Given the challenges related to it, biostorage is not likely to replace other types of storage devices any time soon.[258] Nonetheless, the technology is very promising, especially if the cost and speed of reading and writing DNA would improve and if a DNA language were developed.[259]

Biological computers represent another field that is quickly evolving and gaining relevance. Researchers have developed transistors made of DNA and RNA that operate from inside a living cell. They have the ability to reply in a true or false manner to almost any biological question that could emerge within a cell. They could detect the presence of a substance and communicate their findings by changing colour or smell. Similarly, such cell computers could identify how many times a cell divides. This would be important for cancer treatments, as cells that replicate at an abnormal rate could indicate the presence of a tumour.[260] According to *Endy*, these cellular computers will "*work in places where we don't have computing now*" and "*in places where silicon would never work*".[261]

---

[253]Limbachiya and Gupta (2015, pp. 1–2).

[254]Zakeri and Lu (2015, p. 10).

[255]Limbachiya and Gupta (2015, pp. 2–3).

[256]Limbachiya and Gupta (2015, p. 3) and Brown (2015).

[257]There are also storing systems where the information is expressed in DNA outside and independently of any biological system.

[258]Limbachiya and Gupta (2015, p. 14).

[259]Zakeri and Lu (2015, p. 10).

[260]Krieger (2013b) and Myers (2013).

[261]Krieger (2013b).

Notwithstanding the leaps made by researchers in this field, no highly functional biological computers are expected in the near future. Still, simpler biosensors that detect and record changes in a cell will be available in the short term.[262]

#### 2.2.4.6 Others

Synthetic biology could be used for a variety of applications that go beyond the ones listed above. In particular, synthetic biology has been employed in the cosmetic and fragrance sector, for fabrics and textiles, for the production of rubber, for mining and is even being considered for space missions.

Cosmetics, creams and detergents were amongst the first SynBio products to be placed on the market. This included the aroma of oranges and grapefruits used for perfumes and cosmetics, as well as the moisturising agent squalene which was obtained via engineered algae.[263] Laundry detergents have also been listed amongst the goods that contain synthetic biology products. For example, a company manufactured a detergent containing oil obtained from engineered algae. The use of this oil, which was a replacement for palm oil, sparked controversy not only because it was labelled as natural, but also because this was the first time that a company publicly confirmed the use of synthetic biology for the manufacturing of a specific product.[264]

Synthetic biology is also being used to manufacture textiles. Projects for the development of spider silk, which would be a very strong and elastic fabric, have been initiated and carpets made via bacteria have already entered the market.[265] Synthetic biology companies are also heavily investing in the production of rubber for tyres.[266]

The extraction of copper from mines is also being addressed by synthetic biology. Traditional technologies require considerable amounts of energy and chemicals and cannot reach all copper in a mine. To counter these problems, synthetic biologists are designing microorganisms that boost the solubility and extraction of this metal and thus allow extractions that were either not possible before or not commercially feasible.[267]

Futuristic and speculative uses of synthetic biology for space exploration have also been imagined. Scientists noted that during long space missions it would be

---

[262] Anthony (2013) and Sankin (2013).

[263] Check Hayden (2014, p. 598), Strom (2014), Janicki (2014) and Synthetic Biology Project (n.d.-b, pp. 1–2).

[264] De Nieuwe Band (2015) and Strom (2014). The oil itself can be considered natural, as it was not genetically engineered. It is instead the organism that produced the oil that was engineered (Strom 2014).

[265] Leproust (2015), OECD Development (2011, p. 80) and European Group on Ethics in Science and New Technologies to the European Commission (2009, p. 21).

[266] ETC Group (2014d, p. 1) and OECD Development (2011, pp. 76–77).

[267] Synthetic Biology Project (n.d.-c, p. 1).

ideal to have renewable sources of fuel, food, and medicine, which could save costs as well as improve the well-being of astronauts.[268]

### 2.2.5 Technical Background: The Chemical and Informational Content of DNA

Human genes are sections of DNA contained in chromosomes in the nucleus of a cell.[269] Genes contain the instructions to create the amino acids that are used by the body to produce proteins.[270] Each gene can produce more than one protein and have multiple functions.[271] Interestingly, a gene does not present differences based on its natural or synthetic origin.[272] Structurally, the gene is composed of two sections: introns and exons. Exons correspond to the sections that code for amino acids and are separated by introns, which instead do not perform any known coding activity. Approximately 98% of DNA does not encode proteins and was thus referred to as "junk DNA". However, scientists noted that parts of it are still transcribed into RNA; the reason behind this structure is still unknown. Similarly, RNA performs a number of functions that are still unclear and that are unrelated to the coding of proteins.[273]

DNA found in nature is called genomic DNA or gDNA.[274] gDNA contains both exons and introns. cDNA is complementary DNA that can be manufactured in the lab. It contains only the relevant exons and no introns. cDNA is particularly useful in the production of proteins (via the insertion into bacteria) or for preparing diagnostic probes and it can perform its functions even in the absence of the introns found in gDNA.[275] Generally, cDNA does not occur in nature.[276]

---

[268]Yarris (2014).

[269]There exists no unique definition of "gene", as its meaning changed over time and varies according to the discipline involved (Lesser 2011, pp. 328–329; Calvert and Joly 2011, p. 166). As noted by *Burk*, the notion of gene is a product of human classification (Burk 2013a, p. 95). Similarly, it was held that scientific facts are essentially judgements about what should be considered relevant in a certain situation (Burk 2013a, p. 96).

[270]More specifically, they contain instructions on the order in which amino acids should be linked to produce a functional protein. The creation (or synthesis) of amino acids is regulated on various other levels.

[271]Calvert and Joly (2011, p. 166) and Dutfield and Suthersanen (2008, p. 304).

[272]Falcone (2012, p. 5).

[273]Dutfield and Suthersanen (2008, pp. 303–304).

[274]Schertenleib (2003, p. 125).

[275]Authors held that the issue of function should be addressed with the requirement of industrial application rather than in connection with the notion of invention (Bostyn 2004, p. 42).

[276]"Brief for the United States as amicus curiae in support of neither party in Association for Molecular Pathology, et al. v. Myriad Genetics, Inc., et al. before the U.S. Supreme Court (Myriad)" (2013, p. 9).

Related to cDNA is messenger RNA (mRNA), which naturally conveys genetic information from the DNA to the ribosomes. There, the mRNA sequence is translated into the amino acid sequence needed to produce functional proteins. mRNA does not contain all the sequences that are present in gDNA and, therefore, the two are not identical.[277] cDNA is reverse-transcribed from mRNA via a number of man-made techniques.[278] Therefore, since cDNA derives from mRNA, the former will not be identical to the DNA found in the human genome. This latter point is the reason why cDNA has been considered as non-naturally occurring by some commentators.[279] Those who oppose this conclusion argue that the "*genetic structure of the human body remains part of it even after the reconstruction as a cDNA sequence*" and that cDNA maintains the same coding information found in gDNA.[280]

The status of cloned genes has also been extensively debated. The process employed to clone genes is complex and the final result is not identical to the naturally occurring gene. Indeed, such gene has been copied several times. Hence, it has been argued that cloned genes could be considered as translations or versions of a text rather than a photocopy of it.[281] Cloned genes can be of genomic DNA (i.e. gDNA) or of complementary DNA (i.e. cDNA).

DNA is isolated from its surroundings by cleaving the covalent bonds that are found at each end of a DNA sequence. Despite this cleaving, the resulting DNA fragment maintains its genetic information. The severing of the covalent bonds is performed via a routine and conventional process that creates a molecule that does not normally exist within the cell. Commentators held that isolated DNA is not markedly different from gDNA, considering that they have an identical sequence and genetic information.[282] In particular, "*isolating or removing the gene from the body does not chemically alter the gene... the chemical information, compared to DNA in its natural state, is identical*".[283] Additionally, experiments have shown that if isolated DNA is reinserted into the cell, it continues to function as it did before.[284]

One of the most controversial points regarding the patentability of DNA concerns the nature itself of this molecule. Specifically, whether DNA should be considered a chemical substance (the notation ATCG of DNA is actually a representation of its

[277]DNA sequences undergo significant alterations in their transcription into mRNA, to the point that this creates a: "*Substance that is new and different*" ("Brief for the United States as amicus curiae in support of neither party in Association for Molecular Pathology, et al. v. Myriad Genetics, Inc., et al. before the U.S. Supreme Court (Myriad)" 2013, p. 18).

[278]Schertenleib (2003, p. 125).

[279]Taliadoros and Muratore (2000, p. 125).

[280]Cornish et al. (2019, p. 909); "Brief for the United States as amicus curiae in support of neither party in Association for Molecular Pathology, et al. v. Myriad Genetics, Inc., et al. before the U.S. Supreme Court (Myriad)" (2013, p. 19).

[281]Resnik (2002, pp. 150–151).

[282]Farias-Eisner (2014, p. 15).

[283]Farias-Eisner (2014, p. 15).

[284]Farias-Eisner (2014, p. 16).

chemical structure[285]) or a carrier of information.[286] The approach adopted on this point is particularly significant, as it leads to different attitudes on ownership and patentability.[287] Equally, a patentee might highlight one of the two aspects in order to facilitate patent eligibility in that specific case.[288] No clarifications are generally offered as to why a subject matter was characterised in a specific way.[289]

Genes have generally been construed as chemicals in order to fit them into the pre-existing categories of patentable subject matter. This approach was first adopted in the USA and later reached other jurisdictions. In Europe, genes and chemical substances have been ranked side by side, despite the fact that genes are more complex and dynamic than chemicals and that there is only a limited number of genes possible.[290] While some authors stated that there was no clear evidence that this approach was not valid, other scholars considered this perspective as "*fundamentally flawed or at least as partial and one-sided*".[291] Specifically, it was held that this constitutes an ontological and legal reduction, which is no longer appropriate given the current attitude of the scientific community.[292] For example, Nobel laureate *John Sulston* argued that "*the essence of a gene is the information – the sequence*".[293] Equally, *Rai* criticised the qualification of DNA as a mere chemical molecule as "*fundamentally misconceived*" and stated that "*although DNA is, obviously enough, a chemical compound, it is more fundamentally a carrier of information*".[294]

From an historical perspective, molecular biologists have often thought about DNA in informational terms.[295] Examples of this can be found in the language

---

[285]Burk (2013b, p. 748).

[286]Viewing DNA as information will have an impact of the scope of protection accorded to it (Lesser 2011, p. 362).

[287]Committee on Science, Technology, and Law of the National Academies of Science, Engineering and Medicine (2013, p. 3).

[288]The problem with this is that, once a patent on the product has been issued, the inventor is able to claim also the other use (e.g. genetic, even though he previously emphasised the chemical traits of the product) (Sherman 2015, p. 1220).

[289]Sherman (2015, p. 1225).

[290]Jacobs and Van Overwalle (2001, p. 505).

[291]van den Belt (2009b, p. 1326).

[292]A similar point was made concerning the reduction of living organisms to mere compositions of matter (Calvert and Joly 2011, p. 167).

[293]Sulston and Ferry (2002, p. 269).

[294]Rai (1999, pp. 835–836).

[295]The heated debate on whether DNA could be seen as a chemical or as an information carrier started only in recent times. *Eisenberg* noted that, at the beginning, DNA patenting was not controversial. This is possibly attributable to the fact that patenting genes was considered like patenting drugs, given that many claims were directed to tangible materials used in the pharmaceutical sector ("Molecules vs. information: Should patents protect both?" 2002, pp. 191–192). The situation changed when the informational value contained in the sequences gained more relevance that the molecule itself. Hence, she argued that: "*That shift in perceptions of where the value lies is leading to new intellectual property strategies for appropriating that value. The patent system has*

employed in this field, which refers to code, translation, transcription, editing, expression and messenger.[296]

In spite of the above, the majority of businesses and patent practitioners share the idea that DNA is a chemical.[297] Equally, the informational view of DNA is not universally accepted in biology.[298] Authors who criticised the centrality of the concept of information in biotechnologies did so by stating that:

---

*not come close to digesting this shift. I think forward-thinking patent lawyers are digesting this shift, but the patent office has not really grappled with it yet*" ("Molecules vs. information: Should patents protect both?" 2002, p. 197). A number of strategies to protect the chemical and informational value of genes have been considered. It has been argued that patent protection might be appropriate for the chemical molecule, but inadequate for its information value. Indeed, once the patent has been issued, the information regarding the DNA will become freely available and it would not be possible to prevent others from using this information. Since this approach seems problematic, other strategies suggested restricting access to databases containing such DNA information. Others advised instead to claim sequences that are stored in a computer-readable medium ("Molecules vs. information: Should patents protect both?" 2002, pp. 198–199). However, it is debatable whether this latter solution would be appropriate from a patentability perspective ("Molecules vs. information: Should patents protect both?" 2002, pp. 199–200). An extreme consequence of this would be that patents would be infringed merely by storing, retrieving and analysing information rather than by the production and sale of the molecules (Eisenberg 2002, p. 6). Lastly, it would raise questions concerning which IP rights are appropriate for the protection of information. On this, *Eisenberg* affirmed that patents might be: "*A very dangerous form of intellectual property rights for information because there are so few safety valves built into the patent system that constrain the rights of patent holders*" ("Molecules vs. information: Should patents protect both?" 2002, p. 201). Another approach suggests that the DNA molecule itself could be seen as a tangible medium for the storage of the biological information. This would lead to question why information is patentable when stored in the gene, which is readable by living cells, and not when it is stored in a computer, which is readable by humans. *Eisenberg* replied to this by pointing out that the information stored in a computer is read by man and, by tying it up, it would undermine patent disclosure and, consequently, the patent bargain (Eisenberg 2002, pp. 8–9). Similarly, other scholars have argued that the genetic code itself could be considered a law of nature, with obvious repercussions for its patentability (Kane 2004, p. 751). The characterisation of DNA as an information carrier has an impact on the fundamental patent bargain set by the law. Time-limited exclusivity is conferred to the patentee by the law, but, in exchange for that, the inventor must disclose his invention so that others can build upon it. However, if patents are granted on DNA and DNA is classified as information, this system might break down (Andrews 2014, p. 555). In fact, in this case, the patentee could prevent others from using the information disclosed in the patent, thus blocking their possibility to build upon it. The pre-emptive effect of these patents can thus be substantial, given that further innovation could be dis-incentivised and that genes cannot be invented around (Andrews 2014, p. 555). It has been argued that this conclusion does not apply to cDNA ("Brief for the United States as amicus curiae in support of neither party in Association for Molecular Pathology, et al. v. Myriad Genetics, Inc., et al. before the U.S. Supreme Court (Myriad)" 2013, p. 10).

[296]For example, *Sulston* wrote that: "*It makes more sense to think of genes as software rather than chemical entities. The information can just as well be held in a computer or written in a book; composition of matter is irrelevant, because the conversion of the information from one form to another is unsurprising*" (Sulston 2006, p. 412).

[297]Dutfield and Suthersanen (2008, p. 303).

[298]Griffiths (2001) and Sarkar (1996).

> The whole idea of the 'informatisation' of life or the 'digitising' of biology would be no more than a rhetorical construction or, insofar as it is actually believed by many involved, a collective delusion. Sure, bioinformatics plays such a prominent role in modern biological research, because there is a huge amount of data to be analysed. But these vast amounts of information about genes should not... be assimilated with information encoded in genes.[299]

The introduction of novel technologies, such as microarrays and gene chips, poses new challenges. In particular, it was noted that the difference between the computer-readable and molecular version of a DNA sequence is becoming increasingly complex to determine. Indeed, those new technologies enable the use of computers:

> To perceive information stored in DNA molecules... when contemporary technology blurs the distinction between computer-readable and molecular forms of DNA, what logic is there to drawing this distinction in determining the patent rights of DNA sequencers[300]?

## 2.3 Market and Players

### *2.3.1 Market*

No univocal and precise estimates are available to determine the size of the synthetic biology market. The discrepancies detected in several studies could derive from difficulties to determine exactly which products belong to synthetic biology. This is aggravated by ongoing doubts about the definition of synthetic biology and by the reluctance of companies to explicitly mention if they manufacture goods using this technology.

In 2012, the global market for synthetic biology was valued at approximately $1.8 billion, with the largest market share by sector being healthcare.[301] Other studies placed instead the value of the market for 2013 at $3 billion and pinned biofuels to be the most promising field for investment given their growth potential.[302] Indeed, studies showed that, while in 2016 the healthcare segment is bound to retain its predominance, the energy sector is the one displaying the highest compound annual growth rate.[303]

By 2020, the size of the synthetic biology market should have reached $38.7 billion.[304] Geographically, Europe is considered to be the largest market, while the Asia-Pacific region is forecasted to be the fastest growing one.[305] Other prognoses

---

[299] van den Belt (2009b, pp. 1310–1311).

[300] Eisenberg (2002, p. 10).

[301] Industry Today (2015).

[302] Allied Market Research (2014).

[303] SCHER, SCENIHR, SCCS Scientific Committees (2015a, p. 12).

[304] Rizzuto (2014).

[305] Global Industry Analysts, Inc. (2015).

predict that the global synthetic biology market will reach the $100 billion mark in 2025.[306]

## 2.3.2 *Players*

### 2.3.2.1 Governments

Since the inception of synthetic biology in the early 2000s, governments have invested heavily in this field to support research on fuels, medicaments and, overall, sustainable development.[307] So far, States have been the main contributors to the funding of synthetic biology.[308]

European institutions were slower than their US counterparts in embracing this discipline, but several steps have been taken in the past years to invert this trend and to take advantage of the pool of expertise available in the old continent.[309] Yet, the amounts invested in Europe are considerably smaller than in the US (10s of millions instead of 100s of millions).[310] Over time, the European Commission has supported synthetic biology through its framework programmes (FP) for research and technological development, albeit with slightly different approaches.

The first FP to address synthetic biology was FP6, which spanned from 2002 to 2006. During this time, it funded NEST (New and Emerging Science and Technology) and its projects dedicated to the application of synthetic biology. Everything started in 2003, when synthetic biology was identified as a promising research area,

[306]ERASynBio (2014, p. 8).

[307]For an extensive overview per country and region of the projects, funding and authorities involved in synthetic biology, Kelley (2014, pp. 50–90) and OECD (2014, pp. 133–154).

[308]Calvert and Martin (2009, pp. 201–202).

[309]European Commission, Directorate-General for Research (2005, p. 5). Recently, China has also been investing in synthetic biology and the country is currently setting up a professional committee for this subject (Xinhua 2018).

[310]In the US, no single coordinated funding plan for synthetic biology exists. Numerous agencies are involved. Overall, while no precise estimates are available, it is believed that these sources contributed between $500 million and $1 billion to research on synthetic biology (Kelley 2014, p. 50). The US and Europe have collaborated on several projects (ERASynBio 2014, p. 11) and instituted a joint task force for biotechnology, which has a synthetic biology working group (British Food & Environment Research Agency 2014, p. 35). The legislative branch was also active in this field. In 2015, an Engineering Biology Research and Development Act was presented in order to ensure that America maintains its leadership in this field. The act does not refer to synthetic biology, but rather to engineering biology, and aims to advance this field, increase the number of researchers, accelerate the commercialisation of products, support social sciences related to it and improve coordination between agencies (HR 591) (Johnson 2015). In 2019, a similar act was passed by Congress. Its goal is to promote US national security, sustainability and productivity. It also considers the acceleration of the commercialisation of this type of R&D (Boyd 2019; Johnson 2019).

although no synthetic biology community was yet present in Europe at the time.[311] Soon after, in 2005, a NEST High-Level Expert Group reported on this innovative research area and outlined its challenges, its possible lines of development for the next 10–15 years as well as the actions that could be taken to promote its growth in Europe.[312] As a result, several projects related to synthetic biology were founded via FP6. Amongst those, there were scientific research endeavours on energy, biological computers, and medical applications.[313] Interestingly, these projects were not exclusively dedicated to scientific research, but included also project on policy, strategy and safety.

FP7 began in 2007 and it included a number of specific initiatives targeting synthetic biology. SYBHEL was another project funded via FP7, which was focused on assessing the ethical, legal and policy issues deriving from synthetic biology.[314]

The 8th FP—called Horizon 2020—provided funding for biotechnology and was based on the idea that smart and sustainable growth are pillars for the development of Europe.[315] Specific projects related to synthetic biology have been funded via this mechanism.[316]

European countries are also active individually in the support of synthetic biology, albeit with different approaches. The UK and Switzerland have set up programmes dedicated to synthetic biology, while in France and in Germany synthetic biology projects are funded via general biotechnology programmes.[317]

#### 2.3.2.2 Universities and Research Institutions

The first synthetic biology department was inaugurated in 2003 in Berkeley.[318] Nowadays, more than 40 countries are involved in this area, with almost 700 organisations active in this field, the majority of which are universities and research institutions.[319]

---

[311] Parliamentary Office of Science and Technology (POST) (2008, p. 2).

[312] European Commission, Directorate-General for Research (2005).

[313] European Group on Ethics in Science and New Technologies to the European Commission (2009, pp. 24–26).

[314] SYBHEL (2014, p. 1).

[315] Deutsche Akademie der Naturforscher Leopoldina (2010, p. 28).

[316] Kelley (2014, p. 68). The European Science Foundation (ESF) has also funded projects related to synthetic biology (e.g. EuroSYNBIO) to foster multidisciplinary research and to advance fundamental knowledge (Kelley 2014, pp. 71–72; Deutsche Akademie der Technikwissenschaften et al. 2009, p. 25).

[317] OECD (2014, p. 82).

[318] European Academies Science Advisory Council (EASAC) and Swiss Academies of Arts and Sciences (2011, p. 3).

[319] Conference of the parties to the Convention on Biological Diversity (2014a, p. 10) and ERASynBio (2014, p. 11).

As far as publications are concerned, the majority of papers—including a high proportion of the most significant ones—originates from the US. Nonetheless, the gap between publications in Europe as a whole and in the US is closing.[320] The number of researchers in this area is still limited. These experts operate predominantly in research institutions, with a meagre 7% working in industry. The background of these experts reflects the interdisciplinarity of the field, as 37% of them were biologists, 14% chemists, and about 10% each informatics and engineers.[321] This interdisciplinarity is a challenge to the traditional education system and, therefore, ad hoc classes and programmes for synthetic biology have been introduced at both graduate and undergraduate level.[322]

A key aspect to train young synthetic biologists and to form a community around this subject has been the creation of the international Genetically Engineered Machine (iGEM) competition. The annual iGEM competition, established in 2004, sees the participation of high school, undergraduate and graduate students, who compete in the design and implementation of biological systems.[323] iGEM administers also a Registry of Standard Biological Parts, to which the participating teams shall contribute new parts in case they want to compete for iGEM prizes and medals.[324] Such parts must follow the BioBricks™ assembly standard.[325] The competition attracts an increasing number of teams from all over the world.[326]

Another forum that greatly contributed to the development of this field is the Synthetic Biology x.0 conference series. The BioBricks Foundation organises a series of international conferences addressing both scientific and policy related aspects of synthetic biology.[327] The first gathering was organised in 2004.[328] In the eyes of its proponents, the SB 1.0 conference was "*consciously designed to facilitate the emergence of a tight-knit, cooperative international community of synthetic biology researchers*".[329]

---

[320]ERASynBio (2014, p. 11) and ERASynBio (n.d.).

[321]European Commission's Directorate-General for Health & Consumers (2010, p. 7).

[322]OECD (2014, p. 22) and British Synthetic Biology Centre for Doctoral Training (n.d.).

[323]SCHER, SCENIHR, SCCS Scientific Committees (2014, p. 53).

[324]Frow and Calvert (2013, p. 49). On this, see Sect. 2.2.1.3.

[325]iGEM (n.d.-b). On this, see Sect. 2.2.1.3.

[326]At the beginning, in 2004, 5 teams participated in the competition and submitted approximately 50 parts to the Registry. In 2010, 128 teams took part in iGEM and submitted more than 1800 parts. By 2015, the number of participating teams reached 280 (iGEM n.d.-c).

[327]Bensaude Vincent (2013, pp. 369–371).

[328]BioBricks Foundation (n.d.-b).

[329]Kelley (2014, p. 137).

#### 2.3.2.3 Companies

The companies active in the synthetic biology market operate at different levels of the production chain. Some focus on enabling technologies, while others specialise in enabled products. Amongst the first group are companies manufacturing synthetic genes.[330] A considerable number of companies active in the enabled products sector have been founded by researchers and professors. During early funding rounds between 2005 and 2006, many of these companies were able to raise millions of dollars in capital from wealthy individuals as well as venture capitalists.[331] This led to an increased media attention and to articles placing synthetic biology companies in the list of the hottest start-ups.[332] This trend has continued, as promising start-ups were recently able to obtain several million dollars from venture capitalists.[333]

Venture capital is not the only source of funding of synthetic biology, as some high-profile investors entered this field. Studies show that six of the top ten chemical, energy, and grain corporations and the top seven pharmaceutical companies in the world have all invested in synthetic biology.[334] For example, companies like Bayer, Intel, Microsoft, Total, Exxon, L'Oréal and PepsiCo, just to name a few, are directly investing in synthetic biology or have integrated it into their R&D processes and partnerships.[335] This is a change from the initial reluctance of pharmaceutical companies to invest in this field due to a lack of proof of utility.[336]

Some synthetic biology companies have also gone public during the past few years, taking advantage of its market potential, reducing operating costs and improved business models.[337] Lastly, synthetic biology has been promoted via crowd funding initiatives.[338]

Product-wise, companies are increasingly moving away from the production of fuels and are instead focusing on the manufacturing of chemicals.[339] If at the

[330]Rousseaux (2015) and Presidential Commission for the Study of Bioethical Issues (2010, pp. 40–41).

[331]Balmer and Martin (2008, p. 11). Wealthy individuals, especially tech founders, are still heavily investing in synthetic biology (Cumbers 2019).

[332]Ferry (2015).

[333]Interestingly, venture capitalists are increasingly investing in this field, with $500 millions being raised in the first three quarters of 2015, which is more than the total amount raised in 2013 and 2014 combined (Basulto 2015b). For 2018, Nanalyze (2018).

[334]ETC Group (2012).

[335]ETC Group and Heinrich Böll Stiftung (2015, pp. 10–18), Mistbreaker News (2015), Rousseaux (2015), Ferrari (2014) and Kelley et al. (2014, p. 141).

[336]European Commission's Directorate-General for Health & Consumers (2010, p. 22) and OECD and The Royal Society (2010, p. 24).

[337]Philp (2014, p. 23) and OECD (2014, pp. 88–89).

[338]For example, the Pink Army project to devise personalised medical treatments has appealed to crowd-funding, while the Glowing Plant one initiated a Kickstarter campaign (Hanel 2015; Callaway 2013, p. 15).

[339]van den Belt (2009b, p. 13).

beginning companies pivoted towards lower value chemicals (chemicals that yield between $3–30 for kg), they have now geared towards high value chemicals (spanning between $30–3000 for kg) in order to remain in business. The start-ups of today have learned from the struggles of the first wave of industrialisation of synthetic biology and have steered clear of highly complex and regulated fields (e.g. energy) in order to concentrate their efforts on niche areas, where products could be quickly brought to the market.[340]

#### 2.3.2.4 Public

The success of synthetic biology will depend on if and how this technology will be embraced by the public. After the experience with GMOs, there is a risk of rejection of this technology due to environmental, safety, health and policy concerns.

Until 2010, only 17% of people who participated in a survey commissioned by the European Commission knew of the existence of synthetic biology.[341] Specifically, 16% had either only heard of this discipline or discussed it and researched it occasionally. Only 1% had researched and discussed the topic frequently. With its 17%, synthetic biology had by far the lowest percentage of familiarity amongst the five technologies surveyed by the Commission (i.e. GM food, animal cloning for food production, nanotechnology and biobanks). This trend was consistent all over Europe, albeit with some variations.[342]

EU-wide studies showed that people would like to be more informed about the risks and benefits of this technology. When it comes to the conditions for approving synthetic biology, 17% of interviewees did not approve of synthetic biology under any conditions, while 21% approved in special circumstances and 36% approved only if this discipline were regulated by strict norms.[343] Those approval ratings varied between countries. Overall, the number of disapproving States was basically equal to that of approving ones.[344]

Another report compiled in the UK on the basis of workshops with citizens and interviews with stakeholders showed a number of interesting findings on the perception of synthetic biology. Overall, it was noted that there is:

> Conditional support for synthetic biology – while there was great enthusiasm for the possibilities of the science; there were also fears about control; who benefits; health or environmental impacts; misuse; and how to govern the science under uncertainty.[345]

---

[340] Check Hayden (2015, p. 19).

[341] Bensaude Vincent (2013, p. 369). For US surveys, Pauwels (2013, pp. 81–87).

[342] In Switzerland, the percentage of people claiming familiarity with this discipline peaked at 29%, while in France and in the Czech Republic it sunk at 12% (Gaskell et al. 2010, p. 82).

[343] A high percentage of interviewees (23%) choose not to reply, as they did not have a clear opinion on the matter.

[344] Gaskell et al. (2010, p. 34).

[345] Bhattachary et al. (2010, p. 7).

The participants mentioned the need for scientists to properly take into consideration the implications of their work and believed that citizens should be involved in discussions on synthetic biology and its funding from an early stage. On a more abstract level, they perceived a tension between the concepts of "biological" and "synthetic" and some had the impression that science was transgressing nature. They were also sceptical of the notion that nature can be reduced to parts that can be assembled. Similarly, concerns were expressed over the fast rate of development, unknown long-term impact and possible uncontrolled release in the environment.[346]

Related works showed that participants negatively reacted to the name synthetic biology, so that this term might become a liability for the field. They also indicated that the reaction of the British public towards synthetic biology was more positive than in the US.[347]

All reports examined above detected a connection between the area of application of synthetic biology and its acceptance. For example, health applications were generally seen as more desirable than food-related ones. Indeed, UK studies showed that medical applications were considered to be the most morally acceptable, useful and to be encouraged, while bioremediation and food applications fared more negatively.[348] Other studies have assessed the connection between the acceptance of synthetic biology and religious beliefs, concluding that the latter seems to increase the disapproval for this technology.[349]

To overcome the doubts and fears of the public, numerous approaches have been proposed. Entrepreneurs have argued that the more consumers understand synthetic biology, the better the situation will be.[350] However, the familiarity argument (postulating that public backlash is a result of an informational deficit) has been rebutted by scholars and studies alike.[351]

The media are another factor that will greatly influence the way synthetic biology is perceived. Overall, the media attitude towards synthetic biology has been predominantly positive both in Europe and in the US, with the American press

[346]Bhattachary et al. (2010, pp. 8–13).

[347]OECD and The Royal Society (2010, p. 40).

[348]Philp (2014, p. 30).

[349]Dragojlovic and Einsiedel (2013, p. 869).

[350]Harman (2014).

[351]Pauwels (2013, p. 79).

displaying an even more optimistic view of it.[352] As for their content, articles have often focused on potential future applications rather than on current ones.[353]

Safety concerns related to synthetic biology have also featured prominently in the news. A study focused on English speaking newspapers found that a quarter of the articles mentioning synthetic biology included also references to the words "bio-weapon" or "terror".[354] Older studies, covering the period between 2003 and 2008, showed that the European press focused on ethics as well as biosafety and biosecurity, whereas US newspapers tended to focus mostly on biosecurity issues.[355] Metaphors on "playing Lego", creation, re-designing life and religion were instead very present in the German media portray of synthetic biology.[356]

Stories about scientists "playing God" and the debate between what is natural or not are already recurring in the media and so are associations with previous technologies. For example, studies showed that people tend to associate synthetic biology with other more conventional gene technologies and, in particular, with biotechnology.[357] It was even stated that synthetic biology is "*not perceived as different enough from biotechnology to merit special attention and is supposed to raise the same questions as biotechnology*".[358]

#### 2.3.2.5 DIY Movement

Synthetic biology is not being developed only in institutional settings, but also within a new and heterogenic community known as do-it-yourself biology or biohacker community.

The term biohacker, which should not be read in a negative connotation, refers to individuals who work either alone or in groups to design and manufacture biological systems outside of institutional settings.[359] Participants do not necessarily have a

---

[352]Gschmeidler and Seiringer (2012, p. 164). German-speaking print media presented synthetic biology as a field with great benefits (83% of articles), especially for energy, health and environmental applications. On the other hand, risks were only mentioned in half of the surveyed articles (Gschmeidler and Seiringer 2012, p. 166). Another study showed that only 21% of the examined German articles assessed the pro and contra of this technology and its classification (Lehmkuhl 2011, p. 25). Similar results emerged from an analysis of the Scandinavian press, which showed that the tone of the articles was mostly balanced or positive, with only 13% of publications that expressed a cautious or negative approach (Ancillotti et al. 2015, p. 5).

[353]Cserer and Seiringer (2009, p. 33).

[354]Jefferson et al. (2014a, p. 16).

[355]European Group on Ethics in Science and New Technologies to the European Commission (2009, p. 38). On this, see Sect. 2.4.1.2.

[356]Gschmeidler and Seiringer (2012, p. 169) and Cserer and Seiringer (2009, p. 29).

[357]Gschmeidler and Seiringer (2012, pp. 169–170) and Kronberger et al. (2012, p. 179).

[358]Gschmeidler and Seiringer (2012, p. 170).

[359]Dietrich and Steen (2007, p. 4).

biology background and are inspired by the open-source movement. Hence, they generally judge negatively the use of intellectual property rights in this field.[360]

The number of biohackers is controversial. While some believe that their number is quite limited, other sources argued that they could have hundreds or even thousands of members.[361] Their communities are active around the world, including in less developed countries.[362]

Three factors contributed to the rise of the DIY movement in synthetic biology. First, the reduction of the costs associated with DNA synthesis. Second, the Registry of Standard Biological Parts offered a basis to build new biological devices in an easier and cheaper manner. Third, information on biological devices and their modification is now largely available on the internet.[363]

Biohackers are said to operate in an environment which is almost completely devoid of any regulatory or enforcement oversight.[364] Because of this, biohackers are considered as a possible threat to safety. Reports have described them as a community in which "*lone individuals develop dangerous organisms much as they currently create computer viruses*".[365] By contrast, other commentators pointed out that the skills needed to produce this kind of organisms are much beyond the abilities of most biohackers.[366]

## 2.4 Risks, Concerns and Regulations

### *2.4.1 Risks and Concerns Connected to Synthetic Biology*

#### 2.4.1.1 Ethics

Ethics are an integral part of the debate surrounding synthetic biology. Scholars have considered whether synthetic biology poses new ethical questions. Although it was argued that "*synthetic biology does not create any ethical dilemmas that have not already been raised*", this does not exclude or reduce the complexity of such issues.[367] To address these complex points, studies were undertaken by the European Group on Ethics in Science and New Technologies, the US Presidential

[360] McLennan and Rimmer (2012).

[361] SCHER, SCENIHR, SCCS Scientific Committees (2015a, p. 39), Meyer (2012, pp. 312–314) and Schmidt (2008, p. 2).

[362] Tatalovic (n.d.).

[363] Dietrich and Steen (2007, p. 5).

[364] Rousseaux (2015) and Schmidt (2008, p. 2).

[365] European Commission, Directorate-General for Research (2005, p. 18).

[366] Ledford (2015, pp. 398–399).

[367] SYBHEL (2014, p. 16) and Heyd (2012, p. 581).

Commission for the study of bioethical issues, and the Nuffield Council, just to name a few.[368]

Amongst the most debated issues is the question of whether it would be ethical to produce goods via synthetic biology knowing that this might negatively affect ecosystems and disrupt the livelihood of the farmers that currently grow the natural versions of these goods in some of the poorest regions on the planet.[369] Concerns are not limited to the livelihood of local producers, but extend also to the legal status of the genetic resources found in these countries. The Convention on Biological Diversity supports the equitable sharing of the benefits connected to local knowledge and resources. However, this sharing might be affected by synthetic biology.[370]

From an ethical morality perspective, the two most intricate questions posed by synthetic biology concern the blurring of the line between natural and artificial and claims that scientists are "playing God" and creating life.

On the first issue, *Calvert* argues that, although synthetic is often considered a synonym of artificial, this does not imply that all creations of synthetic biology will be seen as such. The line separating the natural from the artificial could thus become a "*receding horizon*".[371] Still, consensus exist that, independently from their natural or artificial method of creation, these entities shall have equal moral status and rights.[372]

The second issue was often raised in the media, as headlines referred to scientists "playing God" and disrupting the natural course of life.[373] Different opinions have been expressed on whether synthetic biology really invites this accusation more than other technologies and, if so, why. *Campos* pointed out that controversies about the

[368]Nuffield Council on Bioethics (2012), Presidential Commission for the Study of Bioethical Issues (2010) and European Group on Ethics in Science and New Technologies to the European Commission (2009). Papers have also been commissioned by synthetic biologists themselves. For example, in 1999, Craig Venter commissioned an article on the ethical aspects of the synthesis of a minimal genome (Yearley 2009, p. 561; Cho et al. 1999).

[369]Friends of the Earth (n.d., p. 1).

[370]Teller (2015, p. 8) and Conde (2012).

[371]Calvert (2010, p. 108).

[372]Newson (2015, p. 48) and Commissione federale d'etica per la biotecnologia nel settore non umano (CENU) (2010, p. 28).

[373]It could be suggested that this argument is not strictly religious. In a report, the Commission of the Bishops' Conferences of the European Community (COMECE) argued that Craig Venter had not created life, but he rather: "*Obtained a new life form, but to do that he has only exploited, after long and costly efforts, the natural properties of a bacterium which certainly did not owe its existence to him*". Hence, the production of new biological entities does not necessarily amount to "playing God" (Commission of the Bishops' Conferences of the European Community (COMECE) 2016, p. 7). Similar conclusions over the non-religious character of this objection have been raised also by *van den Belt*. He held that the "playing God" argument is generally used by secular organisations and that synthetic biology is seen as an affront to nature rather than to a divine entity (van den Belt 2009a, pp. 263–265).

creation of artificial life and the pushing of its boundaries are not new.[374] On the other hand, it has been argued that synthetic biology pushes these statements even further. Indeed, while "*many a technology has at some time or another been deemed an affront to God... perhaps none invites the accusation as directly as synthetic biology*".[375] According to *Preston*, the construction of organisms from scratch using BioBricks™ severs the connection between the organism and the evolutionary process in a way that prior technologies did not, thus infringing the Darwinian principle of natural selection.[376]

Other arguments criticise the idea of biological building blocks as mechanical entities that can be controlled and dominated, as this would constitute a reductive view of life and would prove that the goal of synthetic biology is not to understand nature, but rather to exploit it.[377] The quest for a minimal genome could also be seen as a reductionist approach to the origin and meaning of life.[378] Equally, authors wondered whether organisms synthesised by men could be considered machines.[379] Relatedly, questions have been raised on which responsibilities mankind has for the repercussions of its synthetic creations.[380]

#### 2.4.1.2 Safety

Synthetic biology postulates the creation of new biological systems that could be used to fulfil tasks set by humans. Although these tasks are usually associated with benign targets, malicious use represents a risk. Those risks can be the result of either wilful or inadvertent dissemination. Given the potential of synthetic biology, these dangers could affect large sections of the population and the environment. Hence, scientists, organisations, governments and companies have developed different strategies to assess and address these threats, keeping in mind that absolute risk removal is generally unattainable.

An open question is whether synthetic biology poses unique safety risks. Some commentators replied in the negative, believing that this discipline is based on the same practices of genetic engineering.[381] Nevertheless, debates have emerged over

[374]In 1905, similar claims were made about an experiment that, by adding radium to a petri dish containing bouillon, resulted in the creation of life-like cell forms or to works done in the sixties on artificial DNA synthesis (Campos 2009, pp. 9–10).

[375]"Meanings of 'life'" (2007).

[376]Preston (2008, p. 35).

[377]Commissione federale d'etica per la biotecnologia nel settore non umano (CENU) (2010, pp. 11–12).

[378]Raho (2012, p. 298).

[379]For example, if machines are intended as devices that are externally controlled and that are used for purposes set by humans, this would raise the question of whether "biological machines" should be considered as such or as living systems (Gregorowius 2012).

[380]Commissione federale d'etica per la biotecnologia nel settore non umano (CENU) (2010, p. 19).

[381]Moe-Behrens et al. (2013, p. 1).

whether current safety procedures and regulations are effective also for synthetic biology. Synthetic biologists and regulators have often argued that they are, especially in the biosafety field.[382]

The level of risk mitigation to be applied to synthetic biology has been a contentious topic. The discussion reverted mostly around whether the precautionary principle or the prudent vigilance one should be applied. The precautionary principle shifts the burden of proof on the possible risks connected to a project from those who are against it to those who are promoting it. In other terms, those who propose a project should be able to show that it does not pose any risks, which—some say—would lead to the paralysis of the project as absence of risk is complex and lengthy to prove.[383] Nonetheless, the Conference of the Parties to the Convention on Biological Diversity urged its parties to adopt it "*when addressing threats of significant reduction or loss of biological diversity posed by organisms… from synthetic biology*".[384] This view was shared by a number of civil society organisations.[385] This led to requests for a "*moratorium on the release and commercial use of synthetic organisms until a thorough study of all the environmental and socioeconomic impacts of this emerging technology has taken place*".[386] Equally, a number of States supported resolutions to prevent the release of synthetic biology products into the wild.[387] By contrast, the US Presidential Commission for the Study of Bioethical Issues believed that the principle of prudent vigilance would be better suited in this case.[388] This position was also supported by proponents of synthetic biology, as the application of the precautionary principle would hinder the development of the field.

Risks in the field of synthetic biology are usually divided into threats to biosecurity and biosafety. Biosafety, which is often at the centre of the debate in Europe and represents a more novel field, targets the "*unintentional exposure to harmful or potentially harmful biological agents and material, or their accidental release*".[389] Conversely, biosecurity, generally a focus in the US and a more established concept, concentrates "*on preventing the misuse through for example loss, theft, diversion or intentional release of harmful or potentially harmful*

[382]OECD and The Royal Society (2010, p. 32).

[383]Kaebnick (2012) and ETC Group (2007, p. 50).

[384]Convention on biological diversity (2018), Conference of the parties to the Convention on Biological Diversity (2014b, p. 6) and *Convention on biological diversity (1760 UNTS 79)* (1992).

[385]Friends of the Earth et al. (2012, p. 3).

[386]Friends of the Earth et al. (2012, p. 1), ETC Group (2012), Sharma (2012a) and Gutmann (2011, p. 21).

[387]During a 2010 meeting of the Conference of the Parties to the Convention on Biodiversity, the Philippines as well as a number of African countries favoured the precautionary principle for the field release of synthetic biology products (Sharma 2012b).

[388]Gutmann (2011, p. 21).

[389]Kelle (2009, p. 85) and Parliamentary Office of Science and Technology (POST) (2008, p. 3).

*biological agents and materials*".[390] In lay terms, they have been described as "*keeping bad bugs from people*" and "*keeping bad people from bugs*" respectively.[391] A third problematic topic is dual-use research, that is, research that could be used for either benign or malevolent purposes. Some research studies have led to the development of pathogens, thus raising concerns that these same research endeavours could be used for malign purposes.

Biosecurity threats could in theory be posed by States, terrorist organisations and lone wolfs alike. Some scholars believe that governmental programmes are the likeliest source, given their resources and history.[392] Still, historical precedents show that even state funded programmes spanning years and involving thousands of scientists encountered many difficulties in trying to develop biological weapons, especially new pathogens. This has been attributed to the difficulty of turning living, mutating organisms that are sensitive to their environment into reliable weapons.[393]

Terrorist organisations have been pinned as possible users of biological weapons developed via synthetic biology. While some authors argue that the employment of such weapons by terrorist groups is inevitable, others have cautioned against the exaggeration of this threat.[394] For example, it has been doubted that terrorists would focus time and resources on new artificial biological agents, when there are plenty of dangerous natural ones available.[395]

The preparation of synthetic biology weapons by terrorist groups or lone wolves has been attributed to the de-skilling of biology, which would allow any layman to design and produce dangerous pathogens and weapons. This idea has been judged as misleading by some commentators. First, it was noted that synthetic biology is not easy and that it would be wrong to believe that anyone can engineer an organism. Also, the skills possessed by DIY-biologists have been considered as grossly overstated. Likewise, despite DNA synthesis becoming cheaper and being outsourced to third companies, it is not easy to assemble a functional genome.[396]

The intentional release of synthetic biology organisms into the environment for benign uses has also raised concerns. First, those organisms could displace already existing ones and become invasive or polluting. Second, they could exchange genetic materials with natural organisms, possibly resulting in unpredictable variations in both the natural and the synthetic organisms. Third, fears over synthetic biology organisms with "*unpredictable and emergent properties*" have also been

[390] Kelle (2009, p. 85) and Parliamentary Office of Science and Technology (POST) (2008, p. 3).

[391] OECD (2014, p. 117).

[392] European Commission's Directorate-General for Health & Consumers (2010, p. 19) and Kelle (2009, p. 86).

[393] Jefferson et al. (2014a, p. 23).

[394] Kelle (2009, p. 86).

[395] European Commission's Directorate-General for Health & Consumers (2010, p. 19).

[396] Marris and Jefferson (n.d., pp. 1–2).

raised.[397] Additionally, worries have emerged over the possibility that synthetic biology organisms could cause allergic reactions in the population.[398] However, for the moment, synthetic biology organisms are generally unfit to survive in the wild, thus decreasing the risk of contamination. Still, evolutionary pressure might change this.[399]

To prevent the accidental spreading of synthetic biology organisms and avoiding unintended contacts with natural environments, containment strategies have been devised. Such containment methods are either physical or biological and, though helpful, cannot guarantee absolute safety. Physical containment strategies include—as the name suggests—physical barriers to the release of the organisms, for instance by restricting use only to laboratories or closed industrial settings. On the other hand, biological containment comprises controls that are intrinsic to the organism; that is, biological systems should be designed with safety in mind.[400] Biological mechanisms are apt for applications that require the spreading of synthetic biology organisms into the wild (e.g. bioremediation). For instance, induced lethality mechanisms, or "kill switches", would cause a cell to self-destruct in certain conditions or when a specific time is reached, thus preventing it from further interacting or spreading.[401] However, evolutionary pressure might disarm those genes or cause them to mutate, thus defeating their purpose. Another possibility is trophic containment. In this case, organisms are designed to make them unable to synthesise a compound needed for their survival. If the organisms are in a safe environment, such compound could be fed to them, thus keeping them alive. Conversely, the release of the organisms in the wild would cause them to die due to their inability to either synthesise or obtain this necessary compound. Still, also this method does not offer absolute reliability. Other strategies rely instead on the prevention of gene transfers. This would block organisms from taking up or inheriting altered genetic materials. Moreover, xenobiology is also being considered to shield the environment from synthetic biology organisms. This strategy, called semantic containment, uses unnatural backbones or nucleotides to prevent communication between natural and synthetic organisms. Research in this field is in its infancy and thus the effect of xenobiology organisms on natural ones is still unclear.[402]

The topic of security has been addressed by the synthetic biology community since its inception. Already at the Synthetic Biology 1.0 conference in 2004, some discussions were hosted to explore biological risks.[403] However, a study carried out in Europe in 2007 regarding the level of risk-knowledge of 20 synthetic biologists

---

[397] Conference of the parties to the Convention on Biological Diversity (2014a, p. 30).

[398] SCHER, SCENIHR, SCCS Scientific Committees (2015b, pp. 9–10).

[399] Parliamentary Office of Science and Technology (POST) (2008, p. 3).

[400] European Commission's Directorate-General for Health & Consumers (2010, p. 21).

[401] Moe-Behrens et al. (2013, p. 3).

[402] Conference of the parties to the Convention on Biological Diversity (2014a, pp. 32–33) and Moe-Behrens et al. (2013, p. 6).

[403] Maurer et al. (2006, p. 4). On safety issues connected to synthetic biology, Maurer (2011).

showed a low to medium awareness.[404] In the DIY field, codes, guidelines and web portals dealing with safety questions have been established.[405]

### 2.4.2 Synthetic Biology in the Normative and Regulatory Framework

Several international conventions and agreements cover the area of synthetic biology.[406] Some of the most relevant focus on the maintenance of biological diversity and on the safety and security issues related to this discipline.

The 1992 Convention on Biological Diversity aims to conserve biological diversity, guarantee sustainable use as well as the fair and equitable sharing of the benefits obtained from genetic resources. The Conference of the Parties to the Convention deemed synthetic biology to fall within the realm of biotechnology as defined by the Convention and recognised that this new discipline could have both positive and negative effects on the conservation and sustainable use of biodiversity.[407] The Convention on Biological Diversity addresses also the use and release in the environment of organisms that have been modified and that might have an adverse impact on biodiversity. The Conference of the Parties to the Convention noted that synthetic biology organisms fall within the definition of "*living modified organisms resulting from biotechnology*" and that the provisions on biosafety of the Convention are thus applicable to them. For the same reason, the norms of the Cartagena Protocol on biosafety are applicable to synthetic biology as well.[408]

The creation of biological weapons is the subject of legislation. The Biological Weapons Convention prevents States from developing and producing biological agents that do not have any justification for peaceful purposes. It covers agents that have been developed via natural or artificial methods, thus including developments in synthetic biology.[409]

---

[404]Kelle (2007, p. 27).

[405]Jefferson et al. (2014b, p. 6).

[406]For an overview of applicable international and European regulations, Falcone (2014, pp. 62–65).

[407]Conference of the parties to the Convention on Biological Diversity (2014a, pp. 8–9). For an overview of the potential implications of synthetic biology on biotrade and access/benefit-sharing, see the study presented by UNCTAD, United Nations Conference on Trade and Development (2019). For an overview of a number of issues raised by synthetic biology within the framework of the Convention on Biological Diversity (Lai et al. 2019).

[408]Conference of the parties to the Convention on Biological Diversity (2014b, p. 6) and *Cartagena Protocol on Biosafety to the 1992 Convention on Biological Diversity (2226 UNTS 208)* (2000).

[409]Conference of the parties to the Convention on Biological Diversity (2014b, p. 7), Schmidt and Giersch (2012, p. 288) and *Convention on the Prohibition of the Development, Production and Stockpiling of Bacteriological (Biological) and Toxin Weapons and on their Destruction (1015 UNTS 163)* (1972).

At EU and US level, synthetic biology is currently regulated by the norms previously set for GMOs.[410] In Europe, Directives 2001/18/EC and 2009/41/EC regulating deliberate release and contained use of GMOs are applicable to synthetic biology.[411] The GMOs directives are based on a case-by-case assessment of risks, on the precautionary principle and adopt a comparative approach.[412] The comparative principle dictates that GMOs are compared to a natural equivalent to assess the possible risks and changes introduced by the mutation. However, this principle could be problematic in synthetic biology, given the growing distance between natural organisms and synthetic biology ones.[413] Also, the high demands and costs associated with existing GMO regulations could discourage innovation, especially by smaller companies. Innovation may thus be redirected towards areas with lower regulatory hurdles (e.g. cosmetics).[414]

These issues raised the question of whether the norms regulating GMOs are sufficient and appropriate for synthetic biology. While this issue is still much debated, commentators noticed that the use of the processes established for GMOs has both positive and negative aspects. On the one hand, it offers a solid background of expertise. On the other, discussions on synthetic biology might be influenced by negative public views associated with GMOs and might be subjected to an onerous regulatory framework.[415] The consensus seems to be that the regulatory and risk assessment regimes for GMOs can be applied to synthetic biology in its current form. However, as the discipline evolves, those regimes might need to be revised.[416]

The commercialisation of synthetic biology products is also raising a number of questions, especially concerning the commercial release and labelling of the

---

[410]The US legislation on GMOs presents some differences in comparison to its European counterpart. For an overview of the regulatory and oversight competences, Presidential Commission for the Study of Bioethical Issues (2010, pp. 80–102).

[411]For an overview of the regulatory framework applicable to specific applications and requirements (e.g. labelling) (British Food & Environment Research Agency 2014, pp. 8–9; OECD 2014, pp. 122–127; SCHER, SCENIHR, SCCS Scientific Committees 2014, pp. 60–63). Directive 2001/18/EC and Directive 2009/41/EC define a Genetically Modified Organism (GMO) and a Genetically Modified Microorganism (GMM) as an: "*Organism in which the genetic material has been altered in a way that does not occur naturally by mating and/or natural recombination*". However, while the Directives apply to: "*Any biological entity capable of replication or of transferring genetic material*" and thus cover most of the current developments of synthetic biology, some doubts have emerged on whether they would also encompass protocells and xenobiology inventions. On this, see Sect. 4.2.2.1.

[412]SCHER, SCENIHR, SCCS Scientific Committees (2014, p. 18).

[413]SCHER, SCENIHR, SCCS Scientific Committees (2015b, pp. 31–60).

[414]Douglas and Stemerding (2014, pp. 8–9).

[415]OECD (2014, p. 127). Other sources doubted whether the GMO regulatory framework is suitable for synthetic biology (SCHER, SCENIHR, SCCS Scientific Committees 2014, p. 8). This idea was shared by the European Group on Ethics, which suggested that the Commission should reassess the legislation applicable to synthetic biology to establish whether it is fit to address the issues raised by this field (SYBHEL 2014, p. 5; European Group on Ethics in Science and New Technologies to the European Commission 2009, p. 53). For further discussions, Seitz (2018).

[416]Conference of the parties to the Convention on Biological Diversity (2014a, p. 35).

products. For instance, commentators in Europe were arguing over whether products might be labelled as natural and whether they should bear a "synthetic biology" tag.[417]

Lastly, questions were raised on the regulatory approach that should be adopted in this discipline. On the one hand, authors advised against the over-regulation of this field as well as regulations that might be too specific and thus not able to cover various methodologies and applications. On the other hand, it was noted that a specific oversight for synthetic biology could reduce problems and consolidate the trust of stakeholders.[418] In its report, SYBHEL warned against the risks of "*exceptionalism*" and argued that "*synthetic biology should be regulated like any other commercial engineering endeavour*".[419] Another debated point concerned the issue of self-regulation. Some argued that self-regulation was appropriate, as the scientific community is able to adhere to voluntary standards, as shown with Asilomar.[420] A proposal for self-regulation along the lines of Asilomar was presented at the SB 2.0 conference in 2006. This proposal was however withdrawn due to the strong criticism it raised amongst civil society organisations.[421] Public response to self-regulation was also sceptical.[422]

## References

Allied Market Research (2014) World synthetic biology market - opportunities and forecasts, 2013 - 2020 [WWW Document]. www.alliedmarketresearch.com. URL https://www.alliedmarketresearch.com/synthetic-biology-market. Accessed 9 Mar 2016

Ancillotti M, Holmberg N, Lindfelt M, Eriksson S (2015) Uncritical and unbalanced coverage of synthetic biology in the Nordic press. Public Underst Sci:1–16

Andrews LB (2014) The progress clause: an empirical analysis based on the constitutional foundation of patent law. N C J Law Technol 15:537–561

---

[417]British Food & Environment Research Agency (2014, p. 15). Questions on commercialisation and labelling have been particularly prominent in the US. The fact that discussions over the adequacy of current regulations are more predominant in the US than in Europe has been attributed to the more restrictive regulatory framework already existing in the old continent (OECD and The Royal Society 2010, p. 37).

[418]Newson (2015, p. 53).

[419]SYBHEL (2014, p. 13).

[420]Maurer et al. (2006, p. 4). Asilomar refers to the International Congress on Recombinant DNA Molecules held in Asilomar in February 1975. In that occasion, scientists evaluated the nature and level of risk deriving from newly discovered recombinant DNA techniques and agreed on guidelines to perform research in this field.

[421]ETC Group (2006, pp. 3–4).

[422]A British study highlighted that: "*There was a strong view that scientists should not be allowed to regulate themselves and people should not be allowed to do synthetic biology in their 'back gardens'. Given the stakes, voluntary standards developed through industry were also not seen as appropriate. A robust and independent regulator was considered to be fundamental in this area*" (Bhattachary et al. 2010, p. 42).

Anthony S (2013) Stanford creates biological transistors, the final step towards computers inside living cells [WWW Document]. www.extremetech.com. URL https://www.extremetech.com/extreme/152074-stanford-creates-biological-transistors-the-final-step-towards-computers-inside-living-cells. Accessed 29 Feb 2020

Ballantyne C (n.d.) Longest piece of synthetic DNA yet [WWW Document]. www.scientificamerican.com. URL https://www.scientificamerican.com/article/longest-piece-of-dna-yet/. Accessed 29 Feb 2020

Balmer A, Martin P (2008) Synthetic biology social and ethical challenges. Biotechnology and Biological Sciences Research Council (BBSRC)

Barclay E (2014) GMOs are old hat. Synthetically modified food is the new frontier [WWW Document]. www.npr.org. URL https://www.npr.org/sections/thesalt/2014/10/03/353024980/gmos-are-old-hat-synthetically-modified-food-is-the-new-frontier?t=1582994572188. Accessed 29 Feb 2020

Basulto D (2015a) Three recent developments in synthetic biology you need to know [WWW Document]. www.washingtonpost.com. URL https://www.washingtonpost.com/news/innovations/wp/2015/05/07/three-recent-developments-in-synthetic-biology-you-need-to-know/. Accessed 27 Feb 2020

Basulto D (2015b) Synthetic biology innovations need a clearer path to market [WWW Document]. www.washingtonpost.com. URL https://www.washingtonpost.com/news/innovations/wp/2015/10/29/synthetic-biology-innovations-need-a-clearer-path-to-market/. Accessed 27 Feb 2020

Bensaude Vincent B (2013) Ethical perspectives on synthetic biology. Biol Theory 8:368–375

Bethge P, Grolle J (2013) Interview with George Church: can Neanderthals be brought back from the dead? [WWW Document]. www.spiegel.de. URL https://www.spiegel.de/international/zeitgeist/george-church-explains-how-dna-will-be-construction-material-of-the-future-a-877634.html. Accessed 29 Feb 2020

Bhattachary D, Pascall Calitz J, Hunter A (2010) Synthetic biology dialogue. Biotechnology and Biological Sciences Research Council (BBSRC)

BioBricks Foundation (n.d.-a) About [WWW Document]. http://biobricks.org. URL https://biobricks.org/biobricks-history/. Accessed 29 Feb 2020

BioBricks Foundation (n.d.-b) SBx.0 Conferences [WWW Document]. https://biobricks.org/. URL https://biobricks.org/sbx-conferences/. Accessed 29 Feb 2020

Boeke JD, Church G, Hessel A, Kelley NJ (2016) HGP-write: testing large genomes in cells - meeting summary. Presented at the HGP-write: testing large genomes in cells workshop, Boston, pp 1–13

Bostyn SJR (2004) Patenting DNA sequences (polynucleotides) and scope of protection in the European Union: an evaluation. Office for Official Publications of the European Communities, Luxembourg

Boyd M (2019) Engineering biology research and development act of 2019 (HR 4373, 116th Congress) [WWW Document]. https://scipol.duke.edu. URL https://scipol.duke.edu/track/hr-4373-engineering-biology-research-and-development-act-2019. Accessed 1 Mar 2020

Brief for the United States as amicus curiae in support of neither party in Association for Molecular Pathology, et al. v. Myriad Genetics, Inc., et al. before the U.S. Supreme Court (Myriad), 2013

British Food & Environment Research Agency (2014) Final report - synthetic biology: prospective products and applications for food and feed - requirements for regulation (No. FS 102068). Food and Environmental Research Agency, York

British Synthetic Biology Centre for Doctoral Training (n.d.) Research themes [WWW Document]. www.synbio-cdt.ac.uk. URL http://www.synbio-cdt.ac.uk/research/research-themes.html. Accessed 29 Feb 2020

Brown N (2015) New storage method allows 70 Million books to be stored in a drop of liquid [WWW Document]. http://futurism.com. URL https://futurism.com/new-storage-method-allows-70-million-books-to-be-stored-in-a-drop-of-liquid. Accessed 29 Feb 2020

Bullis K (2012) Audi backs a biofuels startup [WWW Document]. www.technologyreview.com. URL https://www.technologyreview.com/s/429245/audi-backs-a-biofuels-startup/. Accessed 29 Feb 2020
Burk DL (2013a) Edifying thoughts of a patent watcher: the nature of DNA. UCLA Law Rev Discourse 60:92–102
Burk DL (2013b) Are human genes patentable? Int Rev Intellect Prop Compet Law 44:747–749
Burningham G (2016) Lab-grown beef will save the planet - and be a billion-dollar business [WWW Document]. http://europe.newsweek.com. URL https://www.newsweek.com/lab-grown-beef-will-save-planet-and-be-billion-dollar-business-430980. Accessed 29 Feb 2020
Callaway E (2013) Glowing plants spark debate. Nature 498:15–16
Callaway E (2014) First life with "alien" DNA. https://www.nature.com. https://doi.org/10.1038/nature.2014.15179
Callaway E (2016) Plan to synthesize human genome triggers mixed response. Nat News 534:163
Calvert J (2010) Synthetic biology: constructing nature? Sociol Rev 58:95–112
Calvert J, Joly P-B (2011) How did the gene become a chemical compound? The ontology of the gene and the patenting of DNA. Soc Sci Inf 50:157–177
Calvert J, Martin P (2009) The role of social scientists in synthetic biology. EMBO Rep 10:201–204
Campos L (2009) That was the synthetic biology that was. In: Schmidt M, Kelle A, Ganguli-Mitra A, Vriend H (eds) Synthetic biology. Springer, Dordrecht, pp 5–21
Carlin NF, Wurman I, Zakim T (2014) How to permit your mammoth: some legal implications of "de-extinction". Stanf Environ Law J 33:3–57
Carlson R (2006) Synthetic biology 2.0: Part IV: what's in a name? [WWW Document]. http://www.synthesis.cc. URL http://www.synthesis.cc/synthesis/2006/05/synthetic_biology_20_part_iv_whats_in_a_name. Accessed 29 Feb 2020
Cartagena Protocol on Biosafety to the 1992 Convention on Biological Diversity (2226 UNTS 208), 2000
Carter S, Rodemeyer M, Garfinkel M, Friedman R (2014) Synthetic biology and the U.S. biotechnology regulatory system: challenges and options. J. Craig Venter Institute
Check Hayden E (2014) Synthetic-biology firms shift focus. Nature 505:598–598
Check Hayden E (2015) Synthetic biology lures Silicon Valley investors. Nature 527:19–19
Cho MK, Magnus D, Caplan AL, McGee D (1999) Policy forum: genetics. Ethical considerations in synthesizing a minimal genome. Science 286:2087, 2089–2090
Coghlan A (2016) Small is beautiful: why a synthetic minimal genome is a big deal [WWW Document]. www.newscientist.com. URL https://www.newscientist.com/article/2082313-small-is-beautiful-why-a-synthetic-minimal-genome-is-a-big-deal/. Accessed 29 Feb 2020
Colwell K (2015) Non-GMO project says no to synthetic biology [WWW Document]. https://foe.org. URL https://foe.org/news/2015-06-non-gmo-project-says-no-to-synthetic-biology/. Accessed 27 Feb 2020
Commission of the Bishops' Conferences of the European Community (COMECE) (2016) Opinion of the reflection group on bioethics on synthetic biology. Secretariat of the Commission of the Bishops' Conferences of the European Community
Commissione federale d'etica per la biotecnologia nel settore non umano (CENU) (2010) Biologia sintetica - Riflessioni etiche. Commissione federale d'etica per la biotechnologia nel settore non umano - Svizzera
Committee on Science, Technology, and Law of the National Academies of Science, Engineering and Medicine (2013) Ownership and sharing - Setting the patent framework for innovation in synthetic biology. Committee on Science, Technology, and Law of the National Academies of Science, Engineering and Medicine
Committee on Science, Technology, and Law, Policy and Global Affairs, Board on Life Sciences, Division on Earth and Life Studies, National Academy of Engineering, National Research Council (2013) Positioning synthetic biology to meet the challenges of the 21st century: summary report of a six academies symposium series. National Academies Press, Washington DC

Conde CA (2012) Consecuencias de la biología sintética en los derechos de propiedad intelectual y acceso a los recursos genéticos y distribución de los beneficios. Revista la Propiedad Inmaterial:281–295

Conference of the parties to the Convention on Biological Diversity (2014a) Potential positive and negative impacts of components, organisms and products resulting from synthetic biology techniques on the conservation and sustainable use of biodiversity, and associated social, economic and cultural considerations (No. UNEP/CBD/COP/12/INF/11). Pyeongchang

Conference of the parties to the Convention on Biological Diversity (2014b) Possible gaps and overlaps with the applicable provisions of the convention, its protocols and other relevant agreements related to components, organisms and products resulting from synthetic biology techniques (UNEP/CBD/COP/12/INF/12). Presented at the Conference of the parties to the convention on biological diversity, Pyeongchang

Connor S (2014) Major synthetic life breakthrough as scientists make the first artificial enzymes [WWW Document]. www.independent.co.uk/. URL https://www.independent.co.uk/news/science/major-synthetic-life-breakthrough-as-scientists-make-the-first-artificial-enzymes-9896333.html. Accessed 29 Feb 2020

Convention on biological diversity (1760 UNTS 79), 1992

Convention on biological diversity (2018) Synthetic biology. Draft decision submitted by the Chair of Working Group II (CBD/COP/14/L.31)

Convention on the Prohibition of the Development, Production and Stockpiling of Bacteriological (Biological) and Toxin Weapons and on their Destruction (1015 UNTS 163), 1972

Corbyn Z (2013) Craig Venter: "This isn't a fantasy look at the future. We are doing the future" [WWW Document]. www.theguardian.com. URL https://www.theguardian.com/science/2013/oct/13/craig-ventner-mars. Accessed 29 Feb 2020

Cornish WR, Llewelyn D, Aplin TF (2019) Intellectual property: patents, copyright, trademarks and allied rights, 9th edn. Sweet & Maxwell, London

Cserer A, Seiringer A (2009) Pictures of synthetic biology. Syst Synth Biol 3:27–35

Cumbers J (2019) Meet eight tech titans investing in synthetic biology [WWW Document]. www.forbes.com. URL https://www.forbes.com/sites/johncumbers/2019/09/14/meet-the-8-tech-titans-investing-in-synthetic-biology/. Accessed 1 Mar 2020

Cumbers J (2020) The synthetic biology companies racing to fight coronavirus [WWW Document]. www.forbes.com. URL https://www.forbes.com/sites/johncumbers/2020/02/05/seven-synthetic-biology-companies-in-the-fight-against-coronavirus/#4c132ceb16ef. Accessed 1 Mar 2020

Cynober T (2019) CRISPR: one patent to rule them all [WWW Document]. https://labiotech.eu. URL https://www.labiotech.eu/features/crispr-patent-dispute-licensing/. Accessed 29 Feb 2020

Davies JA (2019) Real-world synthetic biology: is it founded on an engineering approach, and should it be? Life 9:6–20

de Lorenzo V, Danchin A (2008) Synthetic biology: discovering new worlds and new words. The new and not so new aspects of this emerging research field. EMBO Rep 9:822–827

De Nieuwe Band (2015) Dutch wholesale De Nieuwe Band chooses Sonett and Bioclean and stops with Ecover [WWW Document]. www.sonett.eu. URL https://www.sonett.eu/eur/index.php?id=145&lang=nl. Accessed 29 Feb 2020

Dengler R (2019) Scientists have created four new letters of artificial DNA. http://blogs.discovermagazine.com. URL https://www.discovermagazine.com/planet-earth/scientists-have-created-four-new-letters-of-artificial-dna. Accessed 29 Feb 2020

Deutsche Akademie der Naturforscher Leopoldina (ed) (2010) Realising European potential in synthetic biology: scientific opportunities and good governance, EASAC policy report. Halle (Saale)

Deutsche Akademie der Technikwissenschaften, Deutsche Akademie der Naturforscher Leopoldina, Deutsche Forschungsgemeinschaft (ed) (2009) Synthetische Biologie: Stellungnahme. Wiley-VCH-Verlag, Weinheim

Dietrich J, Steen E (2007) Policy initiatives for safe realization of synthetic biology's power

Dillow C (2010) Researchers create the world's first fully synthetic, self-replicating living cell [WWW Document]. www.popsci.com. URL https://www.popsci.com/science/article/2010-05/j-craig-venter-institute-creates-worlds-first-synthetic-cell/. Accessed 29 Feb 2020

Douglas CM, Stemerding D (2014) Challenges for the European governance of synthetic biology for human health. Life Sci Soc Policy 10:1–18

Dragojlovic N, Einsiedel E (2013) Playing God or just unnatural? Religious beliefs and approval of synthetic biology. Public Underst Sci 22:869–885

Dutfield G, Suthersanen U (2008) Global intellectual property law, 1st edn. Edward Elgar, Cheltenham

Dvorsky G (2016) Mad scientists created synthetic bacteria with only 473 genes [WWW Document]. https://gizmodo.com/. URL https://gizmodo.com/mad-scientists-created-synthetic-bacteria-with-only-473-1766686722. Accessed 29 Feb 2020

Eisenberg RS (2002) How can you patent genes? Am J Bioeth 2:3–11

Elowitz MB, Leibler S (2000) A synthetic oscillatory network of transcriptional regulators. Nature 403:335–338

Endy D (2013) Presentation: what & why is synthetic biology? Presented at the U.S. Department of State, Washington DC

ERASynBio (2014) Next steps for European synthetic biology: a strategic vision from ERASynBio. ERASynBio - European Commission

ERASynBio (n.d.) ERASynBio Initiative [WWW Document]. www.erasynbio.eu. URL https://www.cobiotech.eu/about-cobiotech/erasysbio. Accessed 29 Feb 2020

ETC Group (2006) Background document synthetic biology – Global societal review urgent!

ETC Group (2007) Extreme genetic engineering: an introduction to synthetic biology. ETC Group

ETC Group (2012) The world's largest chemical, energy, agribusiness and pharma companies are investing in synthetic biology [WWW Document]. www.etcgroup.org. URL https://www.etcgroup.org/content/worlds-largest-chemical-energy-agribusiness-and-pharma-companies-are-investing-synthetic. Accessed 29 Feb 2020

ETC Group (2014a) Artemisin & synthetic biology: a case study. ETC Group

ETC Group (2014b) Saffron and synthetic biology: a case study. ETC Group

ETC Group (2014c) Case study: cocoa butter and synthetic biology. ETC Group

ETC Group (2014d) Rubber & synthetic biology: a case study. ETC Group

ETC Group, Heinrich Böll Stiftung (2015) Outsmarting nature? Synthetic biology and climate smart agriculture. Heinrich Böll Stiftung - ETC Group

European Academies Science Advisory Council (EASAC), Swiss Academies of Arts and Sciences, 2011. Synthetische Biologie: Eine Einführung

European Commission, Directorate-General for Research (2005) Synthetic biology: applying engineering to biology - Report of a NEST high-level expert group. Office for Official Publications of the European Communities, Luxembourg

European Commission's Directorate-General for Health & Consumers (2010) Synthetic biology from science to governance. European Commission's Directorate-General for Health & Consumers, Brussels

European Group on Ethics in Science and New Technologies to the European Commission, 2009. Opinion 25: ethics of synthetic biology

Falcone A (2012) Biologia di sintesi, bio-bricks e brevetti biotech fra libertà della ricerca scientifica e tutela della salute umana: Alcune considerazioni sui principi. Forum di quaderni costituzionali 9:1–27

Falcone A (2014) Synthetic biology, biotechnology patents and the protection of human health. A consideration of the principles at stake. In: de Miguel Beriain I, Casabona CMR (eds) Synbio and human health. Springer, Dordrecht, pp 55–76

Farias-Eisner A (2014) Patenting human genes violates laws of nature, impedes patient care, stifles research, and inhibits long-term business opportunities. Northwest J Int Law Bus 2014:1A

Ferrari M (2014) Che cos'è la biologia sintetica [WWW Document]. www.focus.it. URL https://www.focus.it/scienza/salute/che-cos-e-la-biologia-sintetica. Accessed 29 Feb 2020

Ferry D (2015) The promises and perils of synthetic biology [WWW Document]. http://europe.newsweek.com. URL https://www.newsweek.com/2015/03/20/promises-and-perils-synthetic-biology-312849.html. Accessed 29 Feb 2020
Friends of the Earth (n.d.) Synthetic biology: GMOs 2.0
Friends of the Earth, International Center for Technology Assessment (CTA), ETC Group (2012) The principles for the oversight of synthetic biology
Frow E, Calvert J (2013) 'Can simple biological systems be built from standardized interchangeable parts?' Negotiating biology and engineering in a synthetic biology competition. Eng Stud 5:42–58
Gaisser S, Reiss T, Lunkes A, Müller K, Bernauer H (2008) TESSY achievements and future perspectives in synthetic biology - Tessy final report
Gaj T, Gersbach CA, Barbas CF (2013) ZFN, TALEN, and CRISPR/Cas-based methods for genome engineering. Trends Biotechnol 31:397–405
Gardner TS, Cantor CR, Collins JJ (2000) Construction of a genetic toggle switch in Escherichia coli. Nature 403:339–342
Garthwaite J (2014) Beyond GMOs: the rise of synthetic biology [WWW Document]. www.theatlantic.com. URL https://www.theatlantic.com/technology/archive/2014/09/beyond-gmos-the-rise-of-synthetic-biology/380770/. Accessed 29 Feb 2020
Gaskell G, Stares S, Allansdottir A, Allum N, Castro P, Esmer Y, Fischler C, Jackson J, Kronberger N, Hampel J, Mejlgaard N, Quintanilha A, Rammer A, Revuelta G, Stoneman P, Torgersen H, Wagner W (2010) Europeans and biotechnology in 2010: winds of change? A report to the European Commission's Directorate-General for Research (No. EUR 24537 EN). Directorate General for Research Science in Society and Food, Agriculture & Fisheries, & Biotechnology, Luxembourg
Gateway to Research - Research Councils UK (n.d.) The Sc2.0 UK genome engineering resource (SUGER) [WWW Document]. https://gtr.ukri.org/. URL https://gtr.ukri.org/projects?ref=BB%2FK019791%2F1. Accessed 29 Feb 2020
GenScript (n.d.) The synthetic yeast genome project - Sc2.0 [WWW Document]. www.genscript.com. URL https://www.genscript.com/synthetic_yeast_genome.html. Accessed 29 Feb 2020
Gibson DG, Glass JI, Lartigue C, Noskov VN, Chuang R-Y, Algire MA, Benders GA, Montague MG, Ma L, Moodie MM, Merryman C, Vashee S, Krishnakumar R, Assad-Garcia N, Andrews-Pfannkoch C, Denisova EA, Young L, Qi Z-Q, Segall-Shapiro TH, Calvey CH, Parmar PP, Hutchison CA, Smith HO, Venter JC (2010) Creation of a bacterial cell controlled by a chemically synthesized genome. Science 329:52–56
Global Industry Analysts, Inc. (2015) The global synthetic biology market [WWW Document]. www.strategyr.com. URL https://www.strategyr.com/market-report-synthetic-biology-forecasts-global-industry-analysts-inc.asp. Accessed 3 Mar 2016
GP Write (2018) Happy 2nd Anniversary GP-write! [WWW Document]. https://engineeringbiologycenter.org. URL https://engineeringbiologycenter.org/2nd-anniversary-gp-write/. Accessed 29 Feb 2020
GP Write (n.d.-a) GP Write: Our Mission [WWW Document]. https://engineeringbiologycenter.org. URL https://engineeringbiologycenter.org/about/. Accessed 29 Feb 2020
GP Write (n.d.-b) GP Write - FAQ [WWW Document]. https://engineeringbiologycenter.org/. URL https://engineeringbiologycenter.org/faq/. Accessed 29 Feb 2020
Greco P (2013) Gli dei in laboratorio: Le frontiere della biologia sintetica [WWW Document]. www.unipd.it. URL https://ilbolive.unipd.it/content/gli-dei-laboratorio-le-frontiere-della-biologia-sintetica. Accessed 29 Feb 2020
Gregorowius D (2012) What does autonomy of organisms mean for the synthesis of life from a philosophical point of view? [WWW Document]. www.sybhel.org. URL http://www.sybhel.org/?p=575. Accessed 29 Feb 2020
Griffiths PE (2001) Genetic information: a metaphor in search of a theory. Philos Sci 68:394–412
Gschmeidler B, Seiringer A (2012) "Knight in shining armour" or "Frankenstein's creation"? The coverage of synthetic biology in German-language media. Public Underst Sci 21:163–173

Gutmann A (2011) The ethics of synthetic biology: guiding principles for emerging technologies. Hastings Cent Rep 41:17–22

Hanel S (2015) The art of the artificial: test-tube organs and organisms [WWW Document]. www.lindau-nobel.org. URL https://www.lindau-nobel.org/the-art-of-the-artificial-test-tube-organs-and-organisms/. Accessed 29 Feb 2020

Harman G (2014) Technology is ready for synthetic foods. Are you? [WWW Document]. www.theguardian.com. URL https://www.theguardian.com/sustainable-business/2014/dec/17/synthetic-foods-technology-labeling-gmo. Accessed 29 Feb 2020

Health Council of the Netherlands (2008) Synthetic biology: creating opportunities (No. 2008/19E). Health Council of the Netherlands, Advisory Council on Health Research, and Royal Netherlands Academy of Arts and Sciences, The Hague

Helman N, Lim W, Peisajovich S, Pincus D, Sommovilla N (2007) Bio building basics: a conceptual instruction manual for synthetic biology

Henry L (2014) Instead of poppies, engineering microbes to make morphine [WWW Document]. http://blogs.discovermagazine.com/. URL https://www.discovermagazine.com/planet-earth/instead-of-poppies-engineering-microbes-to-make-morphine. Accessed 29 Feb 2020

Heyd D (2012) Is there anything unique in the ethics of synthetic biology? Perspect Biol Med 55:581–589

Holman CM (2015) Developments in synthetic biology are altering the IP imperatives of biotechnology. Vand J Entertain Technol Law 17:385–462

iGEM (n.d.-a) Help: an introduction to BioBricks [WWW Document]. https://igem.org. URL http://parts.igem.org/Help:An_Introduction_to_BioBricks. Accessed 29 Feb 2020

iGEM (n.d.-b) Registry [WWW Document]. https://igem.org. URL https://igem.org/Registry. Accessed 29 Feb 2020

iGEM (n.d.-c) Previous iGem competitions [WWW Document]. https://igem.org. URL https://igem.org/Previous_Competitions. Accessed 29 Feb 2020

Industry Today (2015) Global synthetic biology market to expand at 32.60% CAGR till 2019 thanks to growing R&D efforts [WWW Document]. www.industrytoday.co.uk. URL https://industrytoday.co.uk/market-research-industry-today/global-synthetic-biology-market-to-expand-at-3260-cagr-till-2019-thanks-to-growing-rd-efforts. Accessed 29 Feb 2020

Jacobs P, Van Overwalle G (2001) Gene patents: a different approach. Eur Intellect Prop Rev:505–506

Jain A (2015) SynBioBeta SF 2015: synthetic biology unites industry and academia [WWW Document]. http://blogs.plos.org. URL https://blogs.plos.org/synbio/2015/12/04/synbiobeta-sf-2015-synthetic-biology-unites-industry-and-academia/. Accessed 29 Feb 2020

Janicki M (2014) Synthetic groceries: coming soon [WWW Document]. www.syntheticisnotnatural.com. URL http://www.syntheticisnotnatural.com/ecover-petition/ecover-synthetic-biology-natural-food/. Accessed 29 Feb 2020

Jefferson C, Lentzos F, Marris C (2014a) Report: synthetic biology and biosecurity: how scared should we be?

Jefferson C, Lentzos F, Marris C (2014b) Synthetic biology and biosecurity: challenging the "myths". Front Public Health 2:1–15

Johnson N (2014) The next front in the GMO war: synthetic biology [WWW Document]. http://grist.org. URL https://grist.org/business-technology/the-next-front-in-the-gmo-war-synthetic-biology/. Accessed 29 Feb 2020

Johnson EB (2015) H.R. 591 - Engineering biology research and development act of 2015

Johnson EB (2019) Engineering biology research and development act of 2019 - Report to accompany H.R. 4373 (116th Congress - 1st Session No. 116–270)

Joyce GF, Szostak JW (2018) Protocells and RNA self-replication [WWW Document]. https://cshperspectives.cshlp.org. URL https://cshperspectives.cshlp.org/content/10/9/a034801.full. Accessed 29 Feb 2020

Kaebnick G (2012) Carefully precautionary about synthetic biology? [WWW Document]. www.thehastingscenter.org. URL https://www.thehastingscenter.org/carefully-precautionary-about-synthetic-biology/. Accessed 29 Feb 2020

Kane EM (2004) Splitting the gene: DNA patents and the genetic code. Tenn Law Rev 71:707–767

Kelle A (2007) Synthetic biology & biosecurity awareness in Europe (No. 9), Bradford Science and Technology Report

Kelle A (2009) Ensuring the security of synthetic biology - towards a 5P governance strategy. Syst Synth Biol 3:85–90

Keller EF (2002) Making sense of life. Harvard University Press, Cambridge

Kelley NJ (2014) Synberc sustainability initiative: initial findings & recommendations. Synberc - Alfred P. Sloan Foundation, New York

Kelley NJ, Whelan DJ, Kerr E, Apel A, Beliveau R, Scanlon R (2014) Engineering biology to address global problems: synthetic biology markets, needs, and applications. Ind Biotechnol 10:140–149

Kerr D (2016) Lab-grown food: it's what's for dinner! [WWW Document]. www.cnet.com. URL https://www.cnet.com/news/lab-grown-meat-in-vitro-meat-sergey-brin-bill-gates-climate-change-cultured-beef/. Accessed 29 Feb 2020

Krieger LM (2013a) Synthetic biology: Stanford, UC Berkeley engineering a new frontier [WWW Document]. www.mercurynews.com. URL https://www.mercurynews.com/2013/02/15/synthetic-biology-stanford-uc-berkeley-engineering-a-new-frontier/. Accessed 29 Feb 2020

Krieger LM (2013b) Biological computer created at Stanford [WWW Document]. www.mercurynews.com. URL https://www.mercurynews.com/2013/03/28/biological-computer-created-at-stanford/. Accessed 29 Feb 2020

Krieger LM (2015a) Designing life from scratch: a fledgling field is about to take off [WWW Document]. www.mercurynews.com. URL https://www.mercurynews.com/2015/08/08/designing-life-from-scratch-a-fledgling-field-is-about-to-take-off/. Accessed 29 Feb 2020

Krieger LM (2015b) Stanford team turns yeast into painkillers [WWW Document]. www.mercurynews.com. URL https://www.mercurynews.com/2015/08/13/stanford-team-turns-yeast-into-painkillers/. Accessed 1 Mar 2020

Kronberger N (2012) Synthetic biology: taking a look at a field in the making. Public Underst Sci 21:130–133

Kronberger N, Holtz P, Wagner W (2012) Consequences of media information uptake and deliberation: focus groups' symbolic coping with synthetic biology. Public Underst Sci 21:174–187

Krumins A (2015) Startup in biotech adds two base pairs to genetic code — and life on earth may never be the same [WWW Document]. www.extremetech.com. URL https://www.extremetech.com/extreme/212755-startup-in-biotech-adds-two-base-pairs-to-genetic-code-and-life-on-earth-may-never-be-the-same. Accessed 29 Feb 2020

Kuzma J, Tanji T (2010) Unpackaging synthetic biology: identification of policy problems and options. Regul Gov 4:92–112

Lai H-E, Canavan C, Cameron L, Moore S, Danchenko M, Kuiken T, Sekeyová Z, Freemont PS (2019) Synthetic biology and the United Nations. Trends Biotechnol 37:1146–1151

LaMonica M (2014) Why the promise of cheap fuel from super bugs fell short [WWW Document]. www.technologyreview.com. URL https://www.technologyreview.com/s/524011/why-the-promise-of-cheap-fuel-from-super-bugs-fell-short/. Accessed 29 Feb 2020

Ledford H (2015) Biohackers gear up for genome editing. Nature 524:398–399

Leduc S (1912) La biologie synthétique. Poinat, Paris

Lehmkuhl M (2011) Die Repräsentation der synthetischen Biologie in der deutschen Presse

Leproust E (2015) Synthetic biology is not just good, it's good for you [WWW Document]. http://social.techcrunch.com. URL https://techcrunch.com/2015/09/28/synthetic-biology-is-not-just-good-its-good-for-you/?guccounter=1&guce_referrer=aHR0cHM6Ly93d3cuZ29vZ2xlLmNvbS8&guce_referrer_sig=AQAAANVi9Q-CtjxFJInd5YmM9fFfSVBHUdT_X31er_rHcrz-UD6UN0JDVMDUrLQBtqJL-

isnCio7hFAMNU8i5gsOkxD0fYO34Owcx8lcagNaInQilaf8ETAdmvmfp5zgsx7vrl4yXI43fM ogGNoOfdobJz6kbJA5HVqLrG9kQayXBHhk. Accessed 29 Feb 2020
Lesser W (2011) Nature or nurture: is there a case basis for a judicially created product of nature exclusion? Are genes somehow different? John Marshall Rev Intellect Prop Law 11:318–366
Limbachiya D, Gupta MK (2015) Natural data storage: a review on sending information from now to then via nature. ACM J Technol Comput Syst:1–17
Lowe D (2016) The smallest viable genome is very weird [WWW Document]. http://blogs.sciencemag.org. URL https://blogs.sciencemag.org/pipeline/archives/2016/03/28/the-smallest-viable-genome-is-very-weird. Accessed 29 Feb 2020
Mandal A (2014) What is molecular biology? [WWW Document]. www.news-medical.net. URL https://www.news-medical.net/life-sciences/What-is-Molecular-Biology.aspx. Accessed 29 Feb 2020
Marris C (2013) Synthetic biology's malaria promises could backfire [WWW Document]. www.scidev.net. URL https://www.scidev.net/global/biotechnology/opinion/synthetic-biology-s-malaria-promises-could-backfire.html?__cf_chl_jschl_tk__
=ffbf78e172db6d9c741a87361fc52ad515e1eab3-1583000737-0-
AWJM9U4xaf5KaCKLjUw1bQRAQI6BsxKakHJ7ZOLZtDbdu9ihhhZjqxf_
yX5WXaMLfUpz64GHIRr98R39hV5nwodjfovJSX1lbhhXXOsj_
I5VcyXSF6xvXqMdOTinAPzLNfsx1gpQiDqXwAHfz90dEyHiymLkfPgSUP-
S67W8oNvWg9K_
yGBKkhrGBIl1utAkqONZzEN13KsYRUe8KxRDtDMo8j0zjHBKlxRtk88HhCpUeAxEXdq0
ktSLxunFYlYXM9AnDN-dKYXv%2D%2DKLaQwCpP8_F4j5D9Af3jSmEdzo_w-
sCQq0pixBGX7y8y1ngSHxrXFu8r-cC7OT_Tb_
EnSpOR7VAgoVfhL4aTnFbnDxhiep2ez9d3CA0MBWFkqHXf4jPc0zdnYPnCUbcDhFT6_
8b_c. Accessed 29 Feb 2020
Marris C, Jefferson C (n.d.) Synthetic biology and biosecurity: an agenda for social science research
Maurer SM (2011) End of the beginning or beginning of the end? Synthetic biology's stalled security agenda and the prospects for restarting it. Valparaiso Univ Law Rev 45:73–132
Maurer SM, Lucas KV, Terrell S (2006) From understanding to action: community-based options for improving security and safety in synthetic biology
Mayhall TA (2019) The meat of the matter: regulating a laboratory-grown alternative. Food Drug Law J 74:151–169
Maynard A (2016) Three ways synthetic biology could annihilate Zika and other mosquito-borne diseases [WWW Document]. http://theconversation.com. URL https://theconversation.com/three-ways-synthetic-biology-could-annihilate-zika-and-other-mosquito-borne-diseases-54087. Accessed 1 Mar 2020
McEachran R (2015) Creators defend vanilla flavour made using synthetic biology [WWW Document]. www.theguardian.com. URL https://www.theguardian.com/sustainable-business/2015/may/28/creators-defend-vanilla-flavour-made-using-synthetic-biology. Accessed 29 Feb 2020
McLennan A, Rimmer M (2012) Inventing life: patent law and synthetic biology [WWW Document]. http://theconversation.com. URL http://theconversation.com/inventing-life-patent-law-and-synthetic-biology-5178. Accessed 27 Feb 2020
Meanings of "life" (2007) Nature 447:1031–1032
Meyer M (2012) Bricoler, domestiquer et contourner la science: L'essor de la biologie de garage. Réseaux 3:303–328
Mistbreaker News (2015) Synthetic biology is discretely targeted by tech giants [WWW Document]. www.mistbreaker.com. URL http://www.mistbreaker.com/medicine-biotech/synthetic-biology-discretely-targeted-tech-giants/. Accessed 29 Feb 2020
Moe-Behrens GHG, Davis R, Haynes KA (2013) Preparing synthetic biology for the world. Front Microbiol 4:1–9
Molecules vs. information: Should patents protect both? (2002) Boston Univ J Sci Technol Law 8:190–217

Mushegian AR, Koonin EV (1996) A minimal gene set for cellular life derived by comparison of complete bacterial genomes. Proc Natl Acad Sci U S A 93:10268–10273

Myers A (2013) Biological transistor enables computing within living cells, study says [WWW Document]. https://engineering.stanford.edu. URL https://med.stanford.edu/news/all-news/2013/03/biological-transistor-enables-computing-within-living-cells-study-says.html. Accessed 29 Feb 2020

Nanalyze (2018) Most funded synthetic biology companies in 2018 [WWW Document]. www.nanalyze.com/. URL https://www.nanalyze.com/2018/12/synthetic-biology-companies-2018/. Accessed 29 Feb 2020

National Human Genome Research Institute (2014) Frequently asked questions about genetic and genomic science [WWW Document]. www.genome.gov/. URL https://www.genome.gov/human-genome-project/Completion-FAQ. Accessed 29 Feb 2020

National Human Genome Research Institute (n.d.) The human genome project [WWW Document]. www.genome.gov. URL https://www.genome.gov/human-genome-project. Accessed 29 Feb 2020

NewScientist.com (2016) Breakthrough in synthetic biology is far from "playing God" [WWW Document]. www.newscientist.com. URL https://www.newscientist.com/article/mg23030672-700-breakthrough-in-synthetic-biology-is-far-from-playing-god/. Accessed 29 Feb 2020

Newson AJ (2015) Synthetic biology: ethics, exceptionalism and expectations. Macquarie Law J 15:45–58

Nista P (2015) Can synthetic biology create orphan black's clones? [WWW Document]. http://sciencevshollywood.com. URL http://sciencevshollywood.com/can-synthetic-biology-create-orphan-blacks-human-clones/. Accessed 29 Feb 2020

Nowogrodzki A (2018) The automatic-design tools that are changing synthetic biology. Nature 564:291–292

Nuffield Council on Bioethics (2012) Emerging biotechnologies: technology, choice and the public good

O'Malley MA, Powell A, Davies JF, Calvert J (2008) Knowledge-making distinctions in synthetic biology. Bioessays 30:57–65

OECD (2014) Emerging policy issues in synthetic biology. OECD, Paris

OECD Development (2011) Future prospects for industrial biotechnology. OECD, Paris

OECD, The Royal Society (2010) Symposium on opportunities and challenges in the emerging field of synthetic biology: synthesis report. OECD - The Royal Society

Open Science (2012) Synthetische Biologie - Eine Einführung

Palca J, Venter CJ, Norris M, Siegel R, Collins JJ, Kaebnick GE (2010) Interview: scientists reach milestone on way to artificial life [WWW Document]. www.npr.org. URL http://www.npr.org/templates/story/story.php?storyId=127010591. Accessed 23 Feb 2016

Panke S (2008) Synthetic biology - engineering in biotechnology. Swiss Academy of Engineering Sciences, Zurich

Paradise J, Fitzpatrick E (2012) Synthetic biology: does re-writing nature require re-writing regulation? Penn State Law Rev 117:53–88

Parens E, Johnston J, Moses J (2009) Ethical issues in synthetic biology: an overview of the debates. Woodrow Wilson International Center for Scholars - Synthetic Biology Project, Washington DC

Parliamentary Office of Science and Technology (POST) (2008) Synthetic biology (No. Number 298)

Parliamentary Office of Science and Technology (POST) (2015) Regulation of synthetic biology (No. 497), Postnote. Parliamentary Office of Science & Technology (POST) of the House of Parliament, London

Pauwels E (2013) Public understanding of synthetic biology. BioScience 63:79–89

Pei L, Gaisser S, Schmidt M (2012) Synthetic biology in the view of European public funding organisations. Public Underst Sci 21:149–162

Perls D (2014) Biotech industry cooks up PR plans to get us to swallow synthetic biology food [WWW Document]. www.foe.org. URL https://foe.org/2014-05-the-synthetic-biology-industrys-pr-scheme/. Accessed 29 Feb 2020

Philp J (2014) Presentation: emerging policy issues in synthetic biology. Presented at the Ministry of Science and Higher Education - Republic of Poland

Pollack A (2014) Scientists add letters to DNA's alphabet, raising hope and fear [WWW Document]. www.nytimes.com. URL https://www.nytimes.com/2014/05/08/business/researchers-report-breakthrough-in-creating-artificial-genetic-code.html. Accessed 29 Feb 2020

Porcar M, Peretó J (2012) Are we doing synthetic biology? Syst Synth Biol 6:79–83

Porcar M, Danchin A, de Lorenzo V, dos Santos VA, Krasnogor N, Rasmussen S, Moya A (2011) The ten grand challenges of synthetic life. Syst Synth Biol 5:1–9

Powell K (2018) How biologists are creating life-like cells from scratch. Nature 563:171–175

Presidential Commission for the Study of Bioethical Issues (2010) New directions: the ethics of synthetic biology and emerging technologies. Presidential Commission for the Study of Bioethical Issues, Washington DC

Preston CJ (2008) Synthetic biology: drawing a line in Darwin's sand. Environ Values 17:23–39

Preston B (2013) Synthetic biology as red herring. Stud Hist Philos Sci - Part C: Stud Hist Philos Biol Biomed Sci 44:649–659

Raho JA (2012) La metafora morale e le questioni emergenti in bioetica. Il caso della biologia di sintesi. In: Quaderni Della Ricerca. Edizioni ETS, Pisa, pp 287–308

Rai AK (1999) Intellectual property rights in biotechnology: addressing new technology. Wake Forest Law Rev 34:827

Resnik DB (2002) Discoveries, inventions, and gene patents. In: Magnus D, Caplan AL, McGee G (eds) Who owns life? Prometheus Books, Amherst, pp 135–159

Ribarits A, Stepanek W, Wögerbauer M, Peterseil V, Kuffner M, Topitschnig C, Brüller W, Hochegger R, Gansberger M, Widhalm I, Leonhardt C (2014) Synthetic Biology. Bundesministerium für Gesundheit (Österreich)

Rice J (2010) Synthetic genome reboots cell [WWW Document]. www.technologyreview.com/. URL https://www.technologyreview.com/s/418999/synthetic-genome-reboots-cell/. Accessed 29 Feb 2020

Rita V, Novelli G, Redi CA (2014) Interview: Sintesi o creazione? Le nuove frontiere della biologia sintetica. A colloquio con Giuseppe Novelli e Carlo Alberto Redi [WWW Document]. www.quotidianosanita.it. URL https://www.quotidianosanita.it/scienza-e-farmaci/articolo.php?articolo_id=22067. Accessed 1 Mar 2020

Rizzuto P (2014) Synthetic biology products may increasingly fall outside regulatory system, study says [WWW Document]. www.bna.com. URL https://www.bna.com/synthetic-biology-products-n17179890903/. Accessed 23 Mar 2016

Rousseaux A (2015) Synthetic biology: life reconstructed by engineers and multinationals [WWW Document]. www.multinationales.org. URL http://multinationales.org/Synthetic-Biology-Life. Accessed 27 Feb 2020

Royal Academy of Engineering (Great Britain) (2009) Synthetic biology: scope, applications and implications. Royal Academy of Engineering, London

Sankin A (2013) Biological computer: Stanford researchers discover genetic transistors that turn cells into computers [WWW Document]. www.huffingtonpost.com. URL https://guce.huffpost.com/copyConsent?sessionId=3_cc-session_8dd61c45-5a80-4df5-b249-19221fd69940&inline=false&lang=en-us. Accessed 29 Feb 2020

Sarkar S (1996) Decoding "coding": information and DNA. BioScience 46:857–864

Savage N (2007) Making gasoline from bacteria [WWW Document]. www.technologyreview.com. URL https://www.technologyreview.com/s/408334/making-gasoline-from-bacteria/. Accessed 29 Feb 2020

SCHER, SCENIHR, SCCS Scientific Committees (2014) Opinion on synthetic biology I: definition. Scientific Committees of the European Commission

SCHER, SCENIHR, SCCS Scientific Committees (2015a) Opinion on synthetic biology II: risk assessment methodologies and safety aspects. Scientific Committees of the European Commission

SCHER, SCENIHR, SCCS Scientific Committees (2015b) Final opinion on synthetic biology III: risks to the environment and biodiversity related to synthetic biology and research priorities in the field of synthetic biology. Scientific Committees of the European Commission

Schertenleib D (2003) The patentability and protection of DNA-based inventions in the EPO and the European Union. Eur Intellect Prop Rev 25:125–138

Schmidt M (2008) Diffusion of synthetic biology: a challenge to biosafety. Syst Synth Biol 2:1–6

Schmidt M (2010) Xenobiology: a new form of life as the ultimate biosafety tool. Bioessays 32:322–331

Schmidt M, Giersch G (2012) DNA synthesis and security. In: Campbell MJ (ed) DNA microarrays, synthesis, and synthetic DNA. Nova Science, New York, pp 285–300

Schwille P (2015) Jump-starting life? Fundamental aspects of synthetic biology. J Cell Biol 210:687–690

Schwille P, Sundmacher K (n.d.) Synthetic biology

Scudellari M (2015) Inner workings: DNA for data storage and computing. Proc Natl Acad Sci 112:15771–15772

Seitz C (2018) Modifiziert oder nicht? Regulatorische Rechtsfragen zur Genoptimierung durch neue biotechnologische Verfahren. Europäische Zeitschrift für Wirtschaftsrecht 757–764

Sharma Y (2012a) NGOs call for international regulation of synthetic biology [WWW Document]. www.scidev.net. URL https://www.scidev.net/global/genomics/news/ngos-call-for-international-regulation-of-synthetic-biology.html?__cf_chl_jschl_tk__=e5b551e9aaa318c4332585ffff7c033026275570-1583000143-0-AfsvBknFjph9ZcIAB88O-Nif83ddHEDXJgzSysQvSl50jSFi8fNj3Fw_XTPnbwGJhm4RfVlnlBGhkpiNw9BIiM1t1e_fF-W-vZOHr4BXNDvTnOyV33vS7cDYPDhPWnMMDL67Eu2S7scyfbEL4wjhzaIXdpGVkvcXmM9ZEbc8ESeA4JyB3zB9adcBZ1UWp9upXI_ZBh42Me8QcCb-mNWcBRXSqoUxXOkP2RlUFZCSLqEBaxn7Anv0IeIAe5SWT1tbwxd8YUoniFpPGSjv6_27PnITJAE9ZE9eMm6k5N9Okh92osRqPKW87NHJWV35_zLJZ_7aJ201bxS3pgtN0GPasBtycUs_TiGvb2kfL1CTpQ0Yvx4vDrP09MvCnqFGi7E_sK5m4LHNP0L2uBx8y4-SmTU. Accessed 29 Feb 2020

Sharma Y (2012b) Developing countries face up to synthetic biology challenges [WWW Document]. www.scidev.net. URL https://www.scidev.net/global/pollution/feature/developing-countries-face-up-to-synthetic-biology-challenges-1.html. Accessed 29 Feb 2020

Sherman B (2015) What does it mean to invent nature? UC Irvine Law Rev 5:1193

Shivam (2019) Genome project-write & synthetic life | A giant step closer to the reality [WWW Document]. https://labstud.com. URL https://labstud.com/genome-project-write/. Accessed 29 Feb 2020

Simone A (2014) 4 passi nella biologia sintetica [WWW Document]. http://aulascienze.scuola.zanichelli.it. URL https://agricolturadigitale.net/2019/04/15/biologia-sintetica-e-agricoltura/. Accessed 29 Feb 2020

Singer E (2008) Synthesizing a genome from scratch [WWW Document]. www.technologyreview.com/. URL https://www.technologyreview.com/s/409441/synthesizing-a-genome-from-scratch/. Accessed 29 Feb 2020

Specter M (2009) A life of its own [WWW Document]. www.newyorker.com. URL https://www.newyorker.com/magazine/2009/09/28/a-life-of-its-own. Accessed 1 Mar 2020

Spiegel Online (2013) Surrogate mother (not yet) sought for Neanderthal [WWW Document]. www.spiegel.de. URL https://www.spiegel.de/international/spiegel-responds-to-brouhaha-over-neanderthal-clone-interview-a-879311.html. Accessed 29 Feb 2020

Stinson L (2015) To save our ecosystems, will we have to design synthetic creatures? [WWW Document]. www.wired.com. URL https://www.wired.com/2015/01/save-ecosystems-will-design-synthetic-creatures/. Accessed 29 Feb 2020

Strom S (2014) Companies quietly apply biofuel tools to household products [WWW Document]. www.nytimes.com. URL https://www.nytimes.com/2014/05/31/business/biofuel-tools-applied-to-household-soaps.html. Accessed 29 Feb 2020
Sulston J (2006) Staking claims in the biotechnology Klondike. Bull World Health Organ 84:412–413
Sulston J, Ferry G (2002) The common thread: a story of science, politics, ethics, and the human genome. Joseph Henry Press, Washington DC
SYBHEL (2014) Final report summary - SYBHEL (Synthetic Biology for Human Health: Ethical and Legal Issues) (Executive Summary). European Commission
Synthetic Biology Project (n.d.-a) Navigating the regulatory landscape: genetically modified plant case study
Synthetic Biology Project (n.d.-b) Navigating the regulatory landscape: synthetic squalane case study
Synthetic Biology Project (n.d.-c) Navigating the regulatory landscape: biomining case study
Takahashi D (2016) Synthetic biologist discusses the ethics of designing life to save nature [WWW Document]. http://venturebeat.com. URL https://venturebeat.com/2016/01/18/synthetic-biologist-discusses-the-ethics-of-designing-life-to-save-nature/. Accessed 29 Feb 2020
Taliadoros L, Muratore A (2000) Patenting and ownership of genes and life forms - Australian perspective. Int Bus Lawyer 28:117–131
Tatalovic M (n.d.) Poor countries urged to get ready for synthetic biology [WWW Document]. www.scidev.net. URL https://www.scidev.net/global/r-d/news/poor-countries-urged-to-get-ready-for-synthetic-biology-1.html. Accessed 29 Feb 2020
Teller A (2015) Potential impacts of synthetic biology on biological diversity
Tennant J (2015) 75 million-year-old blood cells discovered in dinosaur bones [WWW Document]. http://blogs.discovermagazine.com. URL https://www.discovermagazine.com/planet-earth/75-million-year-old-blood-cells-discovered-in-dinosaur-bones. Accessed 29 Feb 2020
The Economist (2007) Patent pending [WWW Document]. www.economist.com. URL https://www.economist.com/node/9333408/print?story_id=9333408. Accessed 29 Feb 2020
The Economist (2015) Strange medicine [WWW Document]. www.economist.com. URL https://www.economist.com/science-and-technology/2015/02/26/strange-medicine. Accessed 29 Feb 2020
Then C, Hamberger S (2010) Synthetische Biologie - Teil 1: Synthetische Biologie und künstliches Leben - Eine kritische Analyse. Testbiotech e.V. - Institut für unabhängige Folgenabschätzung in der Biotechnologie
Thomas J (2014) What-syn-a-name? [WWW Document]. www.theguardian.com. URL https://www.theguardian.com/science/political-science/2014/jul/08/what-syn-a-name. Accessed 29 Feb 2020
Tinafar A, Jaenes K, Pardee K (2019) Synthetic biology goes cell-free. BMC Biol 17:64. https://doi.org/10.1186/s12915-019-0685-x
Torgersen H, Hampel J (2012) Calling controversy: assessing synthetic biology's conflict potential. Public Underst Sci 21:134–148
Tucker JB, Zilinskas RA (2006) The promise and perils of synthetic biology. New Atlantis:25–45
Twilley N (2016) What's the point of streamlining nature? [WWW Document]. www.newyorker.com. URL https://www.newyorker.com/tech/annals-of-technology/whats-the-point-of-streamlining-nature. Accessed 29 Feb 2020
U.S. Supreme Court (1964) Jacobellis v. Ohio, 378 U.S. 184 (1964)
Ulivieri V (2016) Cereali: In 45 anni produttività crollata del 10% per caldo e siccità [WWW Document]. www.corriere.it. URL https://www.corriere.it/ambiente/16_gennaio_12/cereali-produzione-crollata-caldo-siccita-agricoltura-85bf0b88-b94a-11e5-85a5-46ffd263e960.shtml. Accessed 29 Feb 2020
United Nations Conference on Trade and Development (2019) Synthetic biology and its potential implications for biotrade and access and benefit-sharing (No. UNCTAD/DITC/TED/INF/2019/12). United Nations Conference on Trade and Development

van den Belt H (2009a) Playing God in Frankenstein's footsteps: synthetic biology and the meaning of life. Nanoethics 3:257–268

van den Belt H (2009b) Philosophy of biotechnology. In: Meijers A (ed) Philosophy of technology and engineering sciences. North-Holland, Amsterdam, pp 1301–1340

van der Hoeven D (2015) Synthetic biology in food production [WWW Document]. www.biobasedpress.eu. URL https://www.biobasedpress.eu/2015/03/synthetic-biology-in-food-production/. Accessed 29 Feb 2020

Villa R (2011) La biologia sintetica per creare vaccini universali [WWW Document]. www.corriere.it. URL https://www.corriere.it/salute/11_novembre_20/dossier-genetica-biologia-sintetica_89971a54-1109-11e1-b811-fb0a2ca90bde.shtml. Accessed 29 Feb 2020

Wade N (2007) Scientists transplant genome of bacteria [WWW Document]. www.nytimes.com. URL https://www.nytimes.com/2007/06/29/science/29cells.html. Accessed 29 Feb 2020

Wade N (2010) Researchers say they created a "synthetic cell" [WWW Document]. www.nytimes.com. URL https://www.nytimes.com/2010/05/21/science/21cell.html. Accessed 29 Feb 2020

Watson E (2015a) Lux Research: cost, speed & sustainability benefits of synthetic biology will make it a 'permanent and growing aspect' of flavors market [WWW Document]. www.foodnavigator-usa.com. URL https://www.foodnavigator.com/Article/2015/08/27/Lux-Research-Synthetic-biology-will-drive-flavors-fragrances-market. Accessed 29 Feb 2020

Watson E (2015b) Ingredients made via synthetic biology won't qualify for Non-GMO Project verified stamp [WWW Document]. www.foodnavigator-usa.com. URL https://www.foodnavigator-usa.com/Article/2015/06/16/Synbio-ingredients-won-t-qualify-for-Non-GMO-Project-verified-stamp. Accessed 29 Feb 2020

What's in a name? (2009) Nat Biotech 27:1071–1073

Wilson TV (2008) Can scientists clone dinosaurs? [WWW Document]. http://science.howstuffworks.com. URL https://science.howstuffworks.com/environmental/earth/geology/dinosaur-cloning.htm. Accessed 29 Feb 2020

Xinhua (2018) China to have professional committee of synthetic biology [WWW Document]. www.xinhuanet.com. URL http://www.xinhuanet.com/english/2018-11/14/c_137606119.htm. Accessed 29 Feb 2020

Yarris L (2011) CAD for RNA | Berkeley Lab [WWW Document]. http://newscenter.lbl.gov/. URL https://newscenter.lbl.gov/2011/12/22/cad-for-rna/. Accessed 29 Feb 2020

Yarris L (2014) Synthetic biology for space exploration [WWW Document]. http://newscenter.lbl.gov. URL https://newscenter.lbl.gov/2014/11/05/synthetic-biology-for-space-exploration/. Accessed 29 Feb 2020

Yearley S (2009) The ethical landscape: identifying the right way to think about the ethical and societal aspects of synthetic biology research and products. J R Soc Interface 6:559–564

Yeh BJ, Lim WA (2007) Synthetic biology: lessons from the history of synthetic organic chemistry. Nat Chem Biol 3:521–525

Young Rojahn S (2014) Yeast 2.0 [WWW Document]. www.technologyreview.com. URL https://www.technologyreview.com/s/525946/yeast-20/. Accessed 29 Feb 2020

Zakeri B, Lu TK (2015) DNA nanotechnology: new adventures for an old warhorse. Curr Opin Chem Biol 28:9–14

Zhang S (2015) Cheap DNA sequencing is here. Writing DNA is next [WWW Document]. www.wired.com. URL https://www.wired.com/2015/11/making-dna/. Accessed 29 Feb 2020

Zhang J, Marris C, Rose N (2011) The transnational governance of synthetic biology: scientific uncertainty, cross-borderness and the "art" of governance. BIOS working paper n° 4

# Chapter 3
# Norms and Patents in the Field of Synthetic Biology

## 3.1 Patent Norms Applicable to Synthetic Biology Inventions

In Europe, the patentability of synthetic biology inventions is regulated by a number of legislative instruments.

The primary source of guidance in this field is the European Convention on the Grant of European Patents (also known as the European Patent Convention—EPC[1]). A European patent holder shall have "*the same rights as would be conferred by a national patent granted in that State*".[2] The reference to States derives from the fact that European patents do not constitute a unitary legal title that confers a patent valid in all EPC member States, but rather a bundle of national patent rights that are valid in the EPC member States designated by the patentee. As of now, the EPC is in force in 38 countries.[3] Amongst its members and signatories are both EU and non-EU member States. The EU itself is not a contracting party to the Convention.

Rather than offering a unified legal title, the Convention unifies the patent application process. The EPC assigns to the European Patent Office (EPO) the role of examination authority. The EPO bases its decisions on the EPC, which constitutes "*a body of unified substantive patent law*" and a "*highly autonomous European system of law*".[4] The most relevant substantive patent law provisions are detailed in Articles 52–57 EPC, which present the criteria for patentability. In addition to the

---

[1] *Convention on the grant of European patents of 5 October 1973, as revised on 29 November 2000 (European Patent Convention - EPC)*, 1973. The EPC as revised in November 2000 is also known as EPC 2000 to differentiate it from the original version from 1973 (EPC 1973; *Convention on the grant of European patents of 5 October 1973*, 1973).

[2] Article 64 EPC.

[3] European Patent Office (n.d.).

[4] Kur and Dreier (2013, p. 91) and Singer and Stauder (2003, p. 16).



I. de Lisa, *The Patentability of Synthetic Biology Inventions*,
https://doi.org/10.1007/978-3-030-51206-4_3

requirements of novelty, inventiveness and industrial application, Article 52 EPC establishes that "*European patents shall be granted for any inventions, in all fields of technology*". Hence, the patenting of biological matter is not excluded a priori by the EPC.[5] The current formulation of Article 52 EPC is the result of the amendments introduced by the EPC 2000. In that occasion, the EPC was updated in order to take into account the changes and developments determined by the Agreement on Trade-Related Aspects of Intellectual Property Rights (TRIPS).[6] While the basic structure of the Convention was not affected by this revision, Article 52 EPC was adapted to Article 27 TRIPS by clarifying that patents can be granted in "*all fields of technology*".[7]

TRIPS represents a second source of substantive patent law that influences the patentability of synthetic biology inventions. It constitutes an Annex to the 1994 Marrakesh Agreement establishing the World Trade Organization (WTO). It was entered into by members of the WTO, which include all European countries. The European Union became a party of it on 1st January 1995.[8] TRIPS represents a global "*playing field for standardized intellectual property*".[9] Despite the existence of other international IP agreements and conventions, TRIPS established itself as the "*most comprehensive IP agreement to date, setting minimum standards in almost all areas of IP law*".[10]

In TRIPS, the key norm on patents is Article 27.[11] This provision stipulates that patent protection is available "*for any inventions... in all fields of technology*", as long as the invention is new, inventive and capable of industrial application. Furthermore, it establishes that both patents and patent rights shall be available and enjoyable "*without discrimination as to the... field of technology*". This enshrines the technological neutrality of the patent system.[12] While TRIPS is applicable to biotechnological inventions, it does not offer any specific guidance

[5]Seville (2016, p. 201).

[6]*Agreement on Trade-Related Aspects of Intellectual Property Rights (TRIPS)(1869 U.N.T.S. 299)* (1994).

[7]Kur and Dreier (2013, p. 91). Article 52(4) EPC was affected by the EPC 2000, as this section was relocated from here to Article 53 EPC. This change was implemented to place the exceptions to patentability together in a single article (McDonell et al. 2016, p. 3). The previous formulation of that same article simply explained that: "*Patents shall be granted for any inventions*".

[8]Council of the European Union (1994).

[9]Hashim (2013, p. 657).

[10]Hestermeyer (2013, p. 927).

[11]This norm, which constitutes the *raison d'être* of the entire TRIPS agreement, is aimed at overcoming the discriminations that plagued a number of technological fields in the past, when patent protection for chemicals and pharmaceuticals was often not available (de Carvalho 2018, p. 233). Indeed, by the beginning of the treaty negotiations, patent protection for medicines and other products, such as food and beverages, was not available in approximately 50 countries (Correa 2007, p. 271).

[12]There are authors who consider such technology-neutrality a myth (van den Belt 2014, p. 26).

on this point.[13] Still, "*it is generally accepted that biological material, including human genes, should be regarded as patentable subject matter under TRIPS*".[14] Nevertheless, States have the possibility to deny patent protection for genetic material on ethical grounds.[15]

The relation between TRIPS and the EPC has played a role in the development of these two legislative instruments. For one, the two sets of norms have influenced each other's formulation. As previously mentioned, the EPC 2000 introduced some changes in the text of the EPC in order to comply with TRIPS. However, this did not represent a one-way street, as the opposite was also true, since "*many of the minimal patentability requirements set out in TRIPS were based on the... EPC*".[16] In particular, Article 27(2) and 27(3) TRIPS were heavily influenced by the EPC.[17] Furthermore, TRIPS influences the interpretation of European norms. In fact, the European Court of Justice (ECJ) held that member States should interpret their national norms in light of the wording and purpose of TRIPS.[18]

The third legislative instrument that affects the patentability of synthetic biology in Europe is Directive 98/44, also known as the Biotechnology Directive (Biotech Directive or Directive).[19] The Directive came into force in 1998 after a difficult

[13]Falcone (2014, p. 66) and Ugurlu (2014, p. 18).

[14]Wüger (2008, p. 81).

[15]Van Overwalle (2008, p. 81).

[16]Rudge (2014, p. 45).

[17]Such similarity between the norms contained in the EPC and in TRIPS raised the question of whether the latter could be applied in the context of the Convention and whether it could have any direct effect on it. In assessing these questions, the Board of Appeal considered that the EPO is neither a signatory of TRIPS nor a member of the WTO. Therefore, since TRIPS is binding only on its members, a direct application to the EPC would not be justified. Equally, it highlighted that there was no precise correspondence between the parties to the EPC and TRIPS (Tritton et al. 2018, pp. 85–86; Rudge 2014, pp. 45–46; Technical Board of Appeal of the EPO (TBA) 1999, § 2–3). Nonetheless, despite holding that: "*The only source of substantive patent law for examining European patent applications at this moment is the European Patent Convention*", it believed that: "*Although TRIPS may not be applied directly to the EPC, the Board thinks it appropriate to take it into consideration, since it is aimed at setting common standards and principles concerning the availability, scope and use of trade-related intellectual property rights, and therefore of patent rights. Thus TRIPS gives a clear indication of current trends*" (Tritton et al. 2018, pp. 60–61; Technical Board of Appeal of the EPO (TBA) 1999, § 2.3). Similarly, the Enlarged Board of Appeal, in deciding whether the EPO would be bound by TRIPS given that most of its members are, held that: "*TRIPS provisions, like decisions of the European and International Courts of Justice and national decisions, are elements to be taken into consideration by the boards of appeal but are not binding on them. Whereas it is legitimate for the boards of appeal to use the TRIPS Agreement as a means to interpret provisions of the EPC which admit of different interpretations, specific provisions of TRIPS cannot justify ignoring express and unambiguous provisions of the EPC*" (Enlarged Board of Appeal of the EPO (EBA) 2004, § 8.6).

[18]Tritton et al. (2018, p. 54), European Court of Justice (ECJ) (2010, § 72) and European Court of Justice (ECJ) (2004, § 57).

[19]*Directive 98/44/EC of the European Parliament and of the Council of 6 July 1998 on the legal protection of biotechnological inventions (Biotechnology Directive or Biotech Directive or Directive)*, 1998. The Directive references to TRIPS. In Article 1, it mentions that the Directive: "*Shall be*

normative journey. The Commission had considered the different levels of protection accorded in Europe to biotechnological inventions in comparison to other countries (e.g. USA and Japan). The Directive was thus seen as a way to harmonise the legal panorama in Europe and to increase the protection granted to these inventions.[20] In particular, it was noted that, although national norms were often based on the EPC, their different interpretations could lead to discrepancies affecting the internal market. Hence, it was decided to harmonise this subject via a Directive.[21] Being addressed to the EU member States, the Directive required them to amend their national laws to comply with it.[22] In spite of this, eight States had not implemented the norms in their national laws by the required deadline. This led to challenges before the ECJ.[23] Content-wise, it was argued that the Directive was not meant to be a clarifying law, but rather to implement the policy towards biotechnology jointly expressed in 1988 by the EPO, the United States Patent and Trademark Office (USPTO) and the Japan Patent Office (JPO).[24]

An additional issue stood in the way of European harmonisation in this field: while the Directive was binding on EU member States, it would not have been so for the EPO. However, considering that many patents are granted by the EPO rather than by national offices, the two systems needed to be synchronised in order to achieve the desired degree of harmonisation.[25] This was attained by incorporating the core

*without prejudice to the obligations of the Member States pursuant to international agreements, and in particular the TRIPS Agreement and the Convention on Biological Diversity*". Over time, it has been debated whether some provisions of the Directive are incompatible with TRIPS. This objection was raised before the ECJ in *Monsanto v Cefetra* (European Court of Justice (ECJ) 2010). In that occasion, it was argued that Article 9 of the Directive was incompatible with Article 27 TRIPS. In rejecting this argument, the Court maintained that EU law must be interpreted as much as possible in accordance with TRIPS and that no direct effect is given to TRIPS provisions (European Court of Justice (ECJ) 2010, § 71–72). A second objection was raised in the literature. *Palombi* argued that the Directive violated Article 27(1) TRIPS since the creations described in Article 3 and 5 of the Directive are presumed to be inventions (Palombi 2005, p. 70). In his opinion, this would elevate: "*Biotechnological inventions above the status of other types of technology*", thus infringing Article 27 TRIPS (Palombi 2003, p. 790). He also claimed that the Directive might be inconsistent with Article 52 EPC, as discoveries expressly fall outside the realm of patentable subject matter (Palombi 2016, p. 236). Such arguments dispute the generally held opinion that biological materials produced or isolated via human intervention are patent-eligible (Sommer 2007, p. 40).

[20]Seville (2016, p. 200). Apparently though, the Directive did not facilitate the patenting of biotechnologies, as only 28% of biotechnological patent applications are granted in comparison to 44% in all other sectors (Capocci 2012, p. 89).

[21]Visser (2019, p. 454).

[22]European Patent Office (1999).

[23]Despite this delay, the Directive was transposed in national norms, albeit with some modifications (Falcone 2014, p. 69).

[24]Palombi (2005, pp. 69–70).

[25]On this, the EPO explained that: "*The European Patent Organisation itself is not subject to this formal requirement* [to implement the Directive into its norms]. *However, European patent law does need to be brought into line with the Directive, primarily in order to comply with the*

provisions of the Directive into the Implementing Regulations of the Convention (Implementing Regulations or Rules).[26] The decision to quite literally incorporate the provisions of the Directive into the Implementing Regulations was made in 1999 by the Administrative Council of the European Patent Organisation. The Administrative Council opted for an implementation through the Rules, as an amendment of the EPC would have required a diplomatic conference of all contracting States.[27] Despite being an integral part of the EPC, and thus binding on the EPO and national Courts,[28] in case of conflict between the Implementing Regulations and the Convention, the latter shall prevail.[29]

The provisions of the Directive were inserted in a new chapter of the Implementing Regulations, which entered into force in 1999.[30] Rule 26(1) established that the "*relevant provisions of the Convention shall be applied and interpreted in accordance with the provisions of this Chapter*". In the opinion of the EPO, this would "*prevent the creation of a separate body of law*".[31] Rule 26 (1) proclaimed that the Directive should be used as "*supplementary means of interpretation*". The function of this norm was "*to supply provisions for the application and interpretation of the provisions already contained in the EPC… the reference to the Directive aims to ensure that the recitals preceding the provisions of the Directive are also taken into account*".[32]

The EPO viewed the provisions of the Directive as "*essentially based on the case law of the Boards of Appeal and its interpretation of the EPC*".[33] Despite this, the Convention and the Directive are the expression of two separate institutional, normative and judicial frameworks, as the EPC is a "*sui generis regional European treaty which has no legal connection with the EU*".[34] Nevertheless, a

---

*requirement for uniformity in harmonised European patent law*" (European Patent Office 1999, § 3).

[26]Kur and Dreier (2013, p. 125); *Implementing regulations to the convention on the grant of European patents of 5 October 1973, as last amended*, 1973.

[27]Aerts (2014a, p. 89).

[28]European Patent Office (1999).

[29]Article 164 EPC.

[30]Decision of the Administrative Council of 16th June 1999, entered into force on 1st September 1999 (OJ 1999, 437).

[31]European Patent Office (1999, § 13).

[32]Visser (2019, p. 454).

[33]Tritton et al. (2018, p. 135) and European Patent Office (1999, § 8).

[34]Tritton et al. (2018, pp. 86–87). The introduction of a European patent with unitary effect is eroding the separation between the EPC and EU law (Tritton et al. 2018, p. 87). The agreement on the Unified Patent Court (UPC) mentions that the: "*Unified Patent Court must respect and apply Union law and, in collaboration with the Court of Justice of the European Union as guardian of Union law, ensure its correct application and uniform interpretation; the Unified Patent Court must in particular cooperate with the Court of Justice of the European Union in properly interpreting Union law by relying on the latter's case law and by requesting preliminary rulings in accordance with Article 267 TFEU*" (Council of the European Union 2013, § 175/2). This is significant considering that the UPC agreement is an intergovernmental treaty and not an EU norm, despite

case recently brought before the EPO Enlarged Board of Appeal showed possible tensions between the two systems and highlighted just how intertwined they really are. Following the EPO decision in the *Tomatoes II/Broccoli II* cases admitting the patentability of products of essentially biological processes, the EU Commission issued a non-binding notice to clarify that, under the Directive, such products would not be patentable. As a consequence, the EPO harmonised its law to conform with the Directive by amending Rules 27 and 28 EPC to clarify that, under Article 53 (b) EPC, products obtained exclusively via essentially biological processes are excluded from patentability. The EPO thus overturned its previous practice.[35] This change is particularly interesting since, from a purely normative perspective, there was no technical conflict between the two approaches, given that notices of the EU Commission are not binding on the EPO. Yet, the EPO felt obliged to change its course and follow the line set by the Directive and the EU Commission. According to *Tritton*, this "*illustrates the uneasy relationship between the EPC and EU law in the field of patents where there is an overlap... it also undermines the EPO's initial view that the Directive is aligned with the case law of the EPO*".[36]

From an institutional perspective, the EPO and the EU are two distinct bodies with sovereign rights. As such, they give rise to two separate and autonomous legal systems. This independence affects the judicial bodies established under the two systems, which have a distinct personality pursuant to international law.[37] This has a number of repercussions. The first procedural implication concerns the timing and judicial powers to decide on a patent. Pursuant to the EPC, decisions over the patentability of biotechnological inventions lay with the EPO and its judicial bodies both before the grant of a patent and after it, in case an opposition is lodged. In case a patent is granted (and is not revoked during the opposition phase), the matter can be brought to national Courts and the ECJ.[38]

A second repercussion pertains to the applicable law and its correct interpretation. In their decisions, the Boards of Appeal apply the EPC and its Rules and can interpret them in light of the Directive. Despite the similar formulations of the two norms on biotechnological inventions, an interpretation of the Directive given by the EPO is not authoritative for the sake of EU law.[39] This is because the sole power to

the fact that only EU member States can become contracting parties of this agreement (Tritton et al. 2018, p. 87).

[35]Tritton et al. (2018, p. 87) and Enlarged Board of Appeal of the EPO (EBA) (2015). On this, see Sect. 4.1.5.2. For an overview of the consequences of such hybrid structure on the legal certainty for patents on biotechnological inventions, Aerts (2019).

[36]Tritton et al. (2018, pp. 135–136).

[37]Aerts (2014b, p. 585).

[38]Aerts (2014a, p. 90).

[39]Notwithstanding their similar formulation, EU law has not been implemented in the EPC. Indeed, the Convention has no link to the EU legal order and, being an international agreement, it does not constitute the right instrument to implement EU law with a binding force (Aerts 2014a, p. 92).

interpret and apply an EU Directive lays with the national Courts of EU member States and the Court of Justice of the European Union (CJEU).[40]

Furthermore, the EPO and its judicial bodies cannot take advantage of the referral mechanism to the ECJ established by Article 19 of the Treaty on the European Union (TEU) and Article 267 of the Treaty on the Functioning of the European Union (TFEU). In *WARF*,[41] the Enlarged Board of Appeal noted that the EPC does not provide a power of referral, despite the similarities between the Rules and the Directive.[42] This decision is sensible, as otherwise the ECJ could affect non-EU member States with its rulings.[43] Conversely, the decisions of the ECJ on the Directive are relevant for the interpretation of the EPC and its Rules. However, such judgements are not binding.[44] Similar conclusions can be reached for the relationship between the EPC and national laws. As those represent two distinct legal bodies, national Courts can choose not to adopt the interpretation of the EPO when applying the Convention.[45]

While all three sets of norms seem in principle to be applicable, either directly or indirectly, to synthetic biology, the reality might be different. Indeed, as it will be discussed in detail in the following chapters, the application of the Directive and the corresponding EPC Rules to a number of SynBio inventions is uncertain.[46]

## 3.2 The Patenting Landscape in the Field of Synthetic Biology

Patent protection is being sought for a number of synthetic biology inventions. However, specific figures and trends are difficult to determine. This is mostly due to the difficulty of univocally identifying what constitutes synthetic biology and, as a result, a synthetic biology patent. Furthermore, given that the dividing line between this discipline and other biotechnologies is blurred and considering that patents for nucleic acids, microorganisms and proteins are commonly issued, it is particularly

[40]Aerts (2014a, p. 93) and European Court of Justice (ECJ) (2011). The Court of Justice of the European Union (CJEU) ensures that EU law is interpreted and applied in the same way in every EU country and ensures that States and EU institutions abide by EU law. It is divided in two Courts: the Court of Justice (ECJ) and the General Court (European Union 2016).

[41]Technical Board of Appeal of the EPO (TBA) (2006, § 2–11). On this, see Sect. 5.1.3.3.

[42]Visser (2019, p. 454).

[43]Tritton et al. (2018, p. 86).

[44]Tritton et al. (2018, p. 135).

[45]Minssen and Nilsson (2012, p. 691).

[46]On this, see Sect. 4.2.2.1.

difficult to provide an estimate of the number of synthetic biology patents and filings.[47]

Patent studies conducted so far show an increasing—yet inconsistent—amount of patent activity in synthetic biology. Earlier studies pointed to the existence of approximately 3000 applications worldwide, 437 of which were filed in Europe. According to this calculation, the EPO had granted 44 patents in this field until 2009.[48] Another source indicated instead a total of about 1200 applications until 2013.[49] Other studies showed yet other figures.[50] More recent works counted instead between 7000 and 24,000 SynBio patent families worldwide.[51]

Such notable inconsistencies show just how much the definition of synthetic biology and its boundaries can impact the analysis of its patent panorama. Matters are made even more complex by the absence of a specific patent classification class for synthetic biology within the International Patent Classification (IPC) system as well as by the interdisciplinary nature of this field, which could lead to the use of an array of IPC classes.[52]

Countrywise, the United States, China, Europe and Japan showed significant patent activities.[53] The majority of filings were made by companies, yet universities and research institutes were also particularly active (with an estimated 25% to 40% of filings, depending on the study).[54] In fact, the most active applicant was not a corporation, but the University of California.[55] Synthetic biology patent applications focused predominantly on industrial biotechnology applications. The medicine, energy, chemical, fuel and nanotechnology fields encompassed approximately 10% of patent applications each.[56] The applications encompassed products,

[47]Oldham and Hall (2018) and Rutz (2009a, p. 15). The same author noted that, *de facto*, synthetic biology patents exist since years (Rutz 2010a, pp. 11–12). For a case study on the values connected to synthetic biology patents (Ribeiro and Shapira 2020).

[48]Rutz (2009b, p. 14).

[49]van Doren et al. (2013, p. 213).

[50]A study shows that until April 2008 only 11 patent applications containing the words “synthetic biology” were filed in the US. At that time, no patent containing the words “synthetic biology” either in the specifications or claims had been issued in America (Mohan-Ram and Waxman 2008, p. 1). An analysis from June 2012 displays instead about 6250 applications containing the term “synthetic biology” in their full-text. This includes same applications filed across a range of jurisdictions as well as members of the same patent family (Schneider 2014, p. 155).

[51]Oldham and Hall (2018, p. 8) and Olotu (2018). However, those same articles report similar amounts of total patents or total patent family members (ca. 70,000) (Oldham and Hall 2018, p. 8; Olotu 2018).

[52]van Doren et al. (2013, pp. 210, 217) and Reiss and van Doren (2013, p. 5).

[53]Oldham and Hall (2018, pp. 16–22) and Shapira and Kwon (2018, pp. 14–16).

[54]Olotu (2018) and van Doren et al. (2013).

[55]A study from 2018 showed that the University of California owned a portfolio of 472 SynBio patents. Other universities with vast patent portfolios in the subject were Harvard University, MIT, and the California Institute of Technology (Olotu 2018). Other studies seem to rebuff this statement (Oldham and Hall 2018, pp. 22–27).

[56]van Doren et al. (2013, pp. 213–216).

methods, uses as well as apparatuses claims.[57] Many filings covered basic principles and research activities.[58] Such focus on basic research has been interpreted as an indication of the "*prematurity of synthetic biology in terms of patenting activity*".[59]

Patents for synthetic biology can be divided into two main categories; on the one hand, biotechnological products, processes and tools and, on the other, computer programs to design and analyse biological devices.[60] Examples of patent applications of the first type span from xenobiology and non-standard amino acids, to memory and text encoding, minimal genomes, engineered microorganisms for enhanced substance production as well as manipulation and installation of genomes.[61] Applications of the second type concern instead computerised simulation processes to assess cellular interactions and dynamics as well as software to analyse and visualise molecular information.

These filings show the current state of research in synthetic biology and offer a realistic idea of its present achievements. Indeed, an analysis of the patent panorama in synthetic biology should not focus exclusively on the revolutionary innovations that this discipline is expected to deliver in the future, but should address its current status as well. Indeed, it was pointed out that, "*although popular perception may be that scientists... are creating new and entirely artificial life forms, the progress towards this possibility has been more subtle and pragmatic*".[62]

## References

Aerts RJ (2014a) The patenting of biotechnological inventions in the EU, the judicial bodies involved, and the objectives of the EU legislator. Eur Intellect Prop Rev 36:88–94

Aerts RJ (2014b) The unitary patent and the Biotechnology Directive: is uniform protection of biotechnological inventions ensured? Eur Intellect Prop Rev 36:584–587

---

[57] Rutz (2010b, p. 9).

[58] Reiss and van Doren (2013, pp. 17–18).

[59] van Doren et al. (2013, p. 217).

[60] McLennan and Rimmer (2012). The coming together of two very contentious patent topics (i.e. biotech and software) led commentators to describe synthetic biology as a "perfect storm" from a patent perspective (Rai and Boyle 2007, p. 389).

[61] This brief patent overview is based on a search conducted in the EPO (Espacenet), USPTO, WIPO (Patentscope) free online databases. Patents containing the words "synthetic biology" in either the title or the patent abstract were selected. Similarly, patents mentioned in both scientific and legal literature on synthetic biology were taken into consideration. This search presents only a small variety of synthetic biology patents and is not meant to offer a precise and thorough analysis of synthetic biology patent applications filed so far. For further synthetic biology patents and analysis, "Recent patents in synthetic biology" (2017), "Gen9 awarded four U.S. patents for innovation in synthetic biology" (2016), ETC Group (2014a, p. 4), ETC Group (2014b, p. 3), ETC Group (2014c, p. 2), "Recent patent applications in synthetic biology" (2011), "Recent patent applications in synthetic biology" (2009), Mohan-Ram and Waxman (2008, p. 2) and ETC Group (2007, p. 35).

[62] Chan and Sulston (2010, p. 1315).

Aerts RJ (2019) A switch on a switch on a switch: the status of harmonisation of biotech patent law in Europe. Eur Intellect Prop Rev 41:541–545

Agreement on Trade-Related Aspects of Intellectual Property Rights (TRIPS)(1869 U.N.T.S. 299), 1994

Capocci A (2012) Il brevetto. Ediesse, Roma

Chan S, Sulston J (2010) Patents in synthetic biology. BMJ 340:c2984

Convention on the grant of European patents of 5 October 1973, 1973

Convention on the grant of European patents of 5 October 1973, as revised on 29 November 2000 (European Patent Convention - EPC), 1973

Correa CM (2007) Trade related aspects of intellectual property rights: a commentary on the TRIPS agreement. Oxford University Press, Oxford

Council of the European Union (1994) Council decision of 22 December 1994 concerning the conclusion on behalf of the European Community, as regards matters within its competence, of the agreements reached in the Uruguay Round multilateral negotiations (1986–1994) 94/800/EC

Council of the European Union (2013) Council agreement on a Unified Patent Court

de Carvalho NP (2018) The TRIPS regime of patents and test data, 5th edn. Kluwer Law International, Alphen aan den Rijn

Directive 98/44/EC of the European Parliament and of the Council of 6 July 1998 on the legal protection of biotechnological inventions (Biotechnology Directive or Biotech Directive or Directive), 1998. OJ L 213, 30.7.1998, pp 13–21

Enlarged Board of Appeal of the EPO (EBA), 2004. G 0002/02 (Priorities from India/Astrazeneca)

Enlarged Board of Appeal of the EPO (EBA), 2015. G 0002/13 (Broccoli II)

ETC Group (2007) Extreme genetic engineering: an introduction to synthetic biology. ETC Group

ETC Group (2014a) Artemisin & synthetic biology: a case study. ETC Group

ETC Group (2014b) Rubber & synthetic biology: a case study. ETC Group

ETC Group (2014c) Saffron and synthetic biology: a case study. ETC Group

European Court of Justice (ECJ), 2004. (C 245/02) Anheuser-Busch Inc. v Budějovický Budvar, národní podnik

European Court of Justice (ECJ), 2010. (C 428/08) Monsanto Technology LLC v Cefetra BV and Others

European Court of Justice (ECJ), 2011. Opinion 1/09

European Patent Office (1999) Official Journal EPO 8-9/1999 - Notice dated 1 July 1999 concerning the amendment of the implementing regulations to the convention on the grant of European patents - the new provisions relating to biotechnological inventions

European Patent Office (n.d.) Member states of the European patent organisation [WWW Document]. www.epo.org. URL https://www.epo.org/about-us/foundation/member-states.html. Accessed 28 Feb 2020

European Union (2016) Court of Justice of the European Union (CJEU) [WWW Document]. https://europa.eu. URL https://europa.eu/european-union/about-eu/institutions-bodies/court-justice_en. Accessed 28 Feb 2020

Falcone A (2014) Synthetic biology, biotechnology patents and the protection of human health. A consideration of the principles at stake. In: de Miguel Beriain I, Casabona CMR (eds) Synbio and human health. Springer, Dordrecht, pp 55–76

Gen9 awarded four U.S. patents for innovation in synthetic biology [WWW Document], 2016. www.businesswire.com. URL https://www.businesswire.com/news/home/20160107006262/en/Gen9-Awarded-U.S.-Patents-Innovation-Synthetic-Biology. Accessed 27 Feb 2020

Hashim MR (2013) International influence – TRIPS and patentable subject-matter. IIC 44:656–670

Hestermeyer HP (2013) The notion of "trade-related" aspects of intellectual property rights: from world trade to EU law – and back again. IIC 44:925–931

Implementing regulations to the convention on the grant of European patents of 5 October 1973, as last amended, 1973

Kur A, Dreier T (2013) European intellectual property law: text, cases and materials. Edward Elgar, Cheltenham

McDonell LA, Haley JF, Hosoda Y, Jaenichen H-R, Meier J (2016) From clones to claims: an encyclopedia of the European Patent Office's case law on the patentability of biotechnology inventions with a comparison to the United States and Japanese practice, 6th edn. Carl Heymanns Verlag, Cologne
McLennan A, Rimmer M (2012) Inventing life: patent law and synthetic biology [WWW Document]. http://theconversation.com. URL http://theconversation.com/inventing-life-patent-law-and-synthetic-biology-5178. Accessed 27 Feb 2020
Minssen T, Nilsson D (2012) The industrial application requirement for biotech inventions in light of recent EPO & UK case law: a plausible approach or a mere "hunting license"? Eur Intellect Prop Rev 34:689–703
Mohan-Ram VS, Waxman MJ (2008) Synthetic biology patent applications expected to present new challenges. Life Sci Law Ind 2:1–3
Oldham P, Hall S (2018) Synthetic biology - mapping the patent landscape. bioRxiv 483826. https://doi.org/10.1101/483826
Olotu T (2018) Synthetic biology research reveals key trends and organisations [WWW Document]. www.patsnap.com/blog. URL https://www.patsnap.com/blog/synthetic-biology-research-major-trends-and-organisations?utm_campaign=Q2%202018%20Synthetic%20Biology%20Campaign&utm_content=75886280&utm_medium=social&utm_source=facebook. Accessed 26 June 2019
Palombi L (2003) Patentable subject matter, TRIPS and the European Biotechnology Directive: Australia and patenting human genes. Univ N S W Law J 26:782–792
Palombi L (2005) The impact of TRIPS on the validity of the European Biotechnology Directive. J Int Biotechnol Law 2:62–70
Palombi L (2016) Association for Molecular Pathology v Myriad Genetics (US) and D'Arcy v Myriad Genetics (AU): are gene patents in Europe a threatened species? Eur Intellect Prop Rev 38:231–236
Rai A, Boyle J (2007) Synthetic biology: caught between property rights, the public domain, and the commons. PLOS Biol 5:389–393
Recent patent applications in synthetic biology, 2009. Nat Biotechnol 27:1127
Recent patent applications in synthetic biology, 2011. Nat Biotechnol 29:134
Recent patents in synthetic biology, 2017. Nat Biotechnol 35:719
Reiss T, van Doren D (2013) Presentation: trends in synthetic biology based on patent data. Presented at the EU workshop on synthetic biology - IP, standards and regulatory issues, Royal Society of Chemistry, London
Ribeiro B, Shapira P (2020) Private and public values of innovation: a patent analysis of synthetic biology. Res Policy 49:1–11. https://doi.org/10.1016/j.respol.2019.103875
Rudge A (2014) Guide to European patents. Thomson West, Eagan
Rutz B (2009a) Synthetic biology and patents. A European perspective. EMBO Rep 10:S14–S17
Rutz B (2009b) Presentation: synthetic biology and patents. Presented at the Patenting synthetic biology? A transatlantic perspective workshop, Washington DC
Rutz B (2010a) Patentability of synthetic biology inventions in Europe. Biotechnol J 5:11–13
Rutz B (2010b) Presentation: patent issues in SynBio applications. Presented at the Synthetic biology: from science to governance workshop, Brussels
Schneider I (2014) Exclusions and exceptions to patent eligibility revisited: examining the political functions of the "discovery" and "ordre public" clauses in the European Patent Convention and the arenas of negotiation. In: de Miguel Beriain I, Casabona CMR (eds) Synbio and human health. Springer, Dordrecht, pp 145–173
Seville C (2016) EU intellectual property law and policy. Edward Elgar, Cheltenham
Shapira P, Kwon S (2018) Synthetic biology. Research and innovation profile 2018: publications and patents. bioRxiv. https://doi.org/10.1101/485805
Singer M, Stauder D (2003) European Patent Convention: a commentary, 3rd edn. Sweet & Maxwell, Köln

Sommer T (2007) The scope of gene patent protection and the TRIPS agreement. Int Rev Intellect Prop Compet Law 38:30–51

Technical Board of Appeal of the EPO (TBA), 1999. T 0935/97 (Computer program product II/IBM)

Technical Board of Appeal of the EPO (TBA), 2006. T 1374/04 (Stem cells/WARF)

Tritton G, Davis R, St. Quintin T (2018) Tritton on intellectual property in Europe, 5th edn. Sweet & Maxwell, London

Ugurlu AS (2014) Bioethics and the patent eligibility of human embryonic stem cells-related inventions in Europe. Nomos, Baden-Baden

van den Belt H (2014) Synthetic biology and global health in the age of intellectual property. In: de Miguel Beriain I, Casabona CMR (eds) Synbio and human health. Springer, Dordrecht, pp 19–43

van Doren D, Koenigstein S, Reiss T (2013) The development of synthetic biology: a patent analysis. Syst Synth Biol 7:209–220

Van Overwalle G (2008) Biotechnology and patents: global standards, European approaches and national accents. In: Wüger D, Cottier T (eds) Genetic engineering and the world trade system. Cambridge University Press, Cambridge, pp 77–108

Visser D (2019) Visser's annotated European Patent Convention, 26th edn. Wolters Kluwer, Alphen aan den Rijn

Wüger D (ed) (2008) Genetic engineering and the world trade system: world trade forum. Cambridge University Press, Cambridge

# Chapter 4
# The Patent Eligibility of Synthetic Biology Inventions

## 4.1 Patentable Subject Matter

The subject matter eligibility of synthetic biology inventions is a complex, yet often disregarded, matter.

It is generally assumed that synthetic biology inventions will not be problematic from a patentability perspective. This assumption is based on a variety of reasons. First, synthetic biology is often seen as an incremental technology.[1] This would mean that current norms and approaches to biotechnological patents would be applicable to it as well. For example, *Rutz* argued that synthetic biology is not likely to herald anything new from a patent perspective, since synthetic biology patents are likely to claim molecules, organisms, uses and methods for their production, which are "*mostly indistinguishable from applications in other areas of biotechnology*".[2] Therefore, since the Directive has settled questions on the patentability of biotechnological inventions, no specific issues should arise for synthetic biology either. Second, Europe has adopted a permissive patentability standard; applications are rarely challenged from an Article 52 EPC perspective, whereas objections are often raised based on other requirements.[3] Since most inventions pass the patentability hurdle, it is expected that synthetic biology inventions will do the same. Because of

[1] Authors believed that: "*95-98% of what is declared as synthetic biology is just a direct continuation of modern molecular biology, genetic research or genetic engineering*" (Robienski and Simon 2014, p. 123).

[2] European Commission's Directorate-General for Health & Consumers (2010, p. 24) and Rutz (2009, p. 15). Over time, patents have been granted for a wide range of biotechnological products, including isolated DNA and RNA, genes, polypeptides (e.g. hormones), antibodies and microorganisms, as well as their uses and methods of preparation (Macchia 2012, p. 38; Vogt 2010, p. 12).

[3] It was noted that: "*The approach of the EPO to the exclusion of discoveries, following the EBD, is logically untenable and boils down to an emasculation of the exclusion*" (Sterckx and Cockbain 2012, p. 134).



I. de Lisa, *The Patentability of Synthetic Biology Inventions*,
https://doi.org/10.1007/978-3-030-51206-4_4

this, it has been "*foreseen that the new field of synthetic biology will not raise questions that cannot be answered by present legislation and interpretation of the law by the Boards of Appeal*".[4]

Despite those reasonable points, this approach might be reductive.

First, the conclusion that synthetic biology will not raise subject matter issues is often no more than an assumption, since this point is very rarely investigated in detail. Therefore, to establish whether this will actually be the case, a more thorough analysis is needed. This is particularly important if one considers that "*it is in the interest of practitioners to argue that synthetic biology is no more than a cumulative advance on what is already being done*".[5] Indeed, given the unclear boundaries of this discipline and its contentious nature, it is obvious to try to avoid questions on its subject matter patentability, especially since they might stir social opposition and normative objections.

Moreover, these questions need to be examined in the wake of the recent American and Australian decisions on subject matter eligibility.

In order to achieve this, the following sections will examine the norms, decisions and doctrinal orientations that could be relevant for a patentability assessment of synthetic biology inventions. However, those apply only indirectly to synthetic biology, as no decisions have been issued so far in this field in Europe. Yet, the directions and approaches that emerged in them might be applicable by extension to synthetic biology and offer an idea of how its patentability will be addressed. The analysis will focus on the patent panorama in Europe, but references will be made to the US and other jurisdictions in so far as those could affect the assessment. Subsequently, a theoretical examination of the patentability of synthetic biology inventions will be performed. This analysis will be integrated by a review of a number of SynBio patent applications presented before the EPO and the USPTO.

### 4.1.1 Introduction

The patent analysis of synthetic biology inventions must start by determining whether those inventions fall within the realm of patentable subject matter.[6]

---

[4]Fernandez y Branas (2014, p. 187).

[5]Dutfield (2012, p. 119).

[6]On whether questions on subject matter patentability have any practical relevance (Pila 2011, p. 59; UK House of Lords (HL) 1996, §5). The analysis of the subject matter requirement precedes the one of the other patent criteria (i.e. novelty, inventiveness and industrial application) and must be conducted separately (EPO 2018, § G.II.2; EPO 2013, p. 17; Macchia 2012, p. 43; Technical Board of Appeal of the EPO (TBA) 2000a, p. 14, § 6). The question of whether an invention exists is addressed as first also in the US. In *Parker v. Flook*, the Supreme Court held that: "*The obligation to determine what type of discovery is sought to be patented must precede the determination of whether that discovery is, in fact, new or obvious*" (U.S. Supreme Court 1978, p. 437). Analogous comments on the priority to be afforded to the invention requirement can be read in Article 27

Eligibility has been the subject of much case law and constitutes a particularly problematic analysis, as it has both legal and policy repercussions.[7] Indeed, "*whoever controls the meaning of 'invention' controls what can be patented and hence an important aspect of industrial policy*".[8] Limits to what can and should be protected by patent law have been established over time. Patent offices and Courts are confronted with applying such boundaries to ever evolving technologies, which blurry lines that have developed during another technological era.

Given the complexity of the issues involved, the question of patentable subject matter has repeatedly been underplayed or even sidestepped by deciding organs. Equally, litigants have been reluctant to raise this issue, as decisions finding the ineligibility of a specific subject matter could have far reaching consequences for both parties involved.[9]

The downplaying of this requirement or its confusion with others has been considered problematic. For example, complications emerge if the focus is shifted away from the subject matter eligibility analysis to the inventive step one. This approach, which is popular in Europe but rejected in America, is based on the assumption that those two criteria share a common ground. However, there is no specific symmetry between them. In particular, in the case of biotechnological inventions, this reasoning would not be applicable, since eligibility is often at the core of the dispute.[10]

## 4.1.2 *Inventions*

### 4.1.2.1 Overview

The EPC does not define the concept of invention.[11] Rather, Article 52(2) EPC contains a non-exhaustive list of matters that are not considered inventions. Such exclusions cover:

TRIPS (Palombi 2003, p. 282). However, it has also been noted that this examination could occur at the end rather than at the beginning of the patent assessment (Warren-Jones 2001, p. 78). Subject matter patentability has been subject to academic debate. For example, *Pila* held that: "*Patent eligibility is a widely contested and misunderstood aspect of patent law. It is frequently confused with the secondary patentability requirements, including those of novelty and inventive step, as well as with the public policy exclusions... At the heart of this contestation is whether it can be completely separated from these requirements and exclusions... My own view is that it cannot be; a view which Sichelman disagrees*" (Pila 2014a, p. 395).

[7]Setting the boundaries of patentable subject matter has an impact also on other intellectual property rights (e.g. copyright).

[8]UK Patents Court (2005, §10).

[9]Kane (2010, p. 4) and Kane (2004, p. 726).

[10]Sherman (2015, pp. 1228–1229).

[11]Both TRIPS and the EPC do not provide a legal definition of the concept of invention, with some authors doubting whether finding an agreement on an unambiguous definition would have ever

(a) *Discoveries, scientific theories and mathematical methods;*
(b) *Aesthetic creations;*
(c) *Schemes, rules and methods for performing mental acts, playing games or doing business, and programs for computer;*
(d) *Presentations of information.*

According to the EPO, these items have been excluded because they are either abstract or non-technical.[12] From this, it follows that an invention should have both technical and concrete character.[13] Their underlying rationales have been discussed by Courts and authors alike.[14] For example, the rationale behind the exclusion of discoveries and mathematical methods is seen to reside in their purely abstract and intellectual nature, their lack of technical effect and in the need for core abstract

---

been possible and whether that definition would have been practical, especially considering that it would need to be applicable to all fields of technology (Singer and Stauder 2003, pp. 68–69; Westerlund 2002, p. 25). *Westerlund* provided a definition of invention designed to fit in also biotechnological inventions. The definition stated that: "*The concept of 'invention' as used in patent law means a technical solution, which – as part of its inventive kernel – utilises a law of nature or controllable natural forces which cannot as such be obtained without the interference of human intelligence. The concept includes substances properly characterized not present per se in nature and for which a function is indicated. A precondition is that – as reproducible product or process – it can be used commercially*" (Westerlund 2002, p. 56). Only a limited number of countries offer a legal definition of invention. The UK Patents Act of 1949 defined invention as: "*Any manner of new manufacture the subject of letters patent and grant of privilege within section six of the Statute of Monopolies and any new method or process of testing applicable to the improvement or control of manufacture, and includes an alleged invention*"(*UK Patent Act - The Patents Act 1949 (Chapter 87)* 1949). § 101 of the US Patent Act states that: "*Whoever invents or discovers any new and useful process, machine, manufacture, or composition of matter, or any new and useful improvement thereof, may obtain a patent*" (*U.S. Patent Act - United States Code Title 35 - Patents* 1952). Commentators have objected to the existence of a definition of invention under US law (Bostyn 2004, p. 13).

[12]The idea that the exclusions in Article 52(2) EPC are based on the abstract nature of such items has been contested, in particular for computer programs (Pila 2011, p. 62). Also, it has been argued that only a few of those exclusions have an obvious rationale (Pila 2010, p. 911).

[13]Sommer (2013, p. 268).

[14]For a discussion on the point, see the *Aerotel* case. There, *Jacob LJ* held that: "*i) . . . There is no evident underlying purpose lying behind the provisions as a group. . . ii) One cannot form an overall approach to the categories. They form a disparate group – no common, overarching concept. . . iii) Some categories are given protection by other intellectual property laws. . . iv) Further, some categories are so abstract that they are unnecessary or meaningless. . . What does emerge is that the various categories are the result of various compromises and distinct discussions about each of them. So one can at least find confirmation that no overarching principle was intended. What was done was to formulate the language of each of the categories independently of one another, add the 'as such' rider to all of them and leave it to the EPO and European patent judges to work out the detail*" (Court of Appeal of England and Wales (EWCA) 2006, § 9–11). In the *CFPH* case, the policy objectives underlying these exclusions were assessed and they were divided between soft and hard exclusions. According to *Prescott QC*: "*The harder the exclusion, the more it is the policy of the law to insist that the use of the information be not foreclosed under patent law*" (UK Patents Court 2005, §9). For a scholarly analysis, Pila (2011, pp. 59–69).

scientific knowledge to remain in the public domain.[15] This should foster further scientific progress, which may instead be hampered by the patenting of basic research principles.[16] Generally, those exclusions have been construed narrowly, taking into consideration that Article 52(1) EPC seems to offer a general entitlement to patent protection.[17] This led to a *de minimis* reading of this norm.[18] In relation to these exclusions, Article 52(3) EPC states that they apply "*only to the extent to which a... patent relates to such subject-matter or activities as such*".

Objections to subject matter patentability in synthetic biology are likely to arise under the discovery and mathematical methods exclusions, given that the majority of synthetic biology patent filings are directed to biological matters or to programs to design and simulate biological circuits. Further objections could also be raised under Article 52(2)(d) EPC concerning presentations of information.

The following sections will thus focus on the exclusions found in Article 52 (2) EPC and centre on the issue of patentable subject matter in Europe. The assessment will be integrated by an analysis of recent trends that have emerged globally in the product and laws of nature doctrines, in order to determine whether they have any influence on the European exclusions.

The objective is to determine if the exclusions from patentability will be applied differently in the field of synthetic biology compared to other biotechnologies. This is fundamental since, as mentioned by *van den Belt*:

> When a new field of technology emerges, patent law does not provide a list of ready-made criteria by which the technical accomplishments in the new field can be judged as patentable inventions. Instead, the conditions of patentability have first to be worked out and elaborated vis-à-vis the new technology, if only because the notion of 'invention' is not strictly and universally defined but open to historically variable interpretation.[19]

This approach is confirmed by the findings of the SYBHEL project, which suggested a revision of the patent criteria for SynBio inventions, including the subject matter one.[20]

---

[15]EPO (2018, § G.II. 3.3), EPO (2016, § I.A.2.2.1) and Schneider (2014, p. 149).

[16]Schneider (2014, p. 150).

[17]Bergia (2013, p. 634) and Singer and Stauder (2003, p. 70). It was held that: "*The EPO has been singularly successful in giving a narrow reading to the limits on invention and patentability contained in Articles 52 and 53 of the EPC*" Drahos (1999, p. 442).

[18]Pila (2011, p. 59).

[19]van den Belt (2014, p. 27).

[20]SYBHEL (2014, p. 14).

#### 4.1.2.2 Technical Requirement

European Normative Overview

In order to qualify as an invention, an item shall display technical character. This essential requirement was viewed by the Boards of Appeal of the EPO as an implicit *condition sine qua non* rooted in the European legal tradition.[21] Examiners are required to take it into consideration by the Guidelines for Examination of the EPO, which refer to this requirement as "*implicitly contained in the EPC*" and affirm that:

> The invention must be of 'technical character' to the extent that it must relate to a technical field (Rule 42(1)(a)), must be concerned with a technical problem (Rule 42(1)(c), and must have technical features in terms of which the matter for which protection is sought can be defined in the claim (Rule 43(1)).[22]

It is complex to define the term "technical". Some described it as "*extremely opaque*" and pointed out that its content varies amongst countries.[23] According to the EPO, technical character corresponds to a technical teaching, which is an "*instruction addressed to a skilled person as to how to solve a particular technical problem using particular technical means*".[24] Commentators criticised such definition, as it uses the very term that is attempting to define as a core part of its

[21]Early discussions of the EEC Working Party for the establishment of a Common Market Patent in the 1960s show various attitudes towards this requirement. Some argued that inserting a requirement for a technical process would be redundant, as it would repeat what was already inherent in the requirement of invention. Conversely, others pointed out that national patent laws showed divergent understandings of the concept of technical progress. Specifically, the UK delegation argued that the inclusion of such a requirement would mean that the UK: "*Would not be able to make any contribution to unification of law in this particular aspect*", since it would set a higher patentability standard compared to the one in use in the country (Pila 2005a, p. 760).

[22]EPO (2018, § G.I.2). Before the EPC 2000 amendments, this requirement was read into Article 52 (1)(2)(3) EPC, which referred to the term invention and which presented exclusions that could be seen as lacking technical character. Some authors were not convinced by this argument. For example, *Pila* held that: "*While the interpretation of Art. 52(2) and (3) may be an argument for resolving those provisions to a requirement for technical character, it is not an argument that is easily derived from the terms of Art. 52(2) and (3) themselves* (Pila 2010, p. 911). This requirement has also been traced back to the non-binding EPO Guidelines for Examination of 1985 (Pila 2005b, p. 184). The "as such" proviso of Article 52 EPC was meant to offer patent protection to activities having a technical character, even if connected to the list of excluded items (EPO 2016, § I.A.1.1). Nowadays, the formulation introduced by the EPC 2000 in Article 52(1) EPC explicitly mentions that patent protection is granted for inventions in a technical field (EPO 2016, § I.A.1). The reference to "technical field" is also opaque, as it includes areas that are not usually considered as industrial, such as agriculture (Westerlund 2002, pp. 36–37). Equally, Rules 42 and 43 EPC confirm this trend by requiring that an invention displays technical features, relates to a technical field and concerns a technical problem (Schwartz and Minssen 2015, p. 231).

[23]Pila (2005b, p. 183) and Westerlund (2002, p. 36).

[24]EPO (2016, § I.D.9.1.1).

explanation.[25] The EPO noted that technical inventions are generally both useful and practical, but pointed out that the opposite might not be true. Indeed, usefulness and practicality cannot be considered as substitutes for the technical character criteria. Despite the connection between the industrial applicability and the technical requirement, the two remain distinct.[26]

The technical character requirement raises three additional issues. The first concerns whether such requirement is a formal or a substantive one. That is, whether it is essential for an invention to be described in technical terms or to use technical means or whether it must also provide a substantive technical advance. The second addresses the connection between the technical character requirement and the other patentability criteria (i.e. novelty, inventiveness and industrial application). The third centres on whether this requirement demands the production of a material object or whether this is not necessary. In all three cases, diverging orientations have emerged in the case law.[27] Consensus seems instead established on the patentability of inventions containing both technical and non-technical features.[28] It is not necessary for technical features to be predominant to obtain a patent.[29] However, the mere possibility of having a technical purpose or solving a technical problem does not suffice to fulfil this requirement and overcome the exclusions set out in Article 52 EPC.[30]

Although the EPO still considers items as inventions only if they are connected to a technical effect, this requirement ceased to be a cornerstone of the analysis undertaken by the EPO to assess subject matter patentability. According to some authors, the discussion on technicality shifted instead towards the inventive step.[31]

### European Case Law Overview

To be technical, an invention should have technical character or provide a technical contribution.[32] Courts have often addressed the question of what constitutes the technical character of an invention.[33] Over time, different approaches emerged within the EPO and in Europe. The two main approaches are the "contribution"

---

[25]Bostyn (2004, p. 12).

[26]EPO (2016, § 1.1).

[27]Pila (2005b, pp. 175–176).

[28]"European Patent Office: European Patent Convention, Arts. 52(1)(2)(3), 54, 56, 57 - 'Auction Method/Hitachi,'" (2005, p. 238).

[29]Pila (2010, p. 907).

[30]Fritz et al. (2015, p. 85).

[31]Fox and Corbett (2014, p. 571).

[32]Technical Board of Appeal of the EPO (TBA) (1998, § VIII). This definitional approach has been criticised for its circular nature (Pila 2005b, p. 185; Bostyn 2004, p. 12).

[33]Recently, the EPO defined for the first time what it is meant by this term. According to *Pila*, it encompasses any subject matter representing: "*The casual, perceivable result of a purposive human action on the physical world*" (Pila 2014b, p. 181; Enlarged Board of Appeal of the EPO (EBA)

and the "any hardware" ones. The first requires a non-conventional result that ought to occur in a field not excluded by Article 52(2) EPC. This approach is intertwined with the novelty and inventiveness requirements.[34] As for the "any hardware" approach, inventions will fall outside the exclusions of Article 52 EPC as long as they embody or are implemented via technical means. This would apply also when the technical means relate to non-technical features and irrespective of their novelty and inventiveness. All is needed is for the invention to embody or use technology or hardware.[35]

The "contribution" approach is exemplified in the *Vicom* case, which related to software inventions.[36] There, an invention was required to present a technical contribution in a field that was not excluded from patentability pursuant to Article 52 EPC in order to qualify as patentable subject matter. Contributions relating to the area of software were thus ignored. Novelty and inventiveness were relevant for this approach, as the invention's technical contribution to the prior art was assessed.[37] However, this approach led to critiques because of this confusion between different patentability requirements.[38]

The "any hardware" approach emerged instead in the *Pension Benefits* case.[39] The controversy concerned a computer-related invention for controlling a pension benefit scheme. In its decision, the Board of Appeal held that the method claim did not fulfil the requirements of Article 52(2) EPC. However, with regards to the apparatus claim, the situation was different. Firstly, the Board held that Article 52 (2) EPC is separate and distinct from the other patentability requirements and criticised the "contribution" approach. The Board focused on the character and kernel of the invention. It held that a computer programmed for use in a specific field is a concrete apparatus, that is, a physical entity that was man-made for utilitarian purposes. This would apply even if the field were an excluded one (in this case, the business field). In conclusion, an "*apparatus constituting a physical entity or concrete product suitable for performing or supporting an economic activity, is an invention within the meaning of Article 52(1) EPC*".[40] In commenting on this decision, scholars pointed out that it shows "*the requirement for an invention to be – despite the Board's implicit protestation to the contrary – a requirement*

2010a, b). This definition is expansive enough to support patents on isolated human genes and is consistent with both the EPO case law and the Biotech Directive (Pila 2014b, pp. 181–182).

[34]Pila (2005b); "European Patent Office: European Patent Convention, Arts. 52(1)(2)(3), 54, 56, 57 - 'Auction Method/Hitachi,'" (2005, pp. 237–238).

[35]Bently et al. (2018, p. 485).

[36]Technical Board of Appeal of the EPO (TBA) (1986).

[37]King et al. (2010, pp. 256–257).

[38]Aplin and Davis (2017, p. 650) and Technical Board of Appeal of the EPO (TBA) (2000a, p. 15, § 6). The technical contribution approach was employed also in *IBM* (Aplin and Davis 2009, p. 492; Technical Board of Appeal of the EPO (TBA) 1989).

[39]Technical Board of Appeal of the EPO (TBA) (2000a).

[40]Bently et al. (2018, pp. 485–487) and Technical Board of Appeal of the EPO (TBA) (2000a, p. 13, § 5).

*entirely of form, that exists independent of the requirement of inventiveness, and that does not depend on the production of any material or physical object*".[41]

The line of decision started in *Pension Benefits* was expanded in the *Hitachi* case.[42] This controversy, which related to an automatic auction method carried out on a server computer, saw the application of the "any hardware" approach to both apparatus and method claims. In describing its approach, the Board pushed its theory even further, as it extended it to activities whose technical character is quite limited (e.g. writing using pen and paper).[43]

Equally, in the *Microsoft* case, it was stated that the "*method was implemented in a computer and this amounted to a technical means sufficient to escape the prohibition in Article 52*".[44] Specifically, a computer program that has been stored on a carrier (for example, a disk) could confer technical character to the invention, simply because the claim concerned a computer-readable medium.[45]

The immediate consequence of the "any hardware" approach is that a higher number of claims will be able to pass the hurdle set by Article 52 EPC.[46] Hence, this requirement seems to have been reduced to a formality, as a mere recitation of "technicality" is sufficient.[47] However, this does not imply that a higher number of patents will be granted. The eligibility assessment is only the first step and the other patentability requirements will also need to be fulfilled.[48]

[41]Pila (2005b, p. 179).

[42]"European Patent Office: European Patent Convention, Arts. 52(1)(2)(3), 54, 56, 57 - 'Auction Method/Hitachi,'" (2005) and Technical Board of Appeal of the EPO (TBA) (2004).

[43]Huttermann and Storz (2009, p. 589) and Technical Board of Appeal of the EPO (TBA) (2004, §4.6).

[44]Bently et al. (2018, p. 488) and Technical Board of Appeal of the EPO (TBA) (2006a, § 5.1). Conversely, *King* argued that the *Microsoft* case represent an example of the "technical character" test (King et al. 2010, pp. 257–259).

[45]Aplin and Davis (2009, p. 500).

[46]Bently et al. (2018, p. 488).

[47]King et al. (2010, p. 261).

[48]In particular, the "any hardware" approach has placed an increased emphasis on the inventiveness requirement. This approach was confirmed by other EPO decisions that focused on the inventive step rather than on the Article 52 EPC requirement (Bently et al. 2018, p. 489; Technical Board of Appeal of the EPO (TBA) 2002a, b). Interestingly, a different approach is adopted in the UK. There, the focus is more on subject matter patentability than on inventiveness. Nevertheless, it has been argued that this difference would not lead to contrasting results between the EPO and the UK (Fox and Corbett 2014, p. 573; UK Court of Appeal (CA) 2008, § 11).

## 4.1.3 *Discoveries*

### 4.1.3.1 European Normative Overview

European Patent Convention

Practically all patent regimes differentiate between inventions and discoveries and afford patent protection only to the former.[49] The line between the two—and thus between patent-eligible inventions and not—was generally evident in classical mechanical technologies. However, the growth of the biotechnological field changed this.

Article 52(2) EPC merely lists discoveries amongst the subject matter that shall not be considered an invention. The norm adds that the exclusion operates for discoveries as such. This dichotomy was well present in the EPC since its inception. The *travaux préparatoire* have shown that discoveries were seen to differ from patentable inventions. Specifically, it had been argued that "*as a general rule, a patent can only protect an invention (a creation) and not a discovery, i.e. the mere becoming aware of a pre-existing reality*".[50]

Despite the absence of a legal definition of the term, discoveries are generally considered as the "*unearthing of causes, properties or phenomena already existing in nature*"[51] or as the recognition of the "*objective world existing in nature*" in contrast to inventions, which "*belong to the domain of improving the objective world by human efforts*".[52] International instruments described it also as the recognition of phenomena, properties or laws of the material universe.[53]

From a scholarly perspective, inventions have been seen as the "*application of such knowledge to the satisfaction of social needs*".[54] Other authors pointed out that every invention contains an element of discovery.[55] The contrary is however not true, as a mere discovery is not enough to constitute an invention.[56] The latter can be distinguished, since it includes a teaching to use the discovery in an industrial field.[57] The distance between the discovery and its practical application is pivotal in this

[49]This distinction has increasingly fallen into oblivion and became practically irrelevant (Schneider 2014, p. 150).

[50]Sterckx and Cockbain (2012, p. 115).

[51]Cornish et al. (2019, pp. 222–223).

[52]Westerlund (2002, p. 49).

[53]World Intellectual Property Organization (WIPO) (1978) Art. 1.

[54]Cornish et al. (2019, p. 223).

[55]Trevisan and Cuonzo (2013, p. 270).

[56]Ludlow (1999, p. 302).

[57]Trevisan and Cuonzo (2013, p. 270).

definition.[58] However, the question then turns to when and how a discovery acquires enough functionality to be considered an invention.[59]

Others imagined discoveries to relate to the acquisition of knowledge via experiments and thinking and highlighted their intellectual nature.[60] In contrast, *Kohler* argued that, because human-made arrangements and experimental interventions might be needed in order to make a discovery, the distinction between the two cannot revolve around the idea that one requires an active intervention, while the other merely involves a passive contemplation.[61] This is confirmed by the fact that the research activities that lead to inventions and discoveries are often identical. If those two activities are similar and require comparable investments, it could also be argued that the support offered to them should be identical.[62]

Another line of thought distinguished between discoveries and inventions by using the criteria of human intention and design. This theory argues that much exists because of human intention, but that human design is also necessary to define something as an invention.[63] Other commentators have instead highlighted the absence of an objective demarcation between the two concepts. *Resnik* noted that both discoveries and inventions originate from a combination of human and natural activities. In particular, he noticed that this distinction depends on how the causal role of humans in generating such product is conceptualised. When humans are the primary cause then an invention exists, while in case they are the secondary cause there would be a discovery. Things are more complex in the field of biotechnology, since the theory of causation is of little help here, given that the decision to consider something as a primary or secondary cause reflects values and goals rather than objective standards. For *Resnik*, substantially modified genes represent inventions, whereas genes that have not been substantially changed or that have been artificially reproduced (e.g. cloned) lay in a grey area between discovery and invention.[64] Different philosophical doctrines may also lead to contrasting views of what constitutes an invention and a discovery, as this could be seen as "*a policy issue and not an entirely legal one*".[65]

To adapt the discovery/invention duality to the biotechnological field, authors suggested that there is an invention if there is a human technical intervention that generates effects not present in nature. According to this interpretation, the discovery of a gene becomes an invention once a practical application that embodies a solution

58 Pizzoferrato (2002, pp. 143–144).

59 Cornish et al. (2019, p. 907).

60 Roberts (2010, p. 36).

61 van den Belt (2009, p. 1322).

62 *Sena* held that: "*The need for legislation that motivates research and guarantees a remuneration*" is the same (Sena 1999, p. 733).

63 Koepsell (2014, p. 151).

64 Resnik (2002, p. 149).

65 Westerlund (2002, p. 34).

to a technical problem is found.[66] This interpretation would be opposed by those who believe that "*the possibility to patent a gene or organism as a result of almost any human intervention*" is in contrast with the notion of invention.[67]

Furthermore, biotechnological patents have raised doubts concerning their justification. Patents grant a monopoly to an inventor in return for the disclosure of his invention, thus rewarding his investment and ingenuity. Again, while this system was smoothly applicable to mechanical inventions, it is far less apt in the biotechnological field, where it is complex to reconcile the idea that human ingenuity and creative efforts have contributed to the creation of genes and are thus deserving of patent protection. From this perspective, genes are seen as ever existing items, which humans can merely discover and analyse, but whose inner workings and functions are determined by nature.[68] At the turn of the twentieth century, similar arguments had been put forward by *Kohler* regarding the patentability of chemistry and biology. According to him, chemical products are either already existing or might be already existing in nature, since many natural substances are still unknown. In both cases, he argued, such substances can be discovered but not invented. His belief did not change for synthetic substances not found in nature, as he argued that in theory such substances could be found to exist in nature in the future. Lastly, he maintained that the way in which chemical substances combine is given by nature and that, consequently, human input consists merely in removing the obstacles that might impede such combinations.[69] Therefore, it was argued that genetic engineers are free-riders, who do not create anything new, but rather work on systems without fully knowing their functioning. From this perspective, biological systems are the opposite of machines, which require human operators for their assembly and direction. While natural creatures modified via human intervention might be seen as artefacts, this is still not enough to argue that they constitute inventions since "*the creativity of life comes from within. No amount of human tinkering can change that*".[70] Nonetheless, it was recognised that this argument would not fully apply where components of living things are modified or improved in their structure or function.[71]

Arguments over the reproduction of naturally occurring structures or the creation of new ones are common in this field. With regard to the former, difficulties exist in fitting those cases into the definition of invention. Conversely, fewer problems arise in fitting into the invention category biological products that have been modified or conceived by man. Indeed, according to *Westerlund*, these cases "*clearly fall under*

[66]Pizzoferrato (2002, p. 146).

[67]Falcone (2014, p. 67).

[68]Dutfield (2010, pp. 533–539) and Radick (2002, p. 75).

[69]At the time, German patent law did not provide patent protection for chemical substances, but only for processes used for their production (van den Belt 2009, pp. 1322–1323; Kohler 1900; *German patent law 1877* 1877).

[70]Dutfield (2010, p. 539).

[71]Dutfield (2010, p. 533).

*the invention concept*" since "*the modification consists in the creation of something which has not existed before in the naturally occurring entities*".[72] Similar views are shared by *Ludlow*. She argued that GMOs and their products should not be considered mere discoveries, even if they do or could exist in nature. In her view, "*the objection to patenting a discovered organism on the basis that it is a discovery it's not because of its subject-matter per se but because of how the claimed invention was arrived at*".[73]

An additional argument focused on the quantity of the modifications. It was noted that the difference between an invention and a discovery is somehow an illusion from a biological perspective. Taking transgenic animals into consideration, it appears that an organism possessing about 100,000 genes and in which only one or few genes are inserted could hardly be called invented, since the vast majority of the animal would not have been conceived by humans. This type of argument has been countered by noting that such animals have qualities and functions that they do not possess in nature.[74]

#### Biotech Directive

The Biotechnology Directive addresses the issue of what constitutes an invention or a discovery in the field of biotechnology.

Firstly, in the Recitals, it expressly states that patent protection shall not be available for mere discoveries. Subsequently, Article 3(1) provides patent protection for inventions that fulfil the patentability requirements, even if they concern biological materials or related processes. This is the first norm in Europe to officially confirm that living things could be subject to patent protection.[75] The second paragraph of this article further clarifies that "*biological material which is isolated from its natural environment or produced by means of a technical process may be the subject of an invention even if it previously occurred in nature*". This indicates that naturally occurring materials are not *per se* excluded from patentability and that they shall be treated as inventions in more traditional fields. With reference to the human body, Article 5(1) asserts that the body itself and the simple discovery of one of its elements do not constitute patentable inventions. This includes both genes and partial gene sequences. Conversely, Article 5(2) expressly mentions that elements isolated from the human body or manufactured via a technical process may be considered a patentable invention, even if their structure is identical to that of a natural element. Also this norm applies to both genes and gene sequences. The approach chosen by the legislator in Article 5 raises a question. How is it possible to consider DNA as a patentable invention, while at the same time arguing that the

---

[72]Westerlund (2002, pp. 48–49).

[73]Ludlow (1999, p. 303).

[74]Westerlund (2002, p. 31).

[75]Van Overwalle (2008, p. 88).

human body cannot be patented? In replying to this, authors argued that the rationale is to be found in the "*precedent set by patents on chemical compounds, where naturally occurring compounds can be patented if they are isolated and purified*".[76]

The formulation of the Directive illustrates the importance of the notion of isolation in this context. Pursuant to Article 5, the isolation process can transform a gene discovery in an invention, even in the absence of functional and chemical changes. Hence, some authors saw the boundary between discovery and invention as "*indistinct, with isolation acting as a transformative process*" and observed that this proviso represents a rather arbitrary exception to the distinction between discovery and invention.[77]

The relevance of this concept and its role in the discovery/invention dichotomy had been addressed by the EPO, USPTO and the JPO in 1988. In a joint statement issued before DNA patenting became widespread, the three offices explained that:

> Purified natural products are not regarded under any of the three laws as products of nature or discoveries because they do not in fact exist in nature in an isolated form. Rather, they are regarded for patent purposes as biologically active substances or chemical compounds and eligible for patenting on the same basis as other chemical compounds.[78]

---

[76]Calvert and Joly (2011, p. 165).

[77]Brabin (2014, p. 688). Other authors argued that the Directive distorts the difference between discoveries and inventions and claimed that this was the result of the pressure exercised by the biotech lobby, which was interested in exploiting patent rights on discoveries (Falcone 2014, p. 68). It is important to note that the lobbying campaign behind the Directive was described as the largest one in the history of the EU (Calvert and Joly 2011, p. 165). A background study of the Commission addressed the topic of isolation with regards to discoveries and inventions. There it was held that: "*When I isolate the DNA sequence from its natural environment, and when I separate the exons from the introns... I have made an invention, since I have isolated via a reproducible technical process the DNA sequences from the human body, and I have made a selection in the sequences, i.e., I have selected those parts of the sequence I am interested in. I will basically also copy that sequence, and then I have made cDNA, which does not occur as such in nature. All these elements make that the DNA sequence I have isolated is not a mere discovery but an invention, which provides a teaching to a methodical action. The isolated sequence is not a product of nature, but a product derived from nature... also the directive makes this distinction*" (Bostyn 2004, p. 41). In other words, patent protection is available for natural materials, as long as they are not in their natural context. Indeed, without human intervention, the element would not exist in that specific form and could thus not be used to solve a certain technical problem (Westerlund 2002, p. 32).

[78]Nuffield Council on Bioethics (2002), p. 26 and EPO, USPTO, JPO (1988). While some scholars have embraced the isolation doctrine (Bostyn 1999, pp. 3–4), other commentators have objected to its centrality in determining whether there is an invention or a discovery. Specifically, *Schertenleib* argued that a distinction based on the concept of isolation ignores the fact that all genes need to be isolated in order to be discovered (Schertenleib 2003, p. 127). Similarly, other authors have not acknowledged the distinction imposed by the Biotech Directive and have argued that genes are not patentable due to ethical, philosophical and religious reasons. This conclusion is not affected by their isolation from their natural environment (Van Overwalle 2007, p. 241). A powerful voice against the isolation doctrine and the consequent thinning of the distinction between discoveries and inventions was expressed by the Danish Council of Ethics. In its 2004 report, it affirmed that: "*It cannot be said with any reasonableness that a sequence or partial sequence of a gene ceases to be part of the human body merely because an identical copy of the sequence is isolated from or produced outside of the human body... Before the biotech industries came along, it was an*

A critical view of the isolation doctrine was expressed by *Palombi.*[79] In his opinion, a gene coding for a specific protein remains substantially the same when isolated. The rationale behind the Biotech Directive is that the process used to isolate or produce the element distinguishes it from its natural equivalent, thus turning it into an artificial man-made technology. Indeed, "*the only way of distinguishing the resulting 'artificial' protein to its natural counterpart is by the method used to produce it; but, while the process may be patentable*" the same may not be said for the protein.[80] Furthermore, he argued that, despite the man-made methods used, the identity of the natural and the isolated element would negate the artificiality of the latter.[81] Furthermore, he noted that Article 27 TRIPS distinguishes between inventions and patentable inventions. Consequently, not everything that "*meets the conditions of patentability is a patentable invention. It must first be an 'invention'*".[82] He then argued that, pursuant to the *Genentech* case and the Guidelines for Examination of the EPO, the requirement for an invention is one of the four criteria that are needed to turn an invention into a patentable invention.[83] However, this requirement is rendered superfluous by the Directive, as it relies on isolation or production via technical methods to turn biological materials into patentable inventions.[84]

In light of the above, it was argued that the Directive contradicts its own Recitals in the part where they state that "*this Directive shall be without prejudice to concepts of invention and discovery, as developed by national, European or international patent law*".[85]

---

*occurrence of extreme rarity, and generally regarded as unacceptable, to award a patent on phenomena and processes occurring in nature. That kind of thing was an 'invention of nature'. Only methods for extracting and isolating natural phenomena and natural processes could be patented. For the emergent and expanding biotech industry, however, this was an unacceptable barrier. And it has since transpired that the legislation on this point has now been tailored to the wishes of industry—there are many examples in the gene field of the original distinction in patent law between inventions and discoveries having been obliterated*" (Danish Council of Ethics 2004, pp. 98–99).

[79]Palombi (2009a, pp. 231–232).

[80]Palombi (2009a, p. 231).

[81]In support of his views he cites, UK House of Lords (HL) (2004). On this, see Sect. 4.1.3.2.

[82]Palombi (2005, p. 63).

[83]UK Court of Appeal (CA) (1989a). On this, see Sect. 4.1.3.2.

[84]Since the Directive came into effect after both the EPC and TRIPS, it creates the presumption that those materials fall within the definition of invention set by TRIPS and the EPC. Because of this, the Directive may not have set this presumption legitimately (Palombi 2005, p. 65).

[85]Recital 34 of the Directive (D'Antonio 2004, p. 92).

#### 4.1.3.2 European Case Law Overview

In multiple occasions, the EPO has taken a position on the norms presented above. These cases will be presented in the following sections. National decisions will also be examined when relevant.

European Patent Convention and the Pre-Directive Framework

Decisions of the EPO on discoveries are scarce.[86]

In determining if a method for removing carcinogenic agents from oils found in engines was patentable, the EPO was confronted with the question of whether this was a mere discovery. In its decision, the Board of Appeal held that this removal constituted a technical effect and feature. Therefore, in contrast to the findings of the Examining Division, it affirmed that the claims did not relate to a mere discovery, as they concerned a technical use of the discovered property.[87]

The issue emerged again in the *Nissan Motor* case relating to a car headlamp.[88] In a communication connected to the summons to oral proceedings, the Board commented on the statement of one of the opponents. BMW had argued that since the patentee "*does not offer an explanation of why this effect occurs and as the effect of the protected combination takes place automatically, the subject-matter… has to be regarded as a discovery*".[89] The Board did not accept this point and held that the claim was directed to a concrete technical teaching and not to a discovery as such; hence, it was patent eligible.[90]

As for biotechnologies, in the *Relaxin* case the EPO addressed the question of whether a DNA strand coding for a substance is a discovery as such or an invention. In its reasoning, it differentiated between substances freely occurring in nature (unpatentable discoveries) and substances isolated from their surroundings. In the latter case, both the process for obtaining it and the substance can be patented, as long as the substance "*can be properly characterized by its structure and it is new in the absolute sense of having no previously recognized existence*".[91] This rejected the argument that genes perform the same function when they are found in the human body and when they are isolated. It also rejected the notion that the claims constituted a discovery of the characteristics of an old substance.[92] In its decision, the

[86]Sterckx and Cockbain (2012, p. 130).

[87]Sterckx and Cockbain (2012, p. 118) and Technical Board of Appeal of the EPO (TBA) (1995a, § 3–4).

[88]Technical Board of Appeal of the EPO (TBA) (1997).

[89]Sterckx and Cockbain (2012, p. 119).

[90]Technical Board of Appeal of the EPO (TBA) (1997, § 2).

[91]Technical Board of Appeal of the EPO (TBA) (2002c, § 5.1).

[92]Sterckx and Cockbain (2012, pp. 120–122).

Board referred to Rule 29 EPC and held that the DNA had been obtained via a technical process and was thus patentable.[93]

On this point, scholars argued that the DNA section that encoded for human Relaxin had not been significantly modified and that artificial Relaxin did not display any utility above and beyond its natural equivalent.[94] Some commentators argued that the logic expressed in *Relaxin* (i.e. isolating substances found in nature grants them technical character and avoids them being considered a discovery) has been implemented in the Biotech Directive.[95]

The kernel of this decision was extended to proteins in the *Multimeric Receptors* case.[96] There, the Board of Appeal found that the protein was not a mere discovery. It based its decision on the fact that the claimed receptors could be subjected to industrial exploitation and were thus more than a description of research findings.[97] Indeed, they could "*clearly be applied in an industrial activity*" and they were not purely intellectual and unable to provide any practical or technical character.[98]

Other European Courts have also addressed the dichotomy between discovery and invention.

In the *Genentech* case of 1989, the British Court of Appeal investigated whether purified human t-PA was an invention or a discovery as such.[99] T-PA, while naturally available, was not so in quantities sufficient for medical purposes. Hence, the discovery of its DNA structure could lead to its artificial synthesis and its production in sufficient quantities and with fewer impurities. The decision, which was taken under the 1977 UK Patents Act, identified the substance as a discovery. However, the reasoning of the Judges differed on the legal grounds of the decision. One Judge believed that the claim failed, as it was too broad, while the other believed that the claim did not present a patentable inventive activity.[100] Specifically, a description of the previously unknown DNA structure of a known protein was not considered fit for patent protection. Conversely, the Court maintained that valid claims exist when the discovery is incorporated into a manufacturing process for the production of the substance.[101]

[93]Sterckx and Cockbain (2012, p. 122) and Technical Board of Appeal of the EPO (TBA) (2002c, § 7–9).

[94]Palombi (2009a, p. 242). Relaxin had already been synthesised via recombinant technologies in the 1970s (Palombi 2009a, p. 242).

[95]Aplin and Davis (2017, p. 675).

[96]Technical Board of Appeal of the EPO (TBA) (2002d).

[97]Zimmer et al. (2015, p. 138) and Technical Board of Appeal of the EPO (TBA) (2002d, § 1–3).

[98]Sterckx and Cockbain (2012, p. 123) and Technical Board of Appeal of the EPO (TBA) (2002d, § 3).

[99]UK Court of Appeal (CA) (1989a).

[100]Warren-Jones (2001, p. 110).

[101]UK Court of Appeal (CA) (1989a). While commenting on *Genentech*, *Palombi* held that purified t-PA was not new. In support of his opinion, he cited the view of *Mustill LJ* on this case and his belief that t-PA produced by recombinant technology was identical to the natural version and that the word "recombinant" did not describe "*The product itself, but its history*" (Palombi 2009a, p. 236; UK Court of Appeal (CA) 1989a, § 262).

The House of Lords reviewed again this topic in the *Kirin-Amgen* dispute concerning Erythropoietin (Epo).[102] Amgen had discovered the gene responsible for the production of Epo. It then isolated it, cloned it and devised a process to manufacture it. In applying the pre-Directive normative framework, the House of Lords revoked the patent and held that a person skilled in the art would consider the patent of Amgen as an attempt to patent the discovery of a protein. The problem was that Amgen, despite having devised a ground-breaking process for the production of Epo, intended to patent also the protein itself. However, even though it was isolated, the protein was not new. A number of grounds were listed for this decision.[103] First, the natural and isolated versions of Epo were indistinguishable physically, biologically and genetically. Second, the Court did not differentiate between Epo as a physical substance and the DNA sequence coding for it. Furthermore, the isolated DNA sequences could have been a discovery. However, this last point was considered in contrast with prior case law, as technologies providing a technical contribution with a practical application had been considered inventions.[104] This apparent inconsistency was cxplained with the fact that prior case law required a technical advance in the form of a new result in order to establish a technical contribution. However, in the case at hand, no new result had been produced, since the substance was indistinguishable from its natural equivalent. Hence, pre-Directive, isolated DNA and the protein it codes for may not have been considered inventions.[105]

German Courts also addressed this dichotomy.[106] In 1972, the Federal Supreme Court confirmed that synthetically produced chemical compounds can be patent protected.[107] In a subsequent decision on synthetically produced chemicals, the German Federal Patent Court clarified that, while products of nature are not patentable, products that are derived from nature are. Specifically, the Judges held that:

> Discovery is the finding of something existing but heretofore unknown; it is therefore purely perception. A discoverer turns into an inventor, however, if he provides – based on his perception – instructions for purposeful industrial action.[108]

---

[102]UK House of Lords (HL) (2004).

[103]Palombi (2005, pp. 68–69).

[104]UK Court of Appeal (CA) (1989b).

[105]Still, it needs to be noted that the Court of Appeal had considered those claims as valid.

[106]Under German case law (*Red Dove* case), inventions differ from discoveries, as the former are a teaching to methodical action ("*Lehre zum planmäßigen Handeln*") (Bostyn 2004, pp. 12–13; German Federal Supreme Court (BGH) 1969). In that ruling, the German Federal Supreme Court was confronted with whether: "*Biological phenomena and forces can be treated the same as those of a technological nature*" ("Bundesgerichtshof (Federal Supreme Court) 27.03.1969 File: X ZB 15/67 Rote Taube (Red Dove)," 1970, p. 138). In replying to this question, the Judges found no reason for such exclusion and held that: "*A teaching to methodically utilize controllable natural forces to achieve a casual, perceivable result could be considered patentable*" ("Bundesgerichtshof (Federal Supreme Court) 27.03.1969 File: X ZB 15/67 Rote Taube (Red Dove)," 1970, p. 138).

[107]German Federal Supreme Court (BGH) (1972).

[108]Bostyn (2004, p. 15) and German Federal Patent Court (BPatG) (1977, § III.1).

German Courts applied this line of interpretation to isolated human DNA sequences and thus treated DNA as a chemical entity.[109]

Biotech Directive

While the Directive led to a number of controversies before national and European Courts, only a limited amount of them concerned the relation between discoveries and inventions.

In addressing the action for annulment of the Biotech Directive brought by the Netherlands, the ECJ discussed subject matter patentability. It held that patents are available for inventions that combine a natural element with a technical process and added that the Directive protects only the results of inventive, scientific and technical work.[110]

The isolation doctrine and its relevance in the distinction between discovery and invention was addressed by the Opposition Division of the EPO. In the *Human Stem Cell/Biocyte* case, it was highlighted that, while stem cells occur naturally, they do not occur in an isolated state separate from the body. Hence, they should be considered inventions.[111]

## *4.1.4 Microorganisms*

### **4.1.4.1 European Normative Overview**

TRIPS, the EPC and the Biotech Directive set rules for the patentability of microorganisms.[112]

Article 27(3) TRIPS gives member States the opportunity to exclude from patentability both plants and animals. However, this possibility does not extend to microorganisms and microbiological processes. In a similar formulation, Article 53 (b) EPC maintains that the exception to patentability related to plants, animal varieties and essentially biological processes "*shall not apply to microbiological processes or the products thereof*". Being formulated as an exception to an exemption, this section should be interpreted broadly and thus extend protection to all matters included within its boundaries.[113] A microbiological process is defined by Rule 26(6) EPC as "*any process involving or performed upon or resulting in*

[109]Schwartz and Minssen (2015, p. 229) and German Federal Supreme Court (BGH) (1995).

[110]Bostyn (2004, p. 41) and European Court of Justice (ECJ) (2001, § 72).

[111]Warren-Jones (2001, p. 111) and Technical Board of Appeal of the EPO (TBA) (1999a).

[112]The formulation of TRIPS and EPC on this matter is borrowed from the 1963 Strasburg Convention (*Convention on the unification of certain points of substantive law on patents for invention (ETS No. 047)*, 1963).

[113]Warren-Jones (2001, p. 123).

*microbiological material*".[114] Lastly, Article 4 of the Directive confirms that the exceptions from patentability listed therein "*shall be without prejudice to the patentability of inventions which concern a microbiological... process or a product obtained by means of such a process*".

From this overview, it clearly emerges that microorganisms and microbiological processes constitute patentable subject matter.[115] The reason behind this favourable treatment has been found in the key role played by such organisms in the pharmaceutical, brewing and baking industry.[116]

Despite the importance of microorganisms, this term is not clearly defined. In scientific terms, their defining property is their microscopic size. The word encompasses organisms that vary in form, mode and life cycle; hence the absence of a term that univocally defines them. This lack of definition could also be attributed to the fact that the science around it is still developing.[117] In this regard, the EPO held that "*it does not seem expedient to introduce such a definition as the rapid evolution in the field of microbiology would necessitate its frequent updating*".[118]

The term "microorganism" is interpreted extensively to encompass biological material that is neither a plant nor an animal and should also include materials that find themselves at the border of such definition.[119] The EPO considered bacteria, yeasts, fungi, algae, plasmids, viruses as well as human, animal and plant cells to belong to this category.[120] It thus included in the scope "*all generally unicellular organisms with dimensions not visible to the naked eye which can be propagated and manipulated in a laboratory*".[121] This covers a scope that is wider than the scientific definition of unicellular organisms capable of independent existence. This may have been done to ensure broader patent protection.[122] Some have criticised such expansive interpretation by arguing that it distorts the normative framework.[123]

Protection is also available to microorganisms as such, despite the fact that no explicit provision in this regard is contained in the EPC.[124] Before the entry into force of the Biotech Directive, it was argued that microorganisms would raise the

[114]This formulation mirrors the one found in Article 2 of the Biotech Directive.

[115]In 1873, Pasteur obtained a US patent covering exclusively a living organism (i.e. a yeast culture). Judicially, between 1975 and 1980, the German Federal Supreme Court and the US Supreme Court expressly affirmed the patentability of microorganisms (U.S. Supreme Court 1980; German Federal Supreme Court (BGH) 1975).

[116]Dutfield and Suthersanen (2008, p. 310) and Adcock and Llewelyn (2000, p. 92).

[117]Adcock and Llewelyn (2000, pp. 94–95).

[118]Council for Trade-Related Aspects of Intellectual Property Rights (2006).

[119]Adcock and Llewelyn (2000, p. 96). For an overview of possible definitions, Adcock and Llewelyn (2000, pp. 96–97).

[120]EPO (2016) and Fernandez y Branas (2014, p. 190).

[121]EPO (2018, § G.II.5.5.1) and EPO (2016, § I.B.3.4.1).

[122]Adcock and Llewelyn (2000, p. 96). DNA manipulation can also constitute a microbiological process (Warren-Jones 2001, p. 123).

[123]D'Antonio (2004, p. 118).

[124]EPO (2018, § G.II.5.5.1) and Zimmer et al. (2015, pp. 219–220).

same problems as naturally occurring chemicals with regards to the invention/discovery dichotomy of Article 52 EPC. In particular, microorganisms found in nature could be considered as non-patentable subject matter, since generally occurring substances should remain freely available to the public.[125] Such concerns were however dispelled over time. As mentioned in the EPO Guidelines, "*the product of a microbiological process may also be patentable per se... propagation of the microorganism itself is to be construed as a microbiological process... consequently, the microorganism can be protected per se as it is a product obtained by a microbiological process*".[126] Genetically modified microorganisms could also be claimed.[127] Microbiological processes are patentable even if they claim an essentially biological process for the production of animals and plants.[128]

Questions arose as to what exactly should fall within the definition of microbiological process and, hence, within the realm of patentability. Indeed, it could be argued that microbiological processes exist in all cases in which a microorganism is involved.[129] The EPO rejected this approach and held that processes that include a microbiological step could not be equated with microbiological processes.[130] Nonetheless, the EPO also clarified that this term may encompass processes including both microbiological and non-microbiological steps.[131]

#### 4.1.4.2 European Case Law Overview

The patentability of microorganisms has been the subject of a number of decisions. While in the US acceptance for this kind of patents was troublesome, European Courts more readily agreed to it.[132] Indeed, even before the entry into force of the Directive, no Court of last resort considered microorganisms as such as patent ineligible.[133]

Even before the entry into force of the EPC, Germany had extended patent protection to microorganisms, as confirmed in the *Bäckerhefe* case of 1975.[134] In

---

[125]Teschemacher (1982, p. 35).

[126]EPO (2018, § G.II.3.1) and Zimmer et al. (2015, p. 220).

[127]Intellectual Property Office (IPO) (2016, p. 47).

[128]Cornish et al. (2019, p. 239).

[129]D'Antonio (2004, p. 132).

[130]EPO (2016, § I.B.3.4.1).

[131]EPO (2018, § G.II.5.5.1). The policy behind patents for microorganism has been highly debated. Specifically, a number of scholars and cases have addressed whether microorganisms could be framed as chemical substances or whether their being living matter impedes this and demands another patentability approach. The equation of microorganisms to chemicals has been subject to criticism on the basis that: "*Life is different from non-life, and that is relevant for patent law*" (Dutfield 2012, p. 202).

[132]On this, see Sect. 4.1.6.4.

[133]Teschemacher (1982, pp. 30–31).

[134]Marterer (1987, p. 669) and German Federal Supreme Court (BGH) (1975).

another case, the German Federal Patent Court clarified that naturally occurring microorganisms are:

> Not eligible for protection, unless the inventor demonstrates a reproducible method whereby naturally occurring micro-organisms may be produced by human means... this opinion... has validity not only in the field of microbiology, but constitutes generally a condition of patentability.[135]

The Court also mentioned that microorganisms are not necessarily excluded from patent protection because of their being living organisms. The German case law showed that the patentability of microorganisms was based on the possibility of equating them with chemical inventions.[136]

## *4.1.5 Technical and Essentially Biological Processes*

### 4.1.5.1 European Normative Overview

Pursuant to Article 3, 4 and 5 of the Biotech Directive, biological material produced via a technical process is patentable. This applies to both the process itself and its products. The fact that the biological material previously occurred in nature or that the element of the human body is structurally identical to a natural one has no bearing on patentability. While the norms do not specifically define what is meant by technical process, Recital 21 of the Directive explains that:

> An element... is not excluded from patentability since it is, for example, the result of technical processes used to identify, purify and classify it and to reproduce it outside the human body, techniques which human beings alone are capable of putting into practice and which nature is incapable of accomplishing by itself.

On a similar note, the Commission, in a report to the European Parliament and Council, clarified that:

> As set out in recital 21, the reasoning is that, to qualify for patentability, an element from the human body, including a sequence or partial sequence of a gene, must... be the result of technical processes which have identified, purified, characterised and multiplied it outside of the human body. Such techniques cannot be found in nature... They would show only natural properties which man alone, through genetic engineering, is capable of exploiting and inserting into a technical process. The well-known distinction in patent law between a discovery and an invention thus applies fully in the field of biotechnology.[137]

Based on this, the concept would seem to encompass techniques that can solely be implemented by humans and which are unachievable by nature (e.g. recombinant DNA techniques). The general formulation adopted by the legislator allows it to adapt to new technological developments. Also, it was argued that this concept

---

[135]Bostyn (2004, p. 15) and German Federal Patent Court (BPatG) (1977, § III.2).

[136]Wegner (1976, pp. 236–239).

[137]European Commission (2002, § 49).

should be construed widely and "*not be restricted to essentially chemical manufacturing processes*".[138]

On the opposite side of the spectrum are instead processes that are essentially biological. Pursuant to Article 27(2)(b) TRIPS and Article 53(b) EPC, essentially biological processes for the production of plants and animals are excluded from patentability. The Biotech Directive follows the same orientation. In Article 4, it expressly states that essentially biological processes shall not be patentable. As for what constitutes this kind of process, Article 2 of the Directive explains that "*a process for the production of plants and animals is essentially biological if it consists entirely of natural phenomena such as crossing or selection*". In the view of the EPO, procedures based on sexual crossing and selection are essentially biological, and thus non patentable, even if they include a human intervention aimed at enabling and assisting the performance of the processes. Conversely, procedures involving genetic engineering will not be considered as such, since they do not rely on the natural mixing of genes. Equally, processes that are not based on sexual crossing and selection would fall outside the realm of this norm and would therefore be patentable.[139]

#### 4.1.5.2 European Case Law Overview

The concept of "technical process" has not been specifically addressed in the case law. In a number of occasions, the Board of Appeal of the EPO held that an invention was obtained via a technical process, but it did not further specify this concept, thus retaining its flexibility.[140] The ECJ followed a similar approach in *Brüstle*. It praised the significant progress of medical treatments and thanked "*medicinal products resulting from technical processes aimed at obtaining elements similar in structure to those existing naturally in the human body*" for this result.[141]

By contrast, substantial case law is available on the definition of what constitutes an "essentially biological process". In the *Hybrid Plants* case, the Board of Appeal of the EPO held that whether a process is essentially biological is to be assessed on the basis of the essence of the invention.[142] Over time, the Board of Appeal of the EPO

[138]Intellectual Property Office (IPO) (2017, § 76A.04).

[139]EPO (2018, § G.II.5.4.2).

[140]For example, it was noted that: "*Technical processes refer to those requiring interference by man, for example, where the sequence of the steps in a process otherwise consisting of natural phenomena is manipulated by man and has a decisive effect on the end product... similarly, an invention consisting of a process containing at least one step which makes use of a technical application, such as a genetic engineering step, is one that concerns a technical process and is patentable*" (Hacon and Pagenberg 2008, p. 440; Technical Board of Appeal of the EPO (TBA) 1995b, 1988).

[141]European Court of Justice (ECJ) (2011, § 17).

[142]Technical Board of Appeal of the EPO (TBA) (1988). Specifically, human intervention needs to be taken into consideration in its entirety and in both quantitative and qualitative terms. Equally, its

moved away from its earlier case law. The main reasons for this change were that earlier cases linked the definition to the assessment of the state of the art and a different reading of the preparatory documents pertaining to Article 53 EPC.

In 2007, a referral was made to the Enlarged Board of Appeal on whether a process involving crossing and selection of broccoli could be excluded from patentability.[143] Soon after, a similar referral was made, this time concerning the crossing and selection of tomatoes.[144] In its decision, the Board noted that the exception would operate even in the presence of a step of technical nature that enabled crossing and selection. The opposite would be true in case the technical step was able on its own to introduce or modify a trait in the genome. For this purpose, "*it is not relevant whether a step... is a new or known measure, whether it is trivial or a fundamental alteration of a known process, whether it does or could occur in nature or whether the essence of the invention lies in it*".[145]

In 2012 and 2013, the EPO received a referral concerning the scope of the "essentially biological processes" exception relating again to the crossing and selection of tomatoes and broccoli. These cases sought to clarify a point that had previously remained open, namely, whether product claims and product-by-process claims resulting from excluded processes could be excluded from patentability pursuant to Article 53(b) EPC. The Board maintained the patentability of plants and seeds obtained through those processes, including if such claims are formulated in a product-by-process format.[146] In response to this decision, the European Commission noted that "*the patentability of such products runs into potential conflict with the legal protection provided to plant varieties under EU plant variety legislation as regards access to genetic resources*" and noted that "*it is questionable whether the same result would have been reached in the EU context*".[147] At the same time, the EU recognised that those decisions of the EPO would conflict with the norms of some member States. For this reason, it requested more clarity on this matter.[148] As a result, the Administrative Council of the EPO amended Rules 27 and 28 EPC to clarify that, under Article 53(b) EPC, products obtained exclusively via

impact on the results achieved is to be evaluated. To overcome the exception set in Article 53 (b) EPC, human intervention shall not be trivial and shall visibly impact the result (Zimmer and Quest 2011, p. 461; Technical Board of Appeal of the EPO (TBA) 1988).

[143]Enlarged Board of Appeal of the EPO (EBA) (2010a).

[144]Enlarged Board of Appeal of the EPO (EBA) (2010b).

[145]Enlarged Board of Appeal of the EPO (EBA) (2010b, § 4) and Enlarged Board of Appeal of the EPO (EBA) (2010a, § 4).

[146]Enlarged Board of Appeal of the EPO (EBA) (2015a, b) and Minssen (2015). In commenting on these cases, authors noted that this approach seems to focus on genetic identity to determine subject matter patentability. According to some authors, this approach can be equated to the one of the US Supreme Court in *Myriad* and of the Federal Circuit Court in *Roslin* (Sherman 2015, pp. 1218–1219). On this, see Sect. 4.1.6.5.

[147]European Commission (2016).

[148]Council of the European Union (2017).

essentially biological processes are excluded from patentability.[149] This approach is however still hotly contested and far from settled, as recent cases and referrals show.[150]

### 4.1.6 *The Laws and Product of Nature Doctrines*

While Europe maintains a permissive and expansive approach to the patentability of biotechnological inventions, other countries follow a different direction. The most notable example of this trend are the United States of America, which radically changed their approach towards gene patentability after the *Myriad* case.[151] This change has been remarkable, if one considers that just a few years ago American commentators declared that "*it is not clear whether the patent system has any subject matter boundaries at all*" and that some Courts avoided relying on the patentability criteria for their decisions.[152] Interestingly, the reasons of this turnaround are not entirely clear.[153]

To examine if and how this change in the US is likely to affect the European approach to patentability, the relevant case law will be examined. The focus will be on the so-called "product of nature" doctrine, which postulates that products that occur in nature in essentially the same form cannot be patented, and on the "laws of nature" one, which holds that scientific theories, laws of physics and other fundamental laws of the natural world should not be subject to patent protection. First, however, the applicable patent norms will be examined.

#### 4.1.6.1 US Patent Norms

The US Constitution explicitly addresses the topic of patent protection. In Article I, Section 8, clause 8, it states that Congress shall have the power to:

> Promote the Progress of Science and useful Arts, by securing for limited Times to Authors and Inventors the exclusive Right to their respective Writings and Discoveries.[154]

While the norm speaks of discoveries, this term should not be read as a synonym to the one used in Article 52 EPC. At the time the Constitution was drafted, the term

---

[149] EPO (2018, § G.II.5.4) and Wilkof (2017).

[150] In 2018, the EPO Technical Board of Appeal held that the amended Rule 28 EPC is invalid due to its conflict with Article 53(b) EPC. The matter has now been referred by the President of the EPO to the Enlarged Board of Appeal (Martinez-Lopez 2020; EPO 2019; Aerts 2019; Bentham et al. 2019; Technical Board of Appeal of the EPO (TBA) 2018).

[151] On this, see Sect. 4.1.6.5.

[152] Eisenberg (2006, p. 357).

[153] Schwartz and Minssen (2015, pp. 192–193).

[154] "Constitution of the United States of America" (1789).

"discovery" represented the creation of something new and would thus be equivalent to the modern notion of invention.[155]

The United States Code Title 35 (35 U.S.C.) sets the norms concerning patent protection in America. The legislation, which was laid out in 1952, establishes a series of requirements for patentability. In addition to utility (§101), novelty (§102), nonobviousness (§103) and sufficiency of disclosure (§112), the Code outlines the requirements for subject matter eligibility in § 101. This article states that:

> Whoever invents or discovers any new and useful process, machine, manufacture, or composition of matter, or any new and useful improvement thereof, may obtain a patent therefor, subject to the conditions and requirements of this title.

Differently from Article 52 EPC, the norm does not set explicit exclusions from patentability.

#### 4.1.6.2 The Development of the Doctrines

Despite the absence of explicit exclusions from patentability in US norms, American Courts have established two exceptions from patentability that share common aspects with the discovery exclusion of Article 52 EPC. Those two exclusions are known as the "product of nature" and "laws of nature" doctrines and find their basis in different lines of authorities.

The origin of the doctrines is controversial, with some authors describing them as "*kaleidoscope of doctrines*" and attributing this to the fact that "*cases come at the question from different angles, in different historical contexts, with premises ranging from pragmatic to formalist and from patent-friendly to fiercely patent-skeptical*".[156] The fact that patentability requirements often overlap and are tangled up together in older Court decisions has affected the structure and soundness of the doctrines.[157] In recent times, the U.S. Supreme Court emphasised the importance of separating the different conditions and of not stripping § 101 of its pivotal role. The Court held, with an approach that seems to be in stark contrast to the one currently adopted by the EPO, that:

> The relevant cases rest their holdings upon section 101, not later sections... to shift the patent-eligibility inquiry entirely to these later sections risks creating significantly greater legal uncertainty, while assuming that those sections can do work that they are not equipped to do.[158]

Those doctrines are particularly problematic in the biotechnological field, as nearly every invention involves a product or law of nature, whether directly or

[155]Studies show that in the late eighteenth century both terms related to the creation of something original (Andrews 2014, p. 547).

[156]Beauchamp (2013, p. 310).

[157]Beauchamp (2013, pp. 264–266).

[158]U.S. Supreme Court (2012, p. 21).

indirectly.[159] Matters are complicated even further by the absence of a univocal notion of what is meant by "nature" in patent law and by the fact that almost all-natural things have been manipulated by humans in one way or another.[160]

Policy plays an important role in these doctrines. Commentators have identified the policy behind the product of nature exclusion in the constitutional principle dictating the exclusion of patents that are non-inventive and that "*remove existent knowledge from the public domain or restrict free access to materials already available*".[161] Conversely, the laws of nature exclusion is based on the idea that future innovation should not be impeded.[162]

It has been argued that cases concerning products of nature should be decided by using the statutory patentability criteria instead of "*adding the complexity of a non-statutory and poorly defined consideration like the judicially-imposed product of nature bar*".[163] This unclear nature is confirmed by the fact that:

> Different tests are used at different times, often interchangeably and without explanation. There is not only diversity between different judgements; there is sometimes even diversity within a single judgement.[164]

#### 4.1.6.3 The Laws of Nature Doctrine

Laws of nature, scientific theories, laws of physics and other fundamental laws shall not be the object of patent protection. This prohibition seems to originate from concerns over patent scope.[165] The idea that an inventor could claim something that belongs to the commons of humanity, that pre-exists mankind and was not created by humans seems unreasonable.

The importance of these laws for scientific research and technological development was emphasised by the Supreme Court, which affirmed that "*phenomena of nature... mental processes, and abstract intellectual concepts are not patentable, as they are the basic tools of scientific and technological work*".[166] In *LabCorp*,[167]

---

[159] Rogers (2013, p. 441) and Schilling (2011, p. 754).

[160] Sherman (2015, pp. 1226–1227).

[161] Schwartz and Minssen (2015, pp. 200–201) and U.S. Supreme Court (1966, § 6).

[162] Schwartz and Minssen (2015, p. 201).

[163] Lesser (2011, p. 354).

[164] Sherman (2015, p. 1222).

[165] Problems with patent scope are a real concern. It was noted that between 25% and 75% of all litigated patents are held invalid, at times because their wide scope ends up covering unpatentable subject matter such as laws and products of nature (Andrews 2014, p. 546).

[166] U.S. Supreme Court (1972, p. 67).

[167] The case concerned a petition for certiorari. However, since the issue of patentability had not been raised in the lower instances and despite the fact that some of the Justices were interested in revisiting this doctrine, the Court dismissed the writ as improvidently granted (Conley 2009, p. 121; U.S. Supreme Court 2006).

before the Supreme Court, one of the Justices took a stand on the policy behind this doctrine and held that the justification for this exclusion:

> Does not lie in any claim that 'laws of nature' are obvious, or that their discovery is easy, or that they are not useful. To the contrary, research into such matters may be costly and time-consuming; monetary incentives may matter; and the fruits of those incentives and that research may prove of great benefit to the human race. Rather, the reason for the exclusion is that sometimes too much patent protection can impede rather than 'promote the Progress of Science and useful Arts'.[168]

A bright line between what is patentable and what is not proved complex to determine in practice. A reason for this is that the notion of "laws of nature" is *per se* generic. A commentator noted how this term, at its minimum, defines the "*invariant relationship that governs the interaction of two or more physical entities*" and held that its operation is universal. Based on this, it was then wondered whether also the genetic code could be seen as a law of nature and thus unpatentable.[169]

In *Le Roy v. Tatham*, the Supreme Court took a stand on this matter and famously held that:

> A principle, in the abstract, is a fundamental truth; an original cause; a motive; these cannot be patented, as no one can claim in either of them an exclusive right.[170]

Soon after, the Supreme Court dealt with another case that addressed the laws of nature doctrine as well as patent scope. In *O'Reilly v. Morse*, a particularly wide claim submitted by Morse was examined. While the Justices validated the patent for a process that employed electromagnetism to produce telegraphic signs, they recognised a fallacy in one of the claims. They objected to it by noting that:

> He claims the exclusive right to every improvement... Nor is this all; while he shuts the door against inventions of other persons, the patentee would be able to avail himself of new discoveries in the properties and powers of electro-magnetism which scientific men might bring to light... In fine, he claims an exclusive right to use a manner and process which he has not described and indeed had not invented... The Court is of opinion that the claim is too broad, and not warranted by law.[171]

Along the same lines, more than a century later, the Supreme Court famously declared that:

> A new mineral discovered in the earth or a new plant found in the wild is not patentable subject matter. Likewise, Einstein could not patent his celebrated law that $E=mc^2$; nor could Newton have patented the law of gravity. Such discoveries are manifestations of... nature, free to all men and reserved exclusively to none.[172]

---

[168]U.S. Supreme Court (2006, § 2922).

[169]Kane (2004, p. 751).

[170]U.S. Supreme Court (1852, p. 175).

[171]U.S. Supreme Court (1853, p. 113).

[172]U.S. Supreme Court (1980, p. 309).

In 2012, the Supreme Court addressed again the laws of nature exclusion.[173] In *Mayo*, the Court dealt with a method to treat autoimmune diseases. Patients metabolise drugs in a different manner and thus it is complex for doctors to establish the right amount of drugs to be delivered to patients without incurring into side effects. The claimed subject matter concerned the correlations identified between metabolites levels and the effectiveness of the administered drug. In a unanimous decision, the Court affirmed that patents that claim laws of nature are invalid. Specifically, they held that the claims recited laws of nature (the connection between concentrations of certain metabolites and the likelihood that a drug dosage will prove ineffective or harmful) and are thus unpatentable. In its ruling, the Court concentrated on whether the patent would inhibit future research.

In *Myriad*, the United States referred in its *amicus curiae* to "*manifestations of laws of nature, free to all men and reserved exclusively to none*" when describing the BRCA gene.[174] In that contest, it was noted that the link between the naturally occurring DNA sequence and the molecule it expresses (i.e. the connection between genotype and phenotype) is a law of nature and that man could not influence the fact that a segment of human genome can code for a specific protein.[175]

#### 4.1.6.4 The Product of Nature Doctrine Pre-*Myriad*

A first example of the product of nature doctrine is identified in the 1874 case *American Wood-Paper* before the U.S. Supreme Court.[176] In this instance, the Court was faced with claims for chemically treated wood pulp. It held that a product is not novel even if it constitutes the further refinement or purification of an already existing one. Given the content of the decision, authors have been critical of the characterisation of this case as belonging to the product of nature doctrine, considering that the patent was rejected for lack of novelty.[177]

Similar arguments were put forward in *Cochrane*.[178] The patent concerned improvements in dyes extracted from anthracine (so-called alizarine) and essentially related to the production of synthetic alizarine.[179] The patent aimed at covering any method for producing artificial alizarine as well as artificial alizarine itself,

[173]U.S. Supreme Court (2012).

[174]"Brief for the United States as amicus curiae in support of neither party - in the US Court of Appeals for the Federal Circuit in Association for Molecular Pathology, et al. v. United States Patent and Trademark Office (Myriad)," (2010, p. 19).

[175]"Brief for the United States as amicus curiae in support of neither party - in the US Court of Appeals for the Federal Circuit in Association for Molecular Pathology, et al. v. United States Patent and Trademark Office (Myriad)," (2010, p. 19).

[176]U.S. Supreme Court (1874).

[177]Beauchamp (2013, pp. 271–272) and Smalley (2013, pp. 404–406).

[178]U.S. Supreme Court (1884).

[179]This artificial dye was produced via synthetic chemistry (Palombi 2009a, p. 209).

independently from how it was produced.[180] In the course of the trial, it was argued that the patent was invalid since alizarine is a natural product with a known structure and given that the chemical formula of synthetic alizarine was identical to that of the natural version. While holding that the process was patentable, the Justices rejected the patentability of the product. To reach this conclusion, they developed the following argument:

> The specification of the original patent... states the invention to be a process for preparing alizarine, not as a new substance prepared for the first time, but as the substance already known as alizarine, to be prepared, however, by the new process... the article produced by the process described was the alizarine of madder, having the chemical formula $C^{14}$ $H^{8}$ $O^{4}$. It was an old article. While a new process for producing it was patentable, the product itself could not be patented, even though it was a product made artificially for the first time... Calling it artificial alizarine did not make it a new composition of matter, and patentable as such, by reason of its having been prepared artificially, for the first time.[181]

Further indications of the product of nature doctrine can be found in the opinion issued in 1889 by the Commissioner of Patents in *Ex parte Latimer.*[182] The patent concerned a fibre extracted from pine needles which was claimed to be a new article of manufacture and which was proving very valuable on the market for bagging materials. Despite this, the patent application was rejected by the Commissioner who held that the fibre was "*a natural product and can no more be the subject of a patent in its natural state when freed from its surroundings than wheat which has been cut by... some new method*".[183] Indeed, the fibre, when freed from the sheath in which was naturally found, was "*nowise changed or different from its natural construction*" and its chemical structure was unaffected by this.[184] The impossibility to distinguish the claimed fibre from its natural counterpart was as central in the decision of the Commissioner as it had been in that of the patent Examiner.[185] Some commentators believe that this decision crystallised the main aspects of the product of nature doctrine, namely that a product whose physical characteristics render it indistinguishable from its natural counterpart cannot be subject to patent protection and that the utility and value of a product do not affect its patentability. Lastly, the inherent unpatentability of a product cannot be overcome by the novelty of the process used to manufacture it or by its being an unprecedented discovery.[186]

Despite those precedents, some Courts held synthetic or purified products of nature as patentable. This was for example the case in 1910 in *Kuehmsted.*[187] During this controversy, relating to a patent for aspirin, it was argued that the product and the

---

[180]Palombi (2009a, p. 210).

[181]U.S. Supreme Court (1884, pp. 308–309).

[182]U.S. Commissioner of Patents (1889).

[183]U.S. Commissioner of Patents (1889, p. 127).

[184]U.S. Commissioner of Patents (1889, p. 126).

[185]Conley and Makowski (2004, p. 14).

[186]Conley and Makowski (2004, p. 15).

[187]U.S. Court of Appeals for the Seventh Circuit (1910).

prior art substance had an identical chemical formula. In spite of this, the Court held that the two were not identical, since the substances were physically and therapeutically distinct due to the differences in their purity levels.[188] The fact that they performed differently trumped the fact that they had the same composition.[189]

Purified adrenalin was the subject of the next case on the product of nature doctrine.[190] In *Parke-Davis*, it was argued that purified adrenalin should not be patentable since it was merely a purified version of something existing in the body. The Judge rejected this construction and instead declared that the inventor "*was the first to make it available for any use... it became for every practical purpose a new thing commercially and therapeutically. That was a good ground for a patent.*[191]

In 1928, this doctrine was addressed again in *General Electric v. De Forest Radio*.[192] The controversy related to an improved version of the chemical element tungsten. The Judges came to the conclusion that the inventor "*cannot have a patent for it because a patent cannot be awarded for a discovery or for a product of nature, or for a chemical element*".[193] Specifically, they argued that the process devised by the inventor basically led to obtain natural tungsten in a pure form. In turn, pure tungsten showed its intrinsic characteristics, which originated in nature rather than being a creation of the inventor. Therefore, the newly found ductility of tungsten was seen as a discovery rather than an invention.[194]

The product of nature doctrine was addressed again, albeit not directly, in *American Fruit Growers* before the U.S. Supreme Court.[195] The Justices denied the patentability of oranges, whose skin was coated with borax to render it resistant to decay. They maintained that patentable subject matter exists where inventions have "*new or distinctive form, quality, or property*" and that there should be a "*change in the name, appearance, or general character of the fruit*", which in this case was not present, despite the fact that the fruit had a mould resistant quality that was not found in the natural product.[196]

The Supreme Court was faced again with a case concerning the product of nature doctrine in 1948.[197] The *Funk Brothers Seed* case revolved around an inoculant package that contained multiple types of bacteria needed to enhance the nitrogen

---

[188]The idea that patents cannot be granted for products that merely display a higher degree of purity in comparison to products manufactured via other methods was stated in *In re Merz* Lesser (2011, p. 350) and U.S. Court of Customs and Patent Appeals (CCPA) (1938, p. 600).

[189]Smalley (2013, pp. 408–409).

[190]U.S. Circuit Court, S.D. New York (1911).

[191]U.S. Circuit Court, S.D. New York (1911, p. 103).

[192]U.S. Circuit Court of Appeals - Third Circuit (1928).

[193]U.S. Circuit Court of Appeals - Third Circuit (1928, p. 642).

[194]Similar conclusions were reached in cases concerning the ductility of uranium and vanadium (U.S. Court of Customs and Patent Appeals (CCPA) 1931). Nonetheless, patents have been granted for elements of the periodic table, for example in the case of Americium (Lesser 2011, p. 346).

[195]U.S. Supreme Court (1931).

[196]U.S. Supreme Court (1931, pp. 282–283).

[197]U.S. Supreme Court (1948).

fixing ability of crops. While single strains of bacteria had been used before for this purpose, the patentee discovered a combination of bacterial strains that would be effective and not inhibit each other, as was instead the case in the past. This would allow farmers to buy just one inoculant for several types of plants instead of having to buy separate ones for each. In invalidating the patent, the Supreme Court held that the inventor:

> Does not create a state of inhibition or of noninhibition in the bacteria. Their qualities are the work of nature. Those qualities are, of course, not patentable. For patents cannot issue for the discovery of the phenomena of nature... The qualities of these bacteria, like the heat of the sun, electricity, or the qualities of metals, are part of the storehouse of knowledge of all men. They are manifestations of laws of nature, free to all men and reserved exclusively to none. He who discovers a hitherto unknown phenomenon of nature has no claim to a monopoly of it which the law recognizes. If there is to be invention from such a discovery, it must come from the application of the law of nature to a new and useful end.[198]

Although the Court recognised the commercial relevance of the package, it argued that "*even though it may have been the product of skill, it certainly was not the product of invention*".[199] With its decision, the Court clarified what is meant by "naturally occurring". This includes a product that has the same effect it always had, does not gain an additional utility through human manipulation and performs in its own natural way.[200]

Isolation and purification were at the centre of the jurisprudential debate in 1970 with the case *In re Bergstrom*. There, the Court reversed the decisions of the patent Examiner and of the USPTO Board of Appeals and held that isolated and purified hormones are not naturally occurring. Hence, those compounds could be considered patentable subject matter. In its reasoning, the Court explained that "*those compounds... do not exist in nature in pure form*".[201] From this decision, it was inferred that a difference in composition between pure and impure substances can confer patentability and novelty to an invention.[202]

A paradigm shifting decision on the product of nature doctrine was delivered by the US Supreme Court in its first biotechnology case, *Diamond v. Chakrabarty*.[203] The case revolved around a bacterium that was capable of breaking down crude oil. This bacterium, which could be used to treat oil spills, had been altered via the insertion of plasmids.[204] In solving the case, two questions were central. First, whether living things could be considered patent eligible and, second, whether this bacterium constituted an invention or a mere interference in natural metabolic

[198] U.S. Supreme Court (1948, p. 130).

[199] U.S. Supreme Court (1948, p. 132).

[200] Arcuri (2011, pp. 745–746).

[201] U.S. Court of Customs and Patent Appeals (CCPA) (1970, § 1401).

[202] Kane (2004, p. 740).

[203] U.S. Supreme Court (1980).

[204] The decision did not concern the question of the patentability of genes and other molecular components (Kane 2004, p. 736).

processes.[205] The USPTO had first rejected the claims for the bacterium, arguing that living organisms were not patent eligible, even if they were man-made and non-naturally occurring. In a close five-to-four decision, the Justices affirmed the patentability of the bacterium, despite the fact that it was a living organism and justified their conclusion by arguing that:

> [The] claim is not to a hitherto unknown natural phenomenon, but to a nonnaturally occurring manufacture or composition of matter - a product of human ingenuity having a distinctive name, character [and] use... The point is underscored dramatically by comparison of the invention here with that in Funk. There, the patentee had discovered that there existed in nature certain species of root nodule bacteria which did not exert a mutually inhibitive effect on each other... Here, by contrast, the patentee has produced a new bacterium with markedly different characteristics from any found in nature, and one having the potential for significant utility. His discovery is not nature's handiwork, but his own; accordingly it is patentable subject matter under § 101.[206]

Also commentators compared this decision with *Funk*. In both cases, one could argue that the inventor had produced an aggregation that did not naturally exist. From this perspective, the major distinction between the two would reside in the techniques adopted to achieve this: the mixing of cells in *Funk* and genetic engineering in *Chakrabarty*.[207] Other scholars found instead that the relevant difference between the two must have laid in the "*markedly different characteristics*" found in *Chakrabarty* and not in *Funk*, since both had significant utility.[208] But what constitutes a "*markedly different characteristic*" for these purposes[209]? *Palombi* believed that the term "markedly" should not be overlooked and held that, in comparison to its natural equivalent, the product should display a function that was unprecedented in nature and present a significant utility.[210]

This decision was also compared to *American Fruit Growers,* given the reference found in both of them to the new name of a product. In *Chakrabarty*, the Justices referred to a product of human ingenuity having a distinctive name. In that case, the bacterium had been renamed *Pseudomonas putida*; a change in name seemed to indicate a change in kind, which led to subject matter patentability. Conversely, in

---

[205]Myszczuk and De Meirelles (2014, p. 79). In *In re Bergy* of 1977, the CCPA addressed the patentability of living matter, namely microorganisms. The Court held that: "*We see no sound reason to refuse patent protection to the microorganisms themselves a kind of tool used by chemists... in much the same way as they use chemical elements... which are not considered to be alive*". U.S. Court of Customs and Patent Appeals (CCPA) (1977, § 1038).

[206]U.S. Supreme Court (1980, pp. 309–310). The *Chakrabarty* decision became famous for its reference to the Committee Reports that accompanied the introduction of the 1952 US Patent Act, where it was stated that: "*Congress intended statutory subject matter to 'include anything under the sun that is made by man'*" (U.S. Supreme Court 1980, p. 309). Despite this wide formulation, a few lines later the Court recognised that §101 has limits and that laws of nature, physical phenomena and abstract ideas are not patentable.

[207]Conley and Makowski (2004, p. 24).

[208]Sarnoff (2012, p. 104).

[209]For an overview, Sherman (2015, pp. 1210–1221).

[210]Palombi (2009a, pp. 228–229).

*American Fruit Growers*, the Court held that the product was not patent eligible and based its decision, amongst others, on the fact there was no change in name.[211] Changing the name of a product was a tactic adopted over time by patentees in order to show the novelty of their invention. However, commentators have been critical of this practice.[212]

The patentability of living matter was addressed again in *Ex Parte Allen.*[213] There, the Board of Patent Appeals and Interferences (BPAI) held that polyploid oysters could be considered patentable subject matter, as they are non-naturally occurring. In its decision, the Board argued that the focus of the analysis should be on whether the oysters are man-made or naturally occurring, independently from their being controlled by laws of nature. Therefore, since the Examiner did not put forward any evidence showing that the product could occur naturally and without human intervention, it was considered patentable.

The product of nature doctrine stepped again into the spotlight with the first gene patent controversies in the 1990s. The 1991 *Amgen* case related to the DNA sequence encoding for human Erythropoietin (Epo).[214] In this occasion, the Judges mentioned that "*it is important to recognize that neither Fritsch nor Lin invented Epo or the Epo gene. The subject matter... was the novel purified and isolated sequence which codes for Epo*".[215] This DNA sequence did not have a precise structural equivalent in the human body. This is because the object of the patent was a cDNA strand.[216] Furthermore, the Judges viewed the gene as "*a chemical compound, albeit a complex one*".[217]

In the meantime, in 2001, the USPTO amended its Guidelines to establish that isolated and purified DNA molecules would be patentable if either isolated by extraction or if synthesised in a pure form.[218] In the first case, the rationale was that the DNA molecule did not naturally occur in that form. Similarly, in the second case, the logic behind it is the difference between the compound in its natural and

[211]Sherman (2015, pp. 1200–1201).

[212]Sherman (2015, p. 1201). First, changing the name is not necessarily connected to a change in kind. Second, it is not a reliable criteria since: "*The process of giving something a new name usually tells us very little about the thing itself*" (Sherman 2015, p. 1202).

[213]U.S. Boards of Patent Interferences and Appeals (1987).

[214]U.S. Court of Appeals for the Federal Circuit (1991).

[215]U.S. Court of Appeals for the Federal Circuit (1991, § 46).

[216]Conley (2009, p. 116).

[217]U.S. Court of Appeals for the Federal Circuit (1991, § 45). Commentators underlined the fact that, when examining products of nature, their source is irrelevant. Indeed, if a product is equivalent to its natural counterpart, it would not pass the patent eligibility test, regardless of its method of production (Conley and Makowski 2004, p. 34).

[218]USPTO (2001, p. 1093). It is worth noting that, amongst the comments invited and received by the USPTO on the Guidelines, there were positions arguing that genes are not inventions, but rather unpatentable discoveries. Other comments expressed doubts on whether genes could be considered compositions of matter, since they already exist in nature. Both those comments were rejected by the USPTO (Conley and Makowski 2004, pp. 29–30; USPTO 2000, § 1–2).

purified form.[219] All these factors led to a new approach to the product of nature doctrine. This exclusion turned into a technical claim-drafting issue, thus moving away from a substantial analysis. *Eisenberg* described the situation in the following terms:

> The standard patent lawyer's response to the 'products of nature' limitation is to treat it as a technical, claim-drafting problem. From this perspective, the prohibition against patenting products of nature only prevents the patenting of DNA sequences in a naturally occurring form that requires no human intervention.[220]

Other authors identified a second problem with patents for isolated genetic materials. The isolation doctrine focused on the structural differences between natural and isolated DNA as the basis for patentability. However, while isolation does create a difference in the chemical structure of the genetic material, its function remains the same. Isolated DNA carries the same information as naturally occurring one and this identity is the reason for its usefulness. Therefore, *Conley* wondered why minor differences in chemical structure should be in the focus, while functional identities are fundamentally ignored, despite being the whole point of the patent.[221] This dichotomy between function and structure became a pivotal point in the patentability of biotechnologies thanks to the *Myriad* case.[222]

#### 4.1.6.5 The Product of Nature Doctrine Post-*Myriad*

The *Myriad* Saga in the US

The picture presented above changed drastically in 2010, as Sweet DJ from the District Court of the Southern District of New York issued his decision in the *Myriad* case.[223]

The background of this controversy dates back to 1990, when a team of scientists linked a region of the human genome with the predisposition to develop breast cancer. Soon after, Myriad Genetics Inc. was founded. In a few years, the company was able to locate and sequence two relevant genes, which then became known under the acronym BRCA1 and BRCA2 (Breast Cancer Susceptibility Gene).[224]

---

[219]Some commentators noted that the mere creation of synthetic equivalents of natural products would not render them patent eligible, at least pursuant to Supreme Court precedents (Sarnoff 2012, p. 106).

[220]Eisenberg (2000, p. 786). In the *Myriad* case, it was argued that scientists: "*Have considered this practice a 'lawyer's trick' that circumvents the prohibitions on the direct patenting of the DNA in our bodies but which, in practice, reaches the same result*" (Beauchamp 2013, p. 267; U.S. District Court Southern District of New York 2010).

[221]Conley (2009, p. 120).

[222]On this, see Sect. 4.1.6.5.

[223]U.S. District Court Southern District of New York (2010).

[224]While the most frequent form of the BRCA1 gene is not correlated to the incidence of breast cancer, some mutations of it are (Orford 2014, p. 558).

Myriad developed and commercialised a number of genetic tests using this information. Patents over these two genes were granted. Those patents, specifically the product claims concerning the isolated and purified DNA molecules coding for the BRCA1 and BRCA2 genes, were the subject of the dispute.[225]

The legal saga initiated in 2010, when Sweet DJ invalidated the contested claims.[226] Regarding product claims, the question brought before the Judge revolved around whether or not claims for isolated DNA containing naturally-occurring sequences would fall within the realm of application of the product of nature exception and thus be unpatentable. To answer this question, the Judge reviewed precedents on this matter and stated that the Supreme Court "*has established [that] products of nature do not constitute patentable subject matter absent a change that results in the creation of a fundamentally new product*".[227] Similarly, he noted that "*the purified product must possess 'markedly different characteristics' in order to satisfy the requirements of... § 101*".[228] The discussion then revolved around whether the difference between the isolated DNA and its natural equivalent was indeed marked.

In rejecting the argument presented by Myriad that those differences were marked, Sweet DJ reviewed the technologies at hand and their characteristics. In particular, the debate focused on the unique features of DNA. Indeed, as held by one of the experts consulted by Myriad, "*genes are of double nature: on the one hand, they are chemical substances or molecules. On the other hand, they are physical carriers of information, i.e., where the actual biological function of this information is coding for proteins*".[229] The Judge shared this view on the peculiarities of DNA and argued that its information quality renders it unique amongst the chemicals found in the body and it represents a physical embodiment of laws of nature. Hence, considering it a simple chemical would be erroneous.[230]

The discussion focused on two aspects: isolated DNA and complementary DNA (i.e. cDNA). On the first, Sweet DJ focused on the function of DNA and maintained that isolating it by merely severing existing chemical bonds does not render it markedly different from naturally occurring DNA. As for cDNA, the Judge held

---

[225]The Supreme Court excluded method claims from its analysis. However, they were discussed in the decision of the District Court and the Court of Appeals. The decision was limited to nucleic acids and did not addressed other types of molecules (e.g. RNA).

[226]A claim would be considered ineligible if it encompasses any patent-ineligible matter. This applies even when a claim includes also some patentable subject matter (Holman 2013, p. 290).

[227]U.S. District Court Southern District of New York (2010, § 114).

[228]U.S. District Court Southern District of New York (2010, § 227).

[229]U.S. District Court Southern District of New York (2010, § 122).

[230]U.S. District Court Southern District of New York (2010). *Rai* maintained that approaching DNA: "*As just another species of chemical compound had substantially diminished the balance between property rights and the public domain achieved by various patentability requirements*" (Rai 2000, p. 203).

that, although it does not occur naturally, its information content does.[231] This conclusion was possible because he viewed DNA as information rather a molecule.[232] Hence, the Judge held that both isolated DNA and cDNA were not patent eligible, as they fell within the realm of the product of nature exclusion.

The case landed then before the Court of Appeals, which largely reversed the findings of the first instance. Regarding isolated DNA, the Judges considered the claims as patentable subject matter, since they covered markedly different molecules compared to the ones existing in nature.[233] Furthermore, the fact that isolated DNA is produced by human intervention either via synthesis or by chemically cleaving it from its surroundings was seen as imparting on it a distinctive chemical identity. The informational content of the isolated DNA and its natural counterpart was not taken into consideration.[234] *A fortiori*, the Court consider cDNA as patent eligible, given that its chemical structure differed from that of naturally occurring DNA.[235]

Interestingly, the United States intervened with an *amicus curiae* brief in support of neither party to discuss the two questions posed by the case; namely, whether isolated and unmodified DNA is patent eligible and, secondly, whether cDNA and other engineered DNAs are also patentable.[236] This constituted the first time that the United Stated expressed its view on isolated DNA in litigation.[237] In its overview, it reviewed the practice of the USPTO, which granted patents for both synthetic DNA (e.g. cDNA) and isolated unaltered genomic DNA.[238]

The United States differentiated between cDNA and naturally occurring DNA (gDNA). In the first case, it criticised the approach adopted by Sweet DJ, given that cDNA and similar compositions are engineered by humans, do not occur in nature and "*are instead the synthetic results of scientists' manipulation of the natural laws of genetics*".[239] Hence, according to the United States, they should qualify as man-made inventions and be patentable. Conversely, regarding gDNA, the United

[231]In rare cases cDNA can occur naturally. For this reason, authors believed that it might be incorrect to consider it an invention (Westerlund 2002, p. 33).

[232]Orford (2014, p. 563).

[233]U.S. Court of Appeals for the Federal Circuit (2011, p. 41).

[234]U.S. Court of Appeals for the Federal Circuit (2011, p. 42). Bryson J partially dissenting and concurring opinion, U.S. Court of Appeals for the Federal Circuit (2011).

[235]U.S. Court of Appeals for the Federal Circuit (2011, p. 47).

[236]"Brief for the United States as amicus curiae in support of neither party - in the US Court of Appeals for the Federal Circuit in Association for Molecular Pathology, et al. v. United States Patent and Trademark Office (Myriad)," (2010, p. 1).

[237]"Brief for the United States as amicus curiae in support of neither party - in the US Court of Appeals for the Federal Circuit in Association for Molecular Pathology, et al. v. United States Patent and Trademark Office (Myriad)," (2010, p. 6).

[238]"Brief for the United States as amicus curiae in support of neither party - in the US Court of Appeals for the Federal Circuit in Association for Molecular Pathology, et al. v. United States Patent and Trademark Office (Myriad)," (2010, pp. 4–6).

[239]"Brief for the United States as amicus curiae in support of neither party - in the US Court of Appeals for the Federal Circuit in Association for Molecular Pathology, et al. v. United States Patent and Trademark Office (Myriad)," (2010, p. 15).

States agreed with the approach of Sweet DJ and considered isolated and unmodified DNA unpatentable. Isolated DNA maintains the same function and chemical structure of naturally occurring DNA, the only difference being the fact that it has been isolated. Therefore, the entire weight of the claim is to be carried by this feature, which seems inadequate to obtain patentability.[240]

The case landed then before the Supreme Court, in what was its first decision on the patentability of genes.[241] In a unanimous decision, the Court held that isolated DNA is not patent eligible, whereas cDNA is. With this decision, the Court contradicted the approach adopted in the 2001 USPTO Utility Guidelines, which had established a relatively clear-cut criterion.[242]

In its reasoning on isolated DNA, the Supreme Court followed the statements of Sweet DJ and his findings that this constituted an unpatentable product of nature, given its insufficient differences compared to gDNA.[243] The fact that the isolated and natural DNA were informationally indistinguishable and shared the same function confirmed this view.[244] This conclusion was upheld despite the artificiality of the product.[245]

On the contrary, regarding cDNA, the Justices agreed with the Court of Appeals. They affirmed that, despite arguments holding that a sequence of cDNA is dictated by nature and therefore it is not patent eligible:

> The lab technician unquestionably creates something new when cDNA is made. cDNA retains the naturally occurring exons of DNA, but it is distinct from the DNA from which it

---

[240]"Brief for the United States as amicus curiae in support of neither party - in the US Court of Appeals for the Federal Circuit in Association for Molecular Pathology, et al. v. United States Patent and Trademark Office (Myriad)," (2010, pp. 14–33). Indeed, this merely embodies a law of nature, namely the relationship between genotype and phenotype. It also pointed out that something does not cease to be a product of nature through mere isolation, just as coal extracted from a cave does not. This approach, by the United States' own admission, contrasted with the longstanding practice of the USPTO and with the approach that identified the words "isolated", "purified" and "synthesised" as powerful tools to overcome the product of nature exclusion ("Brief for the United States as amicus curiae in support of neither party - in the US Court of Appeals for the Federal Circuit in Association for Molecular Pathology, et al. v. United States Patent and Trademark Office (Myriad)," 2010, pp. 10–14; Conley 2009, pp. 116–119). Some commentators even spoke of a "*talismanic status*" of such terms (Conley and Makowski 2004, p. 34).

[241]U.S. Supreme Court (2013). The Supreme Court received a petition from the patent opponents for a writ of certiorari. The Court vacated the decision of the Court of Appeals and instructed it to reconsider it in light of its recent *Mayo* decision. However, the Court of Appeals broadly reaffirmed its prior decision. Afterwards, the case was brought again before the Supreme Court (Torrance and Kahl 2013, p. 223; U.S. Supreme Court 2012). Although the case is centred on human genetic material, nothing prevents it from being applied to things other than humans and DNA (Holman 2013, p. 290).

[242]On this, see Sect. 4.1.6.4.

[243]Pila (2014b, pp. 180–181). Studies tried to determine the impact that the new approach to isolation and purification would have on patentability before the USPTO (Demers 2013).

[244]Burk (2013, p. 748).

[245]Palombi (2016, p. 234).

was derived. As a result, cDNA is not a "product of nature" and is patent eligible under §101.[246]

An exception was made for very short DNA strands that have no "*introns to remove when creating cDNA. In that situation, a short strand of cDNA may be indistinguishable from natural DNA*" and hence fall within the product of nature exclusion.[247]

In its assessment of the patentability of cDNA, the Supreme Court pivoted away from the informational framework that it had applied to isolated DNA, which led some commentators to consider this decision as inconsistent.[248] In particular, while isolated DNA was considered naturally occurring from an information perspective and thus patent ineligible, cDNA was held patentable since it "*does not naturally occur in the molecular sense, despite the information equivalence*".[249] The contradictory approach of the Supreme Court could be considered as informational or functional for gDNA and instead structural for cDNA.[250]

The question of how much change is needed for something to meet the threshold of § 101 was also indirectly addressed by the Court. Despite the fact that the sequence of cDNA is generally dictated by nature, the Justices found that cDNA was patentable as it did not occur as such in nature. This would seem to suggest that all that is needed to fulfil the patentability requirement is "*a change that could not occur but for the patent seeker's intervention... even the slightest variation could meet the standard*".[251]

Questions were raised on which approach would be adopted towards synthetic molecules that have the same or very similar structure compared to naturally occurring ones. Authors noted that considering those synthetic molecules as patent ineligible would have a serious impact on biotechnology. One possible reading of *Myriad* would be not to preclude the patentability of synthetic DNA that has the same sequence of naturally occurring one. This statement could be generalised to cover the patentability of synthetic molecules (and not just cDNA) that share the same structure as their natural equivalent. However, it was also considered that the Court would want to examine the type and relevance of the structural differences between the two molecules as well as their functional consequences. Furthermore, the Court emphasised both the synthetic origin of cDNA as well as its structural differences with gDNA. Hence, while it could be argued that synthesis would be enough for patent eligibility, a more likely scenario is that Courts would require also

[246]U.S. Supreme Court (2013, p. 17).

[247]U.S. Supreme Court (2013, p. 17).

[248]Burk (2013, p. 748).

[249]Orford (2014, p. 573).

[250]Lai (2015, p. 1064). With its decision, the Court seems to have aligned itself to the understanding of the scientific and medical communities, which consider genes as products of nature rather than inventions. In 2002, a Nobel laureate explicitly stated that genome sequences should be seen as discoveries and not as inventions (Andrews 2014, p. 551).

[251]Boguniewicz (2014, p. 46).

structural differences.[252] For this reason, synthetic DNA that is identical to gDNA is likely to be seen as non-patentable.[253] This conclusion is based on the fact that *Myriad* invalidated two claims concerning synthetic DNA that was identical to naturally occurring DNA.[254]

After *Myriad*, the USPTO revised its Guidelines on patent eligibility. In 2014, it issued an interim guidance, which has been updated ever since.[255] According to this document, functional characteristics (and not solely structural ones) can be used to show a marked difference relevant for patent eligibility purposes. This is a change from the previous approach, where only structural changes were taken into consideration. Both types of differences (i.e. functional and structural) cannot be innate in the product or occur independently from the actions of the inventor. In particular, laws of nature are patentable when the claim as a whole consists in much more than the law itself. For products, it is held that, to assess whether they are "markedly different", the following aspects will be taken into consideration: "*biological or pharmacological functions or activities; chemical and physical properties; phenotype, including functional and structural characteristics; and structure and form, whether chemical, genetic or physical*".[256] Isolated products could be patented if they display such differences.[257] Small changes could also result in marked differences.[258] Conversely, the ease or difficulty through which an invention was obtained is irrelevant for the assessment of its patent eligibility.[259]

In a series of examples concerning nature-based products, the USPTO offered further guidance on this elusive criterion. A peptide was considered patent eligible if it contained at least one substitution or modification that did not occur naturally. In particular, it was held that those structural differences would suffice, even if no functional difference had occurred. The rationale behind this was to be found in the fact that such claims would not tie up future uses of the naturally occurring version.

---

[252]Holman (2013, pp. 289–292).

[253]The protection of proteins, especially synthetically produced ones that are indistinguishable from natural proteins, has also been examined by scholars. Firstly, it has been argued that isolated proteins, just like isolated DNA, will not be considered patent eligible due to the interpretation of the product of nature exception given by the Court in *Myriad*. However, the possibility to patent isolated naturally occurring proteins has not been excluded in cases where they are claimed in chemical terms, especially if chemical changes can be shown. However, the panorama in this area is still unclear (Phukan 2014, pp. 640–641).

[254]Phukan (2014, pp. 636–637). While product-by-process claims were not discussed in *Myriad*, it was argued that claims to a synthetic DNA that is indistinguishable from natural one are unlikely to be patentable. This result can be attributed to the fact that the Courts and the USPTO take into consideration the product itself to assess the patentability of those types of claims rather than the process through which it is made (Phukan 2014, p. 638).

[255]USPTO (2014a). The Guidelines were not limited to nucleic acids, which had been the focus of the *Myriad* dispute.

[256]Schwartz and Minssen (2015, p. 224) and USPTO (2014a, p. 74623).

[257]Schwartz and Minssen (2015, p. 224) and USPTO (2014a, p. 74623).

[258]USPTO (2014a, p. 74623).

[259]Lai (2015, p. 1066).

The relevance of this policy consideration was repeatedly mentioned. Indeed, for another product, it was held that "*a unique molecule that is distinct from, and does not prevent others from using, naturally occurring... acid, its different structural characteristic rises to the level of a marked difference. Accordingly... is not a 'product of nature'*".[260] Similar conclusions were reached for genes where one or more bases had been substituted. This applied even if the modifications were silent (i.e. no change occurred in the encoded protein).[261] Lastly, in the 2016 examples on life sciences, the USPTO maintained that the attenuated or removed virulence of a virus would constitute a structural and functional characteristic that does not occur naturally and that could be considered a marked difference.[262] The absence of a natural equivalent of a product was also significant for its patentability. For this analysis, only the actual presence of such product was taken into consideration. The Guideline cited *Myriad* and clarified that "*unless the examiner can show that this particular mixture... exists in nature, this mere possibility does not bar the eligibility of this claim*".[263] Therefore, natural can be read to encompass what certainly exists without human intervention.[264]

Around the same time, Courts were confronted again with the patent eligibility of biotechnologies in the *In re Roslin* case, which assessed the patentability of cloned animals and which was brought before the US Court of Appeals for the Federal

[260]USPTO (2014b, pp. 4, 7).

[261]USPTO (2014b, pp. 9–11).

[262]USPTO (2016, p. 3). Commentators suggested that the USPTO has adopted a mixed structural/functional approach (Lai 2015, p. 1066).

[263]USPTO (2014b, p. 9).

[264]Lai (2015, p. 1064). Reports compiled on the effects of the 2014 Guidelines maintained that this document has reduced the scope of patent eligibility in the biotechnological field (Schwartz and Minssen 2015, p. 237; Noonan 2014). Other studies confronted the results obtained by applying the new and the old Guidelines and showed that the outcomes were diametrically opposed (Mouta and Tridico 2015). In a case before the CAFC, short, synthetic, single-stranded DNA was considered structurally identical and as not performing a significantly different function compared to natural DNA (Willgoos and Gont 2015, p. 4). Hence, it was considered unpatentable. Specifically, the Judges maintained that: "*A DNA structure with a function similar to that found in nature can only be patent eligible... if it has a unique structure, different from anything found in nature*" (U.S. Court of Appeals for the Federal Circuit 2014a, p. 9). This decision was seen as further restricting the patentability of biotechnological inventions. In fact, here synthetic DNA was considered unpatentable (Willgoos and Gont 2015, p. 5). The Court interpreted *Myriad* in such a way that: "*Neither naturally occurring compositions of matter, nor synthetically created compositions that are structurally identical to the naturally occurring compositions, are patent eligible*" (U.S. Court of Appeals for the Federal Circuit 2014a, p. 8). With this decision, the Judges effectively side-stepped the distinction proposed in *Myriad* by the Supreme Court and focused on structural distinctions rather than on their origin, as done in *Mayo* (Landers 2015, p. 529). Interestingly, in *Myriad,* the Supreme Court did not address the conclusions it reached in *Mayo.* Therefore, the relationship between laws of nature and products of nature remains unclear and undefined (Burk 2013, p. 749). The second question concerned how often will there be functional differences unaccompanied by structural differences. On this, it could be argued that structural differences might be necessary, but not sufficient for patent eligibility and that functional differences will be the ones required to show a marked change (Mouta and Tridico 2015).

Circuit (CAFC) after *Myriad* had been decided by the Supreme Court.[265] Given the precedents, the Roslin Institute needed to prove patent-eligibility by showing that the clones were markedly different from the donor animals. To do this, the Institute brought forward a number of arguments, which were rejected by the Court. Firstly, they held that the animals showed phenotypic differences due to the different environments in which they lived in. The Court dismissed this by noting that, even if this were the case, those differences were the work of nature. Secondly, the Institute claimed that the clones were different since they were a time-delayed version of the original animals. This argument was also rejected by holding that this statement is true for all copies of an original. Lastly, the Institute pointed to dissimilarities in the mitochondrial DNA.

The Court held the invention patent ineligible, given that the cloned animal was an exact genetic replica of the donor and was thus "*a natural phenomenon that did not possess 'markedly different characteristics than any found in nature*'".[266] Nonetheless, the Court admitted that "*having the same nuclear DNA as the donor mammal may not necessarily result in patent ineligibility in every case*",[267] as it must be recognised that such organisms do not fall neatly within the product of nature exclusion either, considering that they exist only thanks to human efforts.[268] In their ruling, the Judges focused on the genetic features of the clones and of the original animals.[269] This shows that genetic identity was the cornerstone of this decision and that the labour of the creator was effectively disregarded.[270]

#### The *Myriad* Saga Around the World

The Myriad patents were litigated in a number of jurisdictions.

In Europe, several cases were brought before the EPO, but their legal relevance is limited.[271] The oppositions, which were brought before the cases in the US and Australia were decided, culminated in the granting of a limited number of narrow patents on part of the EPO.[272] The oppositions stemmed from the social unease connected to gene patents and, given the formulation of the EPC and the Biotechnology Directive, the "*EPO's role was one of strict legislative application due to the very clear nature of the pertinent parts of EPO law*".[273] The EPO was faced with

[265]U.S. Court of Appeals for the Federal Circuit (2014b).

[266]U.S. Court of Appeals for the Federal Circuit (2014b, p. 4).

[267]U.S. Court of Appeals for the Federal Circuit (2014b, p. 4).

[268]Swedlow (2015, p. 190).

[269]Sherman (2015, pp. 1217–1218).

[270]Sherman (2015, p. 1223).

[271]Technical Board of Appeal of the EPO (TBA) (2008a, b) and Board of Appeal of the EPO (2007).

[272]The claims of Myriad we restricted mostly due to the lack of priority of some of them (Lai 2015, p. 1063).

[273]Lai (2015, p. 1057).

questions on whether some of the claims of Myriad amounted to unpatentable discoveries. Succinctly, the Board of Appeal rejected this view and argued that the probes were patentable since they were obtained via a technical process and were isolated from the human body. In its decision, it referred to Rule 29(2) EPC.[274] Similarly, in a connected case, the opponent argued that some of the claims of Myriad were directed at discoveries (i.e. discovery of the mutation of a human gene and its relevance for diseases). This argument was rejected by the Board by referring to prior cases.[275] Overall, the EPO was not open to the possibility that isolated DNA might not constitute an invention. This is not surprising, considering that at that time it would also have been considered an invention in the US. Equally, the language of the Directive did not provide much leeway.[276]

The situation was different in Australia, where the Myriad patents were litigated all the way to the High Court. In a decision issued in October 2015, the High Court established that isolated DNA is not patent eligible, just like the US Supreme Court had done before. However, differently from their US counterparts, the Australian Judges considered also cDNA as non-patent eligible.[277] The Judges explicitly referred to the US decision in their reasoning. Also, just like in the US, the High Court decision reversed that of the Appeal Court and set this jurisdiction apart from most other developed countries.[278]

For isolated DNA, the High Court focused on substance over form, which entailed focusing on the information contained in the DNA rather than on the physical medium that embodied it. The High Court applied the same approach also to cDNA. Differently from the American Court, the Australian Judges focused on the information contained in the sequences. By doing this, they held that cDNA contained all the exons necessary to produce the BRCA protein and that the lack of introns connected to the structure of cDNA was effectively irrelevant.[279]

The Australian portion of the *Myriad* litigation raised also the question of the role of DNA as information. There, this aspect was connected to patent infringement and it was argued that a patent on DNA as information "*could never be infringed by someone who merely reproduced a DNA sequence in written or digitalized form*".[280]

---

[274]Lai (2015, p. 1058) and Technical Board of Appeal of the EPO (TBA) (2007a).

[275]Lai (2015, pp. 1060–1061) and Technical Board of Appeal of the EPO (TBA) (2008b).

[276]Lai (2015, p. 1063).

[277]High Court of Australia (2015). It has to be noted that the Australian definition of what constitutes an invention differs from the US and European one. Also, the patents at issue here were different from those litigated in the US (Lai 2016, pp. 539–544).

[278]Obranovich (2016, pp. 8–9).

[279]Obranovich (2016, p. 10).

[280]Federal Court of Australia (2013). In the US prong of the case, it was argued that claims could be circumvented by patenting DNA containing one additional nucleotide pair ("Brief of James D. Watson, Ph.D. as amicus curiae in support of neither party to the Supreme Court of the United States in the Association for Molecular Pathology, et al. v. Myriad Genetics, Inc., et al. (Myriad)," 2013). This approach disregards the fact that, even if the sequence were invented around, the patent might be invalid pursuant to the doctrine of equivalents (Schwartz and Minssen 2015, p. 205).

In response to this decision, the Australian Patent Office issued a proposed Examination Practice in October 2015 for public consultation.[281] In this version, it was proposed that naturally occurring nucleic acid sequences (both human and not) that encode for polypeptides, naturally occurring coding RNA and cDNA would be considered non-patent eligible, independently from whether they had been isolated or synthesised. Soon after, the Patent Office finalised an updated version of the Examination Practice.[282] There it was established that isolated naturally occurring nucleic acid would be excluded from patentability. This would apply both to coding and non-coding material and to human and non-human molecules. Furthermore, the exclusion would cover also cDNA, synthetic nucleic acids, primers and nucleic acids that merely replicate the genetic information of a natural organism.[283]

### Impact of the Myriad Decisions in Europe

The patent panorama drastically changed both in the US and in Australia after the *Myriad* rulings, since both decisions changed an approach that had seemed settled.[284] If the US approach was previously considered more advantageous, this may no longer be the case, as the European one is currently more permissive.[285] Such differences and lack of harmonisation might have a number of repercussions on biotech innovation. Indeed, although the European and American legal systems are not directly connected, pivotal US Court decisions have impacted Europe and other jurisdictions. For instance, it was affirmed that:

> History has shown that groundbreaking US Supreme Court decisions can influence patent law and practice all over the world. An example of this is the Chakrabarty decision, where it has been suggested that in the years following this case, the TRIPS Agreement (1994) sought ways to regularize and internationalize the technological and legal culture that flowed in that decision.[286]

Moreover, the US, Europe and Japan have been reciprocally influencing each other since the establishment of the trilateral cooperation of their patent offices in 1983. Equally, the European Biotech Directive was heavily influenced by this cooperation, as in 1988 the three offices issued a joint statement holding that purified natural products were to be considered patent eligible and not discoveries or products of nature.[287] Indeed, until now, the patent practice of the US and the JPO

---

[281]Australian Government (2016).

[282]Finkel (2016) and Australian Government (2015).

[283]Obranovich (2016, p. 9).

[284]Years after the US Supreme Court decision in *Myriad*, the patent eligibility requirement is still highly controversial, as shown in a number of Federal Circuit decisions (Gerber and Campbell 2019; Marks and Carron 2018).

[285]Schwartz and Minssen (2015, p. 238) and Lai (2015, p. 1043).

[286]Huys et al. (2011, p. 1106).

[287]EPO, USPTO, JPO (1988).

has been considered by the EPO as "*indicative of current jurisprudential trends, the EPO's observance of which will 'contribute to the... highly desirable (world-wide) harmonization of patent law*'".[288] However, the US has now moved away from this position and embraced an approach that is at odds with the European and Japanese ones.[289]

If and how this difference is going to affect Europe is still uncertain.[290] Some authors have argued that the effect of *Myriad* on the Directive might be limited, given that the scope of the Directive is much wider than that of *Myriad*. Also, it was held that the scope of the Supreme Court's decision is still unclear and that, until it remains so, it is unlikely to profoundly affect patentability in Europe.[291] However, this decision might reopen the debate on biotech patents in the old continent. The Biotech Directive raised criticism ever since its inception and the *Myriad* decisions may just reignite this, especially since one of the grounds for passing the norm (i.e. ensure in Europe a level of protection comparable to that offered in the US) is now gone.[292] Still, the political and social tensions that characterised the passing of the Directive in the first place would probably not encourage a revision of the norm, unless it was utterly necessary.

In spite of all this, it is undeniable that *Myriad* does have an impact from a policy and commercial perspective. Now that the "*the Supreme Court decisions in Myriad and Prometheus [Mayo] stand in clear contrast to the current legislation and practice in... Europe*",[293] there could be cases in which an invention is considered patent eligible in Europe and ineligible in the US. As some authors have noted, this situation "*is as doctrinally inconsistent as it is commercially undesirable for bio-industry*".[294] Lastly, uncertainties over the legal and policy reasons behind patent eligibility might render decisions on the patentability of new biotechnologies (e.g. synthetic biology) even more problematic.

The rigidity of the European approach in comparison to the American one is also significant. The provisions of the Biotech Directive clearly establish what is patentable and what is not and leave little leeway. These norms mirror a scientific status

---

[288] Pila (2005b, pp. 178–179) and Technical Board of Appeal of the EPO (TBA) (1999b, § 2.3, 2.6).

[289] Lai (2015, p. 1073).

[290] Other scholars held that, although a direct and far-reaching effect in Europe is unlikely, the *Myriad* case might revive the debate on this topic in the old continent. Developments in patent law: "*Have the potential to migrate in an increasingly global economy and destabilize the objective of an increasingly harmonized and efficient world patent system*" (Schwartz and Minssen 2015, p. 240). Similarly, commentators were convinced that *Myriad* will have repercussions in Europe and argued that it would be: "*Difficult to image how the ripple effect will not reach Europe*" (Palombi 2016, p. 236).

[291] Lai (2015, pp. 1072–1073). It was also argued that the most marked differences between the two jurisdictions might arise in the field of method patents instead of that of product patents (Lai 2015, p. 1069).

[292] Lai (2015, pp. 1072–1075).

[293] Schwartz and Minssen (2015, p. 240).

[294] Odell-West (2011, p. 305).

that is long gone. In the 20 years since the adoption of the Directive and 30 since its inception, the technical panorama has vastly changed. Therefore, some of the provisions of the Directive relate to technologies that are now outdated. However, it is not easy to tailor a new approach. It was correctly argued that "*it is more complex to amend a controversial directive than to follow technological developments on a case-by-case assessment of neutral standards*".[295] The clarity of the European standard is thus the reason for its lack of flexibility and also the reason why, under the current legislation, the direct impact of *Myriad* is bound to be limited.

## 4.1.7 Mathematical Methods

### 4.1.7.1 Overview

The exclusion from patentability of mathematical methods set in Article 52(2)(a) EPC could be relevant in the field of synthetic biology, given the significance of programs to design and simulate biological circuits in this discipline.[296]

In the past decades, biotechnology and informatics have come together in a new discipline called bioinformatics. It uses "*computer systems to store, manage and analyse biological information*" and borrows elements from mathematics, statistics and engineering.[297] This discipline is increasingly needed to map and model genes and proteins and to develop drugs. Similarly, it is used to identify new molecules and their functions.[298] Those tools are necessary, as the sheer number of information involved would be impossible to assess either at all or, at least, as fast and as precisely manually.

This growing industry equips researches with tools such as algorithms and user interfaces.[299] The former are used to search, analyse, integrate and store biological

[295]Sommer (2013, p. 47).

[296]The exclusion of scientific theories will not be analysed, as, according to the EPO, these constitute a: "*More generalised form of discovery*" (EPO 2018, § G.II.3.2).

[297]Rimmer (2003, p. 31).

[298]McBride (2002, pp. 1332–1337).

[299]In this context, it is useful to define what an algorithm is and the difference between it and a mathematical formula. The term algorithm defines: "*Any sequence of simple actions entailed in the performance of some more complex task*" (Oksanen et al. 2011, p. 139). Specifically, in the field of computing, algorithms are seen as defining: "*A set of precise rules for performing a recursive computation for solving a problem in a finite number of steps*" (Oksanen et al. 2011, p. 139). Authors have identified the difference between algorithms and mathematical formulas in their equivalence to nature, or lack thereof. Formulas are seen as an equivalence to what already exists in nature. Under this reading, they mathematically depict something that has always been in nature, even before formulas were developed. Conversely, algorithms are seen as inventive methods for solving a problem. They are based on laws of nature, but are not laws of nature themselves (Koepsell 2011, pp. 131–132). Interestingly, the same algorithm can be expressed in different programming languages, which could have an impact on patent infringement. On this, it was argued

information, while the latter allow user access and visualisation.[300] Over time, bioinformatics software, hardware and databases have been developed.

Algorithms have been devised to choose optimal sequences and to avoid unwanted features.[301] With time, it is expected that those algorithms will be used to devise new biological materials that are tailored to human needs and that can be structured in the most efficient way, without having to follow the constraints of natural structures. Such *ex novo* designs operate on the basis of theoretical biological knowledge.[302]

Software simulation systems provide a virtual representation of the invention and offer the possibility to verify its functioning before production, thus providing a reasonable level of confidence that the system will behave as expected once it is manufactured.[303] It is important to remember that, while those simulations are visual representations of something, they cannot be considered the real thing. However, where the "unreal" simulation helps rendering more effective something produced in the real world, this could be seen to address a technical problem.[304]

The development of such simulation tools is complex and expensive. Both in the development of the new biological system and in the development of the simulation tool itself, "*when moving from the initial modelling oriented phase to the later virtual system producing phase, the level of acquired abstraction decreases and, simultaneously, the level of technical consideration increases*".[305] This could have an impact on their patentability. Later stages of development, which are generally characterised by higher technical content, are more likely to be considered patent eligible in comparison to earlier stages.[306]

Computer-aided biological designs do not employ actual compounds. This might remove the need for companies to obtain a license from the patentee to use a substance in its virtual experimentation. This raises questions on the protection afforded to compounds in their virtual form, on whether they could constitute patentable subject matter and on the possibility of patent infringements.[307]

Companies involved in bioinformatics have applied for a variety of patents in this field.[308] The areas covered span from databases containing genetic information to

that if one code violates a patent, another version of it will infringe the patent as well (Oksanen et al. 2011, p. 143).

[300]Bentz (2000).

[301]U.S. Department of Energy (2004, p. 4).

[302]Watson (2014, p. 27).

[303]Baccelli and Hiratsuka (2009).

[304]Reeve (2015).

[305]Baccelli and Hiratsuka (2009).

[306]Baccelli and Hiratsuka (2009).

[307]Watson (2014, p. 27).

[308]Bioinformatics covers a wide range of products and applications. Hence: "*A wide array of protection might be available, depending on the particular nature of the bioinformatics component and its intended use*" (*Conference proceedings*, 2016).

systems that store and analyse genomic data to computer hardware and software related to this discipline.[309] The majority of the patents concerns method claims.[310] Overall, the number of patents in this field is still limited. Studies show that only 50 software-related patents have been granted by the USPTO to companies in the biotech and pharmaceutical sector between 1996 and 2001. However, the trend shows an increase in the number of applications for these types of patents in the examined timeframe (i.e. 1979–1997).[311] Other reports fix instead the number of bioinformatics patents in the 1990–2000 period at more than 150.[312] The low number of patent applications was attributed to the novelty of this industry, to the difficulties relating to software patents and their infringement, and to the fact that many bioinformatics inventions merely combine already existing sources of data into one single product.[313]

#### 4.1.7.2 European Normative Overview

As for discoveries, Article 52(2)(a) EPC establishes that mathematical methods as such shall not be regarded as inventions. No specific provisions on this point are found in TRIPS or the Biotech Directive.

This exclusion is aimed at hindering the patentability of methods that are abstract or purely intellectual.[314] Such items belong to the realm of human cognitive activities and should thus remain in the commons. Equally, they have no direct practical application and present no technical character.[315] An example of such excluded subject matter would be a method to perform a division. However, a machine that carries out such method would be patentable. Similarly, electrical filters that were designed on the basis of a mathematical method would be considered patentable.[316] Hence, the technical character of an invention plays a pivotal role in this exclusion. For example, methods that are computer-implemented would not fall within the

[309] Rimmer (2003, p. 40).

[310] Contreras and Cuticchia (2013, p. 55).

[311] Rimmer (2003, pp. 41–42). At the USPTO, a specific designated Art Unite (AU) has been established to deal with inventions combining computers and biology. In its realm are included applications for algorithms predicting gene function as well as screening procedures to identify drug candidates. Studies have compared this AU with another one concerning software (AU 2123) and noted that the rate of rejection of those applications on patentability grounds was not significantly different (Vishnubhakat and Rai 2015, pp. 231–232).

[312] Bentz (2000).

[313] Hultquist et al. (2002, p. 517).

[314] EPO (2018, § G.II.3.3).

[315] Stauder et al. (2016, p. 178).

[316] EPO (2018, § G.II.3.3).

realm of the exclusion of Article 52(2) EPC, since they would be considered to employ technical means.[317]

The EPO noted that mathematical methods used in a technical process, which are carried out on a physical entity via technical means and which determine a change in that entity, contribute to the technical character of the invention. Furthermore, the EPO held that claims for technical processes in which mathematical methods are used and which are restricted to a specific application in a technical field would be patentable. Similar conclusions can be reached for methods whose underlying idea resided in a mathematical method, but that, as a whole, are not restricted to a mathematical method as such. Methods of encryption could also qualify as technical methods and thus be patentable, even if they rely on mathematical methods.[318]

Bioinformatics inventions have often been divided into three subsets: algorithms, databases and interfaces. With regards to algorithms, it was noted that, while they are not patentable *per se*, they are patent eligible if they have a practical application.[319] Furthermore, authors argued that innovative bioinformatics tools that are embodied in software and that utilise complex algorithms employing genomic information could be the object of patent protection.[320]

[317]EPO (2018, § G.II.3.3). The approach of the EPO towards computer programs has resulted in their patent eligibility, if they address and solve a technical problem. This has been interpreted to mean that they must make the computer on which the program is running or its network better (Reeve 2015). Computer programs are thus patentable if they are able to provide, when running on a computer, a further technical effect that goes beyond the usual interaction between the program and the hardware on which it operates. Furthermore, the solved problem must be of technical nature, instead of a purely mathematical one (EPO 2013, pp. 13–14). The EPO distinguishes between computer programs and computer-implemented inventions. The latter cover claims involving "*computers, computer networks or other programmable apparatus wherein at least one feature is realised by means of a computer program*" (EPO 2018, § G.II.3.6). The difference between the two thus lies in the fact that computer programs refer to a sequence of computer-executable instructions that specify a method, whereas computer-implemented methods are methods being actually performed on a computer. For this reason, "*claims directed to a computer-implemented method, a computer-readable storage medium or a device cannot be objected to under Article 52 (2) and (3) as any method involving the use of technical means (e.g. a computer) and any technical means itself (e.g. a computer or a computer-readable storage medium) have technical character and thus represent inventions in the sense of Article 52(1)*" (EPO 2018, § G.II.3.6). For this reason, this analysis focuses on mathematical methods rather than computer programs, given that bioinformatics inventions would be considered computer implemented inventions rather than computer programs as such. Other authors considered instead patent applications for software for design of biological devices and genetic circuits, design of synthetic nucleic acids and analysis of biochemical activities within cells to fall within the category of computer programs (McLennan 2017, p. 64).

[318]EPO (2016, § I.A.2.2.2). The EPO maintained that mathematical methods are carried out on numbers and their results are expressed in numerical form. This is independent of what these numbers represent. Those methods can thus be considered as: "*Abstract concept[s] prescribing how to operate on the numbers*" and they do not produce direct technical results (EPO 2016, § I.A.2.2.2).

[319]Hultquist et al. (2002, p. 518).

[320]Purevdorj (2011, p. 25).

With regards to databases, it was assumed that their informational content could not be protected via patents.[321] However, patents could cover the structure of such databases. In light of this and considering that data relating to DNA may not be seen as mere cognitive information, it has been argued that the data structures that comprise such information could be protected via patents.[322]

Lastly, concerning user interfaces, it was argued that they could be subject to patent protection. Specifically, the EPO would probably favour the patentability of interfaces, where information is exchanged about the conditions found in an apparatus or system.[323]

Further patentability issues are raised by computer simulations. It was held that those simulators often do not fall within the realm of patentable subject matter. In spite of the accuracy of the resemblance between the simulation and the real world or the usefulness of its results, the computer itself is not affected by the simulation. In other words, the computer is not transformed into a better computer because of the simulation. In spite of this reasoning, the case law of the EPO shows that exceptions are possible.[324] Additionally, commentators argued that the patentability of simulators depends on what they are used for. For instance, improvements in fields excluded from patentability (e.g. methods of doing business) are likely to be held non-patentable, whereas simulations in industrial fields and connected to highly technical processes have a better chance of being held patent eligible.[325]

#### 4.1.7.3 European Case Law Overview

The case law of the EPO on this exclusion often intertwines with that on computer programs. In *Vicom*, the Board of Appeal, in assessing the patentability of a method to digitally process images and of an apparatus for carrying it out, maintained that such inventions should not be prejudiced against because they are implemented via technical means (e.g. a computer program).[326] The Board described mathematical methods as those carried out on numbers and whose results are expressed in numerical form.

A few years later, in *IBM*, the Board addressed again the patentability of algorithms. It maintained that:

> Mathematical methods as such would be excluded from patentability by Article 52(2)(a) in conjunction with (3) EPC. However, clearly, the claim does not relate to a mathematical

[321] *Directive 96/9/EC of the European Parliament and of the Council of 11 March 1996 on the legal protection of databases* (1996).

[322] Hultquist et al. (2002, pp. 517–518). This approach was seen as a reference to the decision of the EPO in *Philips* (Technical Board of Appeal of the EPO (TBA) 2000b). On this, see Sect. 4.1.7.3.

[323] Hultquist et al. (2002, p. 518).

[324] Reeve (2015).

[325] Reeve (2015).

[326] Shemtov (2009, p. 507) and Technical Board of Appeal of the EPO (TBA) (1986).

> method as such; it does not define any formula according to which the calculations should be performed. Rather, the calculating steps mentioned are only means, or tools, used within the overall method claimed.[327]

The patentability of data structures was considered by the EPO in *Philips.*[328] There, the Board dealt with a picture retrieval system that stored data in a record carrier. In its decision, it stated that there was "*a difference between the functional data, which controlled the technical working of the system, and the cognitive information, which represented the picture*".[329] Hence, it was patent eligible. Some authors applied this statement to data structures pertaining to biological materials to conclude that they are patentable, since they are not mere cognitive information.[330]

The EPO addressed the issue of the patentability of simulation tools in multiple occasions. In one of its earlier cases, the Board examined the patentability of a method to carry out a design.[331] It held that the method could be seen as covering mere designs that do not exists in the real world and that may never exist. The Board cited *Vicom* and argued that where there are no references to the physical entities represented by the data, the method could be seen as abstract and lacking technical character.[332] Similarly, individual designing steps could be seen as performing mental acts and thus be outside the realm of patentability. In light of this, the Board held that some of the previous versions of the method claims presented by the applicant would be rejected, as they did not concern a technical process that made a contribution in a field that was not excluded from patentability.[333]

The Board reached a partially different conclusion for methods of manufacturing physical objects displaying technical features, as they did not include mere designing methods, but also producing steps. In spite of this, the software simulator itself was not protected.[334] Furthermore, it was noted that such claims are not easily enforced, especially since proof of infringement is difficult to obtain.[335]

Designing methods based on design programs were addressed by the EPO in *Philips.*[336] There, the Board rejected the view of the patentee that the claims were

---

[327]Technical Board of Appeal of the EPO (TBA) (1994a, § 3.2).

[328]Technical Board of Appeal of the EPO (TBA) (2000b).

[329]Hultquist et al. (2002, pp. 517–518) and Technical Board of Appeal of the EPO (TBA) (2000b, § 3.3).

[330]Hultquist et al. (2002, p. 518).

[331]Technical Board of Appeal of the EPO (TBA) (1994b).

[332]Technical Board of Appeal of the EPO (TBA) (1994b, § 5) and Technical Board of Appeal of the EPO (TBA) (1986).

[333]Baccelli and Hiratsuka (2009).

[334]*Baccelli* and *Hiratsuka* noted that: "*Such claim does not cover protection for the process of computer-assisted simulation itself. In fact, its restricted scope includes the steps of designing together with the subsequent step of producing the corresponding product. The software simulator alone would not be covered directly by the claim*" (Baccelli and Hiratsuka 2009).

[335]Baccelli and Hiratsuka (2009).

[336]Technical Board of Appeal of the EPO (TBA) (2007b).

directed at a technical method for designing an optical system and held instead that the claims referred to mathematical and optical abstract concepts that did not require physical or technical implementation. In particular, given that neither the designing method nor the resulting designs required a technical activity or entity, the creation concerned the mere design of an optical system. The methods could be seen as purely mental acts or purely mathematical design algorithms, which are excluded from patentability.[337] The Board noted that the invention could be implemented via a physical activity and on a physical entity. However, the clams did not encompass only this type of physical means, but also excluded subject matter. Conversely, claims restricted to methods of using an optical design program were held patentable.[338] The line between patentable and excluded claims relates to whether a method, even if technical, covers also abstract embodiments. If this were the case, the invention would not be patentable. Instead, if those claims were solely implemented and executed on a computer, they would not be considered abstract and would thus be patentable.[339]

In *Infineon*, the EPO examined an application for a computer-implemented simulation method for integrated circuits.[340] The method provided several advantages; for instance, it shortened the time needed to perform certain operations, required less storage space and expanded the range of computers that could be used for it as well as the circuit range that could be simulated. In spite of this, the application had been rejected holding that the simulation method pertained to a mental act or to a mathematical method.[341]

In a decision that signed a departure from its prior case law and was seen as an example of judicial activism,[342] the Board of Appeal reversed this finding and held that computer-implemented methods would not fall within the realm of excluded subject matter. Interestingly, in reaching this conclusion, the Board overlooked the fact that the claims were limited to the mere performance of a calculation and did not comprise interactions with the outside world (e.g. steps to manufacture the circuit).[343] In its decision, the Board held that both apparatus and method claims displayed technical character. The EPO did not consider the simulation method as either a mathematical method or a computer program as such. Indeed, while those were at the basis of the method, the invention itself went beyond them. The simulation presented a realistic prediction of how the circuit would perform, thus

[337]Baccelli and Hiratsuka (2009).

[338]Baccelli and Hiratsuka (2009).

[339]Baccelli and Hiratsuka (2009). Similar conclusions had already been reached by the Board with regards to a method for designing loading arrangements for nuclear reactors. According to this line of reasoning, inventions will be patentable if the claimed subject matter is restricted to technical means. On the other hand, claims that can be exclusively carried out mentally would be excluded (Technical Board of Appeal of the EPO (TBA) 2005).

[340]Technical Board of Appeal of the EPO (TBA) (2006b).

[341]Shemtov (2009, p. 512).

[342]Reeve (2015) and Shemtov (2009, p. 513).

[343]Reeve (2015).

improving its development. The use of such tools in virtual trials had a practical and practice-oriented value and did not represent a purely mathematical or mental act. Hence, all the steps that were relevant for the circuit simulation, including also the mathematical ones, could be seen as contributing to the technical character of the method.[344]

Lastly, the Board referred to an earlier case on chip design, which had been held unpatentable since it did not concern a physical entity.[345] The difference between the two was explained in policy terms. While in the older case a manufacturing step was required to show technical character, this approach may no longer be up-to-date. Indeed, in a number of disciplines, simulations have become essential and should thus be awarded patent protection.[346]

After *Infineon*, the EPO offered further guidance on simulations. In one instance, it maintained that not all kinds of optimisations can be considered technical. In particular, it explained that optimisation methods that use mathematical algorithms and that can be performed by a computer should not be patent eligible in case no technical application is apparent. Furthermore, concerning the computer that performs the algorithm, it was stated that it did not provide a further technical effect beyond what is expected from the internal workings of the computer and that its output amounted to nothing more than a presentation of the calculation.[347]

---

[344]Shemtov (2009, pp. 512–513). In particular, the Board affirmed that: "*Specific technical applications of computer-implemented simulation methods are themselves to be regarded as modern technical methods which form an essential part of the fabrication process and precede actual production, mostly as an intermediate step. In view of this development it must be assumed that the outlay for implementing a technical product will increasingly shift to the numerical simulation phase, while final implementation of the simulation result in the actual manufacture of the product will entail no or only comparatively little extra innovation effort. In that light, such simulation methods cannot be denied a technical effect merely on the ground that they do not yet incorporate the physical end product… A further fundamental change is to be found in the fact that development and production are increasingly separated, materially and geographically… in that light, too, the Board considers specific patent protection to be appropriate for numerical development tools designed for a technical purpose*" (Technical Board of Appeal of the EPO (TBA) 2006b, § 3.4). In its Guidelines, the EPO confirmed this approach, but noted that: "*Meta-specifications of an undefined technical purpose… could not be considered adequate*" (EPO 2018, § G.II.3.3).

[345]Technical Board of Appeal of the EPO (TBA) (1994b).

[346]Shemtov (2009, p. 513). Authors expressed doubts over the approach to be adopted following this case, in particular regarding the technical character of those inventions. The ruling seemed to focus on the specific industry at hand and to tailor the definition of technical character around it, thus creating uncertainty. It was questioned whether the physical element requirement is fulfilled as long as the results of the simulation could be used in the future for a concrete purpose in the real world or whether this would depend on the industry involved and the need to protect such intermediary stages in that field (Shemtov 2009, p. 513). Other commentators noted instead that references to non-defined technical systems would not suffice for patentability, but that simulations on appropriately defined circuits could be seen as technical features and thus patentable (Baccelli and Hiratsuka 2009).

[347]Technical Board of Appeal of the EPO (TBA) (2008c, § 3).

Conversely, in another case, a method to determine genotypes was seen to display technical character and thus to be an invention.[348]

In a case from 2012, the Board, referring to *Infineon*, held that the simulation presented in that case would constitute an example of an "*adequately defined technical purpose for a computer-implemented method, provided that the method is functionally limited to that technical purpose*".[349] The Board re-examined also the *Vicom* case and held that technical processes were considered different from mathematical methods, since the former are carried out on physical entities and provide, as a result, a change in that entity. However, the Board admitted that such a definition would exclude simulations, whose purpose is exactly to replace physical entities with virtual ones. Hence, it stated that *Infineon* "*goes beyond the earlier decision in holding that the simulation of an adequately defined class of technical items could be a functional technical feature*".[350]

National Courts have also issued decisions on the patentability of simulation methods. In 1999, the German Supreme Court in *Logikverifikation* considered patent eligible a method of verifying the layout of integrated circuits via a computer.[351] In doing so, the Judges found that the invention was a step within the process of production of integrated circuits, which is in itself a technical process, and that the purpose of the invention was technical. Furthermore, given the type of products involved, the invention was the result of technical considerations.[352] Interestingly, this decision was cited by the EPO in *Infineon* in support of its conclusions, as the

[348](Technical Board of Appeal of the EPO (TBA) (2010).

[349]Technical Board of Appeal of the EPO (TBA) (2012, § 3).

[350]Technical Board of Appeal of the EPO (TBA) (2012, § 3).

[351]German Federal Supreme Court (BGH) (1999). In 1996, the German Federal Patent Court discussed mathematical methods and the technical requirement in the *Viterbi Algorithm* case (German Federal Patent Court (BPatG) 1996). The Judges stated that a claim cannot be considered technical merely because its field of application is. However, the Court argued that, by indicating the purpose of the patent: "*The algorithm is brought into so close a relationship with technical processes and, as far as its contents are concerned, so much restricted to technical values that this Court considers the teaching... as technical*" (Betten 2000, p. 445). Hence, by stating its purpose, the algorithm acquires technical character and cannot be considered a mathematical method as such. In other words, it is brought so close to a technical process that its teaching can also be considered technical (Betten 2000, pp. 444–445). The same Court issued further guidance on this point in *Suche fehlerhafter Zeichenketten/Tippfehler* (German Federal Supreme Court (BGH) 2001). There the Judges maintained that claims were generally patentable, even if they performed method steps on a computer, if they aimed at solving concrete problems in conventional technical fields. This would include engineering, physics, chemistry and biology (Rummler 2005, p. 228). In 2015, the German Federal Supreme Court considered the exclusion of mathematical methods as such in relation to the technical requirement. In *Flugzeugzustand*, it viewed them as inventions if they solve a concrete technical problem via technical means (German Federal Supreme Court (BGH) 2015). Conversely, such methods will not be technical if they do not concern the targeted use of natural forces. This decision was seen as furthering the possibility of patent protection for inventions that have a mathematical background (Fernolend 2015).

[352]Rummler (2005, p. 226).

Board stated that "*in effect the German Federal Court of Justice ruled in the same way*".[353]

Lastly, an interesting UK case referred to the *Infineon* decision of the EPO. In *Halliburton*, the Patents High Court decided on a method to design a drill.[354] Despite the absence of claims referring to the actual manufacturing of the drill, the Court considered the invention as patentable subject matter.[355] In particular, it was held that:

> Designing drill bits is obviously a highly technical process, capable of being applied industrially... designers are... highly skilled engineers. The detailed problems to be solved... are technical problems with technical solutions. Accordingly finding a better way of designing drilling bits in general is itself a technical problem.[356]

#### 4.1.7.4 American Normative and Case Law Overview

Mathematical methods and algorithms are not mentioned in § 101 on subject matter patentability of the US Patent Act. However, they could fall within the realm of the judicially made exclusion of abstract ideas. Abstract ideas, natural phenomena, and laws of nature have been deemed not eligible for patent protection through a long line of cases.[357] The language used to describe those exceptions has varied depending on the case and has spanned from "physical phenomena", "scientific principles", "systems that depend on human intelligence alone", "mental processes", to "disembodied mathematical algorithms and formulas". The rationale of these exceptions is to be found in the idea that basic scientific tools should not be removed from the public, as they are "*free to all men and reserved exclusively to none*".[358]

The US Supreme Court has dealt multiple times with the patentability of algorithms, especially after the emergence of the software and computer industry. A first example is the *Gottshalk v. Benson*, *Parker v. Flook* and *Diamond v. Diehr* case trilogy between 1972 and 1981.[359]

The first case related to a method of converting binary-coded decimals into pure binary numerals. The claims concerned a method of programming a general-purpose computer to perform such task. Such conversion could also be performed mentally by a person. The claims were not restricted to a particular machinery or end use. The Court rejected the patent and maintained that the claims were so abstract and

[353] Technical Board of Appeal of the EPO (TBA) (2006b, § 3.4.2).

[354] UK High Court of Justice (Patents Court) (2011).

[355] Reeve (2015).

[356] UK High Court of Justice (Patents Court) (2011, § 74).

[357] U.S. Supreme Court (1972, 1978, 1981).

[358] U.S. Supreme Court (1948, p. 332). Interestingly, authors have also suggested the application of the "abstract idea" exception (as interpreted by the US Supreme Court in *Alice Corp. v. CLS Bank International* and its progeny) to gene patents (Ulen 2019).

[359] U.S. Supreme Court (1972, 1978, 1981).

sweeping that they would cover known and unknown uses of this conversion. The patent would wholly pre-empt the mathematical formula and basically patent the algorithm itself.[360]

A patent for a method to update alarm limits in a catalytic conversion process was discussed in *Parker*. This method relied on a mathematical algorithm and was held non patentable by the Court. The claimed alarm limit was a number and the patent sought to protect merely a formula for calculating it.[361]

In *Diamond v. Diehr*, the Court examined the patentability of a process to cure rubber that incorporated the Arrhenius equation. The Justices reversed prior judgements that had rejected the patent and argued that, although the equation could be seen as prior art, the invention as a whole was new. The patent would not foreclose the use of the equation for the public, but would prevent it from realising the invention.[362]

Given the opposite conclusions reached by the Courts, what were the different deciding factors in those cases? The distinction was seen to lay in the fact that *Flook* claimed a method of computing a number, whilst in *Diehr* the patent was directed at a method for curing synthetic rubber. However, it was noted that the main difference between the two might just concern the way in which the claims were drafted.[363]

Since then, a number of cases were brought before the Courts. Those show that the patentability of this type of inventions is far from settled. In 2007, the BPAI stated that a mathematical algorithm was non-eligible, since it covered the algorithm itself. It noted that the invention did not transform any physical matter, did not manipulate numerical values and did not reach a useful, concrete and tangible result.[364]

A somehow similar approach was followed by other US Courts. In *Classen v. Biogen* concerning a claim for an algorithm, the District Court focused on whether there was a transformation or a practical application on a machine.[365] In a later stage of the dispute, before the Federal Circuit, the claims were considered ineligible, since they did not satisfy the machine or transformation test.[366] Also, in *Ex parte Bilski*, before the BPAI, it was held that methods that were not implemented via a machine would be problematic, since they could cover abstract ideas.[367] Obviously, this approach could be problematic for algorithms.[368]

[360]Merges and Duffy (2017, p. 91) and Smalley (2013, pp. 418–419).

[361]Merges and Duffy (2017, pp. 91–92).

[362]Koepsell (2011, p. 132).

[363]Eisenberg (2013, pp. 344–345).

[364]U.S. Boards of Patent Interferences and Appeals (2009) and McGarrigle and Norviel (2008, pp. 321–322).

[365]U.S. District Court District of Maryland (2014) and McGarrigle and Norviel (2008, p. 327).

[366]Sieman (2011).

[367]U.S. Boards of Patent Interferences and Appeals (2006).

[368]McGarrigle and Norviel (2008, p. 311).

The patentability of algorithms landed before the Supreme Court in 2012 and in 2014 with the *Mayo* and *Alice Corp.* cases.[369] The first case concerned patents for the use of drugs for the treatment of autoimmune diseases. The Supreme Court held that the claims did not pertain to patentable subject matter, as they merely recited a natural law and required doctors to engage in routine activities. Following this decision, algorithms for personalised medicine might be considered patentable only with great difficulty.[370] The Court referred to its *Flook* and *Diehr* decisions and stated that *Flook* had merely recited a mathematical formula, followed by the instruction to apply it and it was therefore held unpatentable. By contrast, the claims presented in *Diehr* contained additional steps and "*transformed the process into an inventive application of the formula*".[371] While some were critical of this distinction between *Flook* and *Diehr*, the Court considered the claims in *Mayo* similar to those in *Flook.*[372] Although this decision refers exclusively to method claims, its conclusions could be extended to other types of claims. This could have a major influence on the patent panorama in this field, since *Mayo* could render almost every biotechnology patent suspect.[373]

This line of thought was strengthened in *Alice Corp.*, as the Justices held that abstract inventions (e.g. algorithms) cannot be considered patentable only because they are implemented on a computer.[374]

Since software simulators offer a visual representation of the relevant molecules or biological systems, this circumstance could have an impact on patent infringement. Researches could generate a virtual representation of a compound patented by someone else. The researcher could have based this virtual representation on the information disclosed in the patent, but did not reproduce the actual compound in real life. While in the US it has been argued that a patent on a compound relates to the actual substance and not to its representation, the patentee might still argue that the virtual version of the substance infringes his patent rights. A way to argue this would be to refer to the doctrine of equivalents. This doctrine maintains that "*if two devices do the same work in substantially the same way, and accomplish*

[369]U.S. Supreme Court (2012, 2014).

[370]Price (2015, pp. 1420–1421).

[371]Eisenberg (2013, p. 345) and U.S. Supreme Court (2012, p. 12).

[372]Eisenberg (2013, p. 345) and U.S. Supreme Court (2012, p. 13).

[373]Smalley (2013, pp. 432–439).

[374]Price (2015, p. 1425) and U.S. Supreme Court (2014). Following *Alice Corp.*, the USPTO issued several examination guidances to determine if an invention covers an abstract idea (Konski 2020; USPTO 2018, 2019). Then, in 2020, the USPTO published a report analysing the impact of *Alice Corp.* on its examination outcomes. It noted that: "*The likelihood of receiving a first office action with a rejection for patent-ineligible subject matter increased by 31% in the 18 months following the U.S. Supreme Court decision in Alice Corp. v. CLS Bank International in 33 "Alice-affected" technology areas. For these technologies, uncertainty in patent examination — measured as variability in patent subject matter eligibility determinations across examiners in the first action stage of examination — increased by 26% in the 18 months following the Alice decision*" (Toole and Pairolero 2020; Rose 2020).

*substantially the same result, they are the same, even though they differ in name, form or shape*".[375] Despite possible arguments to the contrary, it was held that Courts are likely to dismiss the presence of an equivalence between the actual compound and its virtual representation. In fact:

> The arrangement and interaction of elements in the virtual compound do not follow the laws of chemistry and physics like the arrangement of elements in the real patented compound. Instead, they interact according to a pre-programmed mathematical algorithm that simulates those laws using today's imperfect models.[376]

### 4.1.8 Presentations of Information

Presentations of information as such are excluded from patentability pursuant to Article 52(2)(d) EPC. This exclusion is justified by their lack of inventive character.[377] The words "presentations of information" refer to the conveying of information to a user and do not encompass technical representations of information directed to a technical system. This latter type of information is therefore not excluded from patentability under Article 52(2)(d) EPC, as it does not concern information as such.[378] Equally, features of data encoding schemes, data structures and electronic communication protocols representing functional data (as opposed to cognitive ones) fall outside of this exclusion.[379]

Since DNA could be seen as both information and a chemical compound, it is important to briefly review the application of this exclusion to biotechnology.

The EPO listed a number of inventions that would not be considered patent eligible pursuant to this norm. Those encompass protein structures and coordinates as well as computer disks comprising information.[380] Equally, DNA and amino acid sequences would also be considered unpatentable presentations of information. By contrast, claims formulated as "*a DNA molecule comprising the nucleotide sequence...*" would be acceptable.[381] Commentators reached a similar conclusion, as it was held that, while biological molecules are patentable, the information contained therein (i.e., the abstract biological sequence itself) is not. In particular, it was argued that:

---

[375]Watson (2014, p. 30) and U.S. Supreme Court (1950, § 608).

[376]Watson (2014, p. 31). Authors argued that there is a fundamental difference between the virtual and the actual compound, since one is composed of atoms and can be subjected to chemical reactions, whereas the other is a representation that can only react to software (Watson 2014, p. 31).

[377]Stauder et al. (2013).

[378]EPO (2018, § G.II.3.7).

[379]EPO (2018, § G.II.3.7).

[380]Vogt (2010, p. 8).

[381]Covone (2015, p. 15).

> It is unlikely that one could patent a biological sequence since it may be characterized as a natural phenomenon. Therefore, patent protection for DNA, RNA, or protein extends only to the physical/biological composition, and not to the abstract biological sequence information that describes the composition. Thus, a patentee could only prevent another from using the composition itself and not the information within the molecule.[382]

This conclusion is supported by the work of other scholars, who suggested that patenting DNA as a composition of matter offers the patentee exclusionary rights on the tangible molecule, but does not block the public from using and analysing the information contained in the sequence.[383]

Concerning the patenting of DNA sequence information stored in a computer-readable medium, authors argued that the decisions issued by US Courts on computer-implemented inventions could offer valuable guidance.[384] In particular, it was affirmed that:

> It is not obvious why DNA sequence information stored in a computer-readable medium – a product that requires human intervention and serves human purposes – would be categorically excluded from patent protection.[385]

However, this opinion has been criticised for failing to comprehend the full implications of this approach.[386]

On the technical requirement, the EPO held that features concerning a presentation of information that are defined only by the content of that information do not have technical character. Additionally, in assessing whether a feature pertaining to the presentation of information is technical, the focus should be on whether it solves a technical problem.[387]

Lastly, it needs to be noted that the exclusion set by Article 52(2)(d) is limited to mere presentations of information, which would leave information *per se* free to be patented. Therefore, "*such information has the potential to provide patentable inventions*" even in the biotechnological field, where already numerous patents cover information rather than more traditional types of inventions.[388]

---

[382]McBride (2002, p. 1342).

[383]On this, see Sect. 2.2.5.

[384]Eisenberg (2000, p. 791).

[385]Eisenberg (2000, p. 792).

[386]Rimmer (2003, pp. 39–40).

[387]EPO (2018, § G.II.3.7).

[388]Baldock et al. (2000, p. 41).

## 4.2 The Patentability of Synthetic Biology Inventions

### *4.2.1 Introduction*

The following analysis will concentrate on the norms applicable to synthetic biology inventions in Europe. It will then assess whether those inventions are patent eligible pursuant to them. The assessment will be integrated with an analysis of the US panorama to evaluate whether the different standards developed on both sides of the Atlantic will have any specific repercussions on the patentability of synthetic biology inventions.[389]

### *4.2.2 General Assessment*

#### 4.2.2.1 Applicable Norms

The provisions of the EPC are applicable to patents in Europe. Hence, the patentability of synthetic biology inventions will be examined pursuant to Article 52 and 53 EPC. Article 27 TRIPS is equally applicable in this field. By contrast, the application of the Biotech Directive and of the related EPC Rules on biotechnological inventions is more problematic and cannot be taken for granted.[390]

The Directive has shaped the patentability of biotechnological inventions in Europe for the past two decades. Hence, authors have argued that it will continue to do so also in the field of synthetic biology,[391] since:

> The present European Patent Convention and the case law issued in the last 30 years by the European Patent Office in the biomedical field constitute a solid legal foundation able to accommodate the new developments of Synthetic Biology and provide for the granting of patents with a high presumption of validity and which are ethically correct from the perspective of the European Patent Convention and the European Directive 44/98/EC of 1998.[392]

Nevertheless, the Directive is a child of its time and does not consider the giant technical leaps that biotechnology made in the past two decades.[393] As a result, some

[389]The boundaries of what research areas belong to synthetic biology are still blurry. For this reason, the choice of the research areas and applications is subject to a certain degree of discretion. Nonetheless, the list includes the major research areas discussed in Chap. 2.

[390]In the following assessment, references to the applicability of the Directive should be considered to include the Implementing Regulations of the EPC on biotechnological inventions.

[391]This conclusion is sustained by the absence of specific patent provisions for synthetic biology and by the circumstance that: "*From the point of view of European patent law, it seems difficult to envisage specific provisions for synthetic biology*" (Rutz 2009, p. 15).

[392]Fernandez y Branas (2014, p. 189).

[393]Already in 2000, it was stated that: "*The Directive is a child of its time, the only problem being that its 'time' was 12 years ago, the very distant past in biotechnology. Thus it reflects the concerns*

synthetic biology inventions do not fit squarely into the realm of application of the Directive and of the EPC Rules on biotechnologies. This could have an impact on the norms applicable to them as well as on the uniformity of the European patent panorama.

### Biological Material

The Directive covers biotechnological inventions, yet it does not define this term. By contrast, the EPC Rules delimit what is meant by it. According to Rule 26 EPC, biotechnological inventions relate to "*a product consisting of or containing biological material or a process by means of which biological material is produced, processed or used*". Pursuant to both Rule 26 EPC and Article 2 of the Directive, biological material encompasses "*any material containing genetic information and capable of reproducing itself or being reproduced in a biological system*".[394]

The problem is that this definition of biological material does not adapt well to a number of synthetic biology inventions, with the consequence that they would fall outside of this concept and, in turn, of the scope of application of the Directive and the EPC Rules on biotechnology.[395]

Some authors argued that this discipline would be regulated by the Directive, as it is a combination of known technologies and given that it is within the realm of biotechnologies.[396] Conversely, others were more sceptical, since SynBio inventions might increase the gap with naturally occurring biological systems to the point that these inventions might be considered as non-biological. For example, *Rutz* raised the question of whether synthetic biology would really fit within the boundaries of Rule 26 EPC. In particular, he inquired whether artificial codons, non-natural

*and misunderstandings of the early days of biotechnology, and its relevance to current issues such as the human genome project is limited*" (Cook 2000, p. 56).

[394]In particular, it was argued that: "*If material not capable of replication is involved, the products and procedures obtained will be subject to the general rules on patenting i.e. will lie outside the area of law relating to biotechnologies... if, on the contrary, material capable of replicating itself and which constitutes some new, simple life form (artificial life) is involved, the invention would be subject to the specific rules intended for this type of organism*" (Casabona 2014, p. 183).

[395]For example, the ability of SynBio inventions to reproduce themselves is pivotal in this assessment, given that products lacking this ability would be patented as inert material (Spanish Bioethics Committee and Portuguese National Ethics Council for the Life Sciences 2011, pp. 21–22). Specifically, it was stated that: "*If material not capable of replication is involved, the products and procedures obtained will be subject to the general rules on patenting, i.e. will lie outside the area of law regulating biotechnologies*" (Spanish Bioethics Committee and Portuguese National Ethics Council for the Life Sciences 2011, p. 24). Still, most synthetic biology inventions would generally fall within the scope of application of the Directive, as they could be considered biological material.

[396]Sommer (2013, pp. 99–100).

amino acids, protocells and other inventions could be considered biological material.[397] In his opinion:

> A tricky legal question might arise should synthetic biology systems become so different from biological systems that they can no longer be considered 'biological material' (e.g. by the use of unnatural nucleic acids or completely artificial coding systems).[398]

Similarly, another EPO employee expressed his personal views on the matter and maintained that:

> Whether synthetic biology departs so much from natural biological systems so as to be considered 'non-biological' is difficult to predict, and future case law may need to be established in this respect.[399]

The concept of biological material is thus pivotal in this analysis. From an historical perspective, the term biological material first appeared in the 1988 proposal for the Directive. At the time, this concept did not play a pivotal role and was not defined.[400] Under the heading "*categories of biotechnological inventions*", biological materials (e.g. plasmids, viruses) were merely listed amongst possible product inventions. The proposal clarified that the natural or artificial origin of the entity was not relevant. Similarly, biological materials were mentioned in describing the categories of biotechnological inventions that were patentable. Specifically, those would include inventions concerning processes for the creation of a living organism or the production of other biological material as well as inventions relating to their use.

The concept of biological material was first defined in 1992, following the amendments introduced by the European Parliament. In replacing the concept of living matter with that of biological material, the draft defined the latter as "*any*

---

[397]Rutz (2010a, p. 13).

[398]Rutz (2010b, p. 12). It could be argued that also proteins cannot be considered biological material *strictu sensu* pursuant to Article 2 Biotech Directive, as they do not directly store genetic information and are not reproduced. In particular, authors viewed critically the tendency to include them within the scope of protection granted for biotechnological inventions, as their patent protection should be aligned instead with that of chemical inventions. From this perspective, the protection granted to chemicals would constitutes a residual approach, which would apply in case an invention could not be considered biotechnological (D'Antonio 2004, pp. 30–31). Nonetheless, proteins are explicitly referred to in the Recitals of the Directive (European Commission 2000 Recital 24). Equally, protein sequences were considered to fall within Rule 23 EPC (now Rule 29 EPC) by employees of the EPO (Friedrich 2007, p. 4). The extension of the Directive to proteins could be considered generally acceptable also from a social and moral perspective. By contrast, the inclusion of some synthetic biology inventions within the scope of the Directive could be more controversial. Taking for example xenobiology systems, it can be noted how profoundly they affect some features that have always been connected to the concept of biological (e.g. four bases and nucleic acids having a deoxyribonucleic backbone). Obviously, this represents conceptually another level of challenge to the notion of biological material in comparison to the one raised by proteins.

[399]Fernandez y Branas (2014, p. 190).

[400]For an overview of the history of the Directive, Porter (2009, pp. 3–17).

*self-replicating living matter and any matter capable of being replicated through a biological system or by any indirect means"*.[401]

This definition is not identical to the one currently used in the Directive. Firstly, it centres on the concept of replication, whether self-occurring, passively occurring through a biological system or via indirect means. By contrast, the current definition encompasses self- and passive reproduction within biological systems, thus restricting its scope. In particular, the previous formulation would seem to hint to the inclusion within the norm of materials that had not been replicated in biological systems, whereas the current one seem to demand that all biological materials are replicating within biological systems. Although the Directive does not define the concept of biological system, the term biological is generally seen as referring to life and living processes.[402] In the case of synthetic biology, this aspect could be pivotal.

Secondly, the two definitions of biological material differ in their reference to genetic information. The 1992 definition does not mention this concept, which is instead pivotal in the 1998 Directive.[403] Despite its relevance, this notion is not defined in the Directive and the definition adopted in other legislative contexts does not seem fully applicable here.[404] However, commentators have considered this notion as equivalent to that of heritable information.[405]

Despite the above uncertainties, the introduction of a definition of biological material was saluted by the Economic and Social Committee. In assessing the draft of Article 2(1) of the Directive, it held that "*this article is new and was not in the previous text. It gives a precise definition of 'biological material'... precise definitions are indeed highly important to avoid misunderstandings*".[406] Conversely, others felt that this definition is "*so inherently lacking in clarity that this must be*

---

[401]Article 2 (European Commission 1992). This implements the changes discussed by the European Parliament on April 8th 1992 and on October 29th 1992 (*Amendments suggested by the European Parliament to the Draft Directive on the legal protection of biotechnological inventions ((1992) OJ C 305/160)*, 1992) (*Amendments suggested by the European Parliament to the Draft Directive on the legal protection of biotechnological inventions ((1992) OJ C 125/112)*, 1992).

[402]Merriam Webster's Collegiate Dictionary (2014).

[403]Still, the use of the notion of genetic information is not new in this field. In the 1988 proposal, the concept of genetic information was already employed, as it was maintained that the Directive would give the possibility to patent products: "*Consisting of or containing... genetic information*" (*Proposal for a Council Directive on the legal protection of biotechnological inventions (COM/88/496FINAL - SYN 159)* 1988) Art. 13 and Explanatory Memorandum). In the 1994 draft version of the Directive, the definition of biological material already included a reference to genetic information, as it considered biological material to mean: "*Any material containing genetic information and capable of self-reproducing or capable of being reproduced in a biological system*" ((*Common position (EC) No 4/94 adopted by the Council with a view to adopting European Parliament and Council Directive 94/. . ./EC of . . . on the legal protection of biotechnological inventions (94/C 101/01)*, 1994) Art. 2). For a comparison of the different formulations of the relevant articles starting from 1995 (European Commission 2016, Annex).

[404]*The US genetic information nondiscrimination Act of 2008* (2008).

[405]Rutz (2010b, p. 11).

[406]Economic and Social Committee (1996).

*assumed to be the consequence of only the most careful drafting. The meaning... is so important but so opaque that it will inevitably be taken to mean whatever the reader wishes, at least until the ECJ rules definitively on it*",[407] which has not happened so far.

In the absence of ECJ decisions and specific guidance, one could wonder whether these terms could be interpreted extensively. If this were the case, the Directive could apply to SynBio inventions, even if they do not squarely fit within the literal definition of biological material.

In its first Directive proposal in 1988, the Commission defined biotechnology as including techniques that use or cause organic changes in biological materials or cause changes in inorganic material by biological means.[408] The techniques listed (e.g. recombinant DNA, gene transfer, cell culture) noticeably belonged to a different technological era. To accommodate future technological developments, it was mentioned that the "*term 'invention' should always be sufficiently broad to include all new developments in biotechnology*".[409] Similarly, with regards to the notion of microorganism, it was stated that this should not be construed narrowly, so that "*future developments in biotechnology in respect of animate matter... can benefit from the principles of the Directive*".[410]

Such statements would point to an expansive interpretation of what constitutes a biotechnological invention.[411] Indeed, the reference to "*all new developments in biotechnology*" is wide enough to encompass anything that pertains to the biotechnological field. Since synthetic biology can be considered a biotechnology, all its research endeavours, both clearly biological and not (i.e. protocells and xenobiology), would be governed by the Directive under this approach. On the other hand, the reference to animate matter, albeit made with regards to

---

[407] Sampson (2002, p. 414).

[408] *Proposal for a Council Directive on the legal protection of biotechnological inventions (COM/88/496FINAL - SYN 159)* (1988) Explanatory Memorandum.

[409] *Proposal for a Council Directive on the legal protection of biotechnological inventions (COM/88/496FINAL - SYN 159)* (1988) Explanatory Memorandum.

[410] *Proposal for a Council Directive on the legal protection of biotechnological inventions (COM/88/496FINAL - SYN 159)* (1988) Explanatory Memorandum.

[411] On this point, it is interesting to see the approach adopted by the ECJ in a case concerning the regulation of GMOs. The case concerned the exclusion of certain techniques from the scope of the Directive on the deliberate release into the environment of genetically modified organisms (Directive 2001/18/EC). In that occasion, the Judges had to examine the applicability of this Directive to some techniques of genetic modification. They noticed that "*a provision derogating from the requirement to subject GMOs to the obligations laid down in Directive 2001/18... must be interpreted strictly... It is necessary to consider not only its wording but also the context in which it occurs and the objectives pursued by the rules of which it is part*" (European Court of Justice (ECJ) 2018, para 41–42). Extending this approach to the Biotech Directive, one could argue in favour of an extensive interpretation of the norm, since its goal is to reach an effective and harmonised protection of biotechnologies throughout Europe, with the consequence that exceptions to this regime should be interpreted narrowly.

microorganisms, could hint to the exclusion of research areas not dealing with living matter or not strictly biological (e.g. protocells).

Obviously, those results are likely to be influenced by the social acceptance of these inventions and by whether the public considers them biological or not. Equally, whether those inventions are legally considered biological or not, could have an impact on their social and moral acceptance and on the possible criticisms moved against them. In practical terms, it is still too early to tell in which direction the public and normative debate will orient themselves, but their relevance cannot be underestimated.

It is not the first time that this type of problem has been raised. In the past, unnatural nucleotides and amino acids generated similar concerns.[412] As for xenobiology, those unnatural nucleotides are unknown in nature and can be recognised only by modified RNA. At the time, patents were filed for those unnatural elements and authors noted that they could raise questions under Article 2 of the Directive.[413] Still, it was argued that the notion of biological material should not be applied following exclusively scientific criteria, as it ultimately concerns a point of law and not of science. For this reason, a purposive interpretation of this term was suggested. This would allow the inclusion of those unnatural DNA bases into the realm of genetic material. Following this interpretation, all oligonucleotides sequences would be considered to contain genetic information and thus to fall within Article 2(1)(a) of the Directive.[414] This solution was suggested because, if one were to exclude sequences with unnatural nucleotides from the realm of what is biological, this could:

> Lead to a somewhat bizarre conclusion, namely that such sequences are not patentable under the definition contained in Article 2(a) of the Directive... while microbiological processes that utilised them would be patentable under Article 2(b).[415]

However, the possibility that they might fall outside this definition was not excluded, as the level of human manipulation might be too extensive for them to be considered biological. Nonetheless, it was observed that even if they would fall outside Article 2(a) of the Directive for not being biological systems, they could still be patentable if integrated in a process claim under Article 2(b) of the Directive.[416]

Another question that might arise in the field of synthetic biology is whether synthetically constructed materials could be considered biological. The Directive is silent on this point. Yet, a combined reading of Rule 26 and 27 EPC would seem to confirm their inclusion within the scope of the norms. Indeed, the latter proviso states expressly that biotechnological inventions can cover both isolated and technically

---

[412] This discipline, which already emerged around the year 2000, is now further developed and merged into the xenobiology area of synthetic biology.

[413] Sampson (2002, p. 411). Some of those applications were however rejected, albeit not on patentability grounds (Sampson 2002, p. 411; Kurzchalia 1995).

[414] Sampson (2002, p. 411).

[415] Sampson (2002, p. 412).

[416] Sampson (2002, p. 412).

produced materials.[417] Since the definition of technical process includes techniques that humans alone are capable of implementing, there is little doubt that synthetic materials could also be considered biological. Furthermore, no provisions differentiate between materials based on their origin (i.e. synthetic or natural). The norms merely refer to materials containing genetic information and capable of reproducing or being reproduced, which could apply to both natural and synthetic biological materials. This would be confirmed also by the provisions of the 1988 Directive proposal, which considered artificial and natural entities as equal.

Hence, the source of the material (i.e. obtained via a natural process or via chemical synthesis) is not relevant for the application of the Directive.[418]

### Consequences of the Inapplicability of the Biotech Directive

In light of the above, it is then necessary to consider if and how the inapplicability of the Directive might impact the patentability of synthetic biology inventions.

If the Directive were not applicable, those inventions would be treated on par with inventions in other technological fields.[419] Hence, they would be subjected solely to the provisions of TRIPS and the EPC. Although those two legislative instruments apply to all fields of technology, they are not tailored to biotechnological inventions. Indeed, even though "*no prohibition or exclusion exists in national or European patent law (Munich Convention) which precludes a priori the patentability of biological matter*",[420] the EPC does not set specific norms for their patentability.

Of course, it would be possible to argue that no difference would arise, for example by stating that the Directive merely codified the orientations that had already emerged in the prior case law. This point was sustained by the EPO itself, as it expressed the view that the Directive was based on its case law.[421] Equally, one could refer to the statements of Advocate General *Jacobs* in the challenge proposed by the Netherlands against the Directive, where he maintained that "*the patentability of living material is not an innovation of the Directive but the recognition of what is actually happening in conformity with national law*".[422]

In spite of this, it is reductive to argue that the inapplicability of the Directive to SynBio inventions would not have an impact on their patent protection.

Firstly, this would affect the possibility of referring questions to the ECJ pursuant to Article 19 TEU and Article 267 TFEU. According to the referral mechanism,

---

417 Fernandez y Branas (2014, p. 190).

418 Rutz (2010b, p. 12).

419 Rutz (2010b).

420 Recital 15 of the Directive.

421 Tritton et al. (2018, p. 195) and European Patent Office (1999, § 8). Yet, this assumption might be undermined by recent events concerning the patentability of products of essentially biological processes. On this, see Sect. 3.1.

422 Advocate General Jacobs (2001, § 67).

national Courts can refer questions to the ECJ on the interpretation of European law. Being limited to European law, this uniformity mechanism could not be applied to non-EU laws (e.g. EPC) and could not be triggered by the EPO. Gone this mechanism, diverging national interpretations could emerge.[423]

States could regulate matters independently and would not be bound by the orientation of the ECJ. Concerns in this sense had already been expressed at the time the Directive was first formulated, as it was held that only the Directive would be capable of securing a harmonised practice in such a policy and politically charged field.[424] If the Directive were not applicable, this could lead to dissimilar understandings and approaches and increase legal uncertainty.[425]

Additionally, it could be objectionable that the Directive represents a mere restatement of both the EPO and national approaches to the patentability of biotechnologies. The heavy criticism drawn by this Directive both during its negotiation and after its passing could be indicative of the fact that its content was not unanimously accepted. This is confirmed by the fact that several States delayed the implementation of this norm into their national laws. Equally, concerning Rules 26–29 EPC, it could be argued that if they were a mere clarification of existing law, it might not have been necessary to amend the Implementing Regulations. To overcome any problems, authors have suggested focusing on the EPC and considering the Directive as a mere confirmation of the patentability of natural genetic sequences and related products and processes, rather than an all-embracing legislation regulating the patentability of all inventions in the biotechnological field.[426]

---

[423]The *Monsanto v. Cefetra* case offers a good example of those possible inconsistencies (European Court of Justice (ECJ) 2010). In discussing the scope of patent protection granted to biotechnologies pursuant to the Directive and national laws, commentators held that: *"The Directive is not a minimum standard of protection, as is evidenced by Recitals 16, 23, 24 and Art. 11. The point of reference for determining the scope of biotechnological patents is thus not the absolute product protection contained in national patent law, but Arts. 8–11 of the Directive... In other words, if a patent claim that is relied upon in an infringement suit falls within one of the categories of Arts. 8 or 9, the scope of protection is exclusively determined by the provisions of the Biotech Directive... To the extent that national law in its scope of protection for biotechnological inventions... differs from the provisions of the Directive, the former should be interpreted in the light of the latter, or be amended by domestic legislation... Which provisions should be applied to patent applications or patents granted prior to the implementation of the Directive (the subject of the third question referred to the ECJ by the Dutch Court) is a matter of national law"* (Heath 2009, pp. 945–947). This means that, had the Directive not existed, national laws would have determined the scope of protection of the patent. This would have likely led to different positions around Europe.

[424]*Proposal for a Council Directive on the legal protection of biotechnological inventions (COM/88/496FINAL - SYN 159)* (1988).

[425]Sampson (2002, p. 412). For instance, this could affect the line between discoveries and inventions in the biotechnological field. Indeed, it needn't be forgotten that, before the entry into force of the Directive, national Courts rejected a number of patent applications for claiming mere discoveries. This was the case in *Genentech* and *Kirin-Amgen* before British Courts. On this, see Sect. 4.1.3.2.

[426]Sampson (2002, p. 412).

Lastly, the application of the Directive would affect the legal grounds for finding an invention patentable. Biological materials obtained via technical processes would be patentable pursuant to the Directive. Once this point is established, no further legal grounds are needed to prove patentability. This proviso is likely to be of pivotal importance in synthetic biology, given that those inventions are exclusively obtained via technical processes and cannot be achieved naturally. However, should the Directive not be applicable, the eligibility analysis could not benefit from this and other legal grounds would need to be found to establish patentability.

The inapplicability of the Directive would also raise a number of policy, moral and social issues. As argued by commentators, "*it would be a question for the legislator and society at large if specific provisions for the patentability of non-biological self-replicable systems were desirable and needed*".[427] So far, the issue has only been seldom raised, since the gap between SynBio inventions and nature is still limited. However, the more complex and different from nature synthetic biology inventions become, the more legislators and society might consider them as non-biological. Still, it is complex to draw a bright line between what is biological and what is not. This would demand a definition of what is biological and to establish how much change must occur for a biological product to become non-biological. Equally, which relevance should be assigned to the fact that synthetic biology inventions are based upon and cannot overcome the general laws of nature? For example, in the future, synthetic biology might create wholly synthetic systems that are solely devised by man on the basis of engineering principles to fulfil human needs. However, even in this case, it could be argued that those systems are based upon natural laws and platforms and that, therefore, they are biological systems. Would this conclusion be satisfactory and defendable?

The problem with those issues is that it is impossible to univocally reply to them and to foresee which approaches will be adopted by patent offices, Courts, legislators and the public. This is especially true considering that those points are not strictly legal, but have a moral and social connotation. Furthermore, this discussion is likely to be influenced by whether synthetic biology will be perceived as an incremental technology or whether it will be considered a ground-breaking change that requires to be addresses specifically.

Given this complexity and the fact that patent applications in problematic fields (e.g. xenobiology) have already been filed, it would be important to clarify those issues soon, before those technologies become more powerful and increasingly challenge the patent system.

Yet, at the moment, this seems unlikely. Despite the serious repercussions connected to the inapplicability of the Directive and the importance of clarifying whether the definition of biological material should be assessed from a scientific or a legal viewpoint, practical reasons suggest that this topic is unlikely to be decided upon.

[427] Rutz (2010b, p. 12).

Before the EPO, industry players are unlikely to raise such issue. The Directive provides clear-cut and favourable criteria for the patentability of synthetic biology inventions. Therefore, companies are unlikely to jeopardise this by raising questions on the applicability of the Directive. Civil society organisations and NGOs are also unlikely to bring up this issue, given that it has no direct patentability consequences (i.e. it would not stop on its own a patent from being granted). Lastly, this legal issue is unlikely to be raised by the EPO, as the Office would probably avoid venturing into such a complex debate of its own volition unless it was utterly necessary.

Similar conclusions could be reached for an analysis before Courts. Here again the question is which party would be interested in raising such an issue. A decision on this does not necessarily lead to the rejection of the patent (which might be a problem for civil society organisations and NGOs) and could open a Pandora's box by retracing the line between points of law and points of science (which might be a problem for industry players). Lastly, to address this topic, Courts would need to analyse in detail a number of complex and rapidly evolving scientific issues.

On the basis of this, it is unlikely that this point will be thoroughly assessed, even though it constitutes a reasonable and interesting question deserving of closer analysis.

#### 4.2.2.2 Subject Matter Patentability

The patentability of synthetic biology inventions has drawn only limited attention. Given the assumption that SynBio systems will introduce human design and engineering into biology, their patentability has often been taken for granted.[428] Indeed, it has been argued that:

> The more 'synthetic' biology becomes, the closer to classical chemical areas and the further away from the controversy raised by the patenting of products which exist in nature it becomes.[429]

[428]The application of the discovery exclusion to synthetic biology may also depend on the *fora* in which its patentability is discussed. It was held that, should the matter be discussed only before patent offices, this issue is not likely to be raised. On the other hand, should it be addressed by the legislative and judicial branches, it is likely to gain a more prominent position. However, those last two *fora* will probably not decide on this matter any time soon (Schneider 2014, p. 170).

[429]Fernandez y Branas (2014, p. 187). The reason for the limited impact of the subject matter inquiry in the field of synthetic biology was attributed to the long patentability tradition of chemical and natural substances, both in Europe and in the US. Hence, the weight of the examination of those patents could fall on Articles 56, 83 and 84 EPC. The advantage of those requirements is that they can be used to fine-tune and regulate the scope of protection, instead of having to use a "either-or" method, as in the case of discoveries (Schneider 2014, pp. 169–170).

Yet, this assumption is not uncontested. For example, *Bagley* held that the full patent eligibility of synthetic biology inventions is "*not free from doubt*".[430] Likewise, *Palombi* expressed his doubts by stating that:

> The implication here is that, like a machine, the synthetic genome, being the sum total of these genetic parts, is patentable subject matter and that to the extent that these 'natural genes… have been redesigned to function more efficiently or… have been designed and synthetized from scratch', they meet the subsidiary thresholds of novelty and inventive step. Their analysis therefore suggests that these synthetic genomes are patentable inventions. Indeed the efficiency of synthetic genomes over generally modified 'natural genes' to express proteins is supposedly an improvement that is useful in that it purports to improve cellular productivity, which in turn makes them valuable. Accordingly they appear to possess the attributes of things that have traditionally met the requirements of invention; but do they qualify?[431]

Those issues cannot be overlooked, as the subject matter eligibility requirement plays a fundamental role in the patent system and given that other requirements are not equipped to do its work.[432] Also, this criterion should not be disregarded, especially now that differences between the EU and the US approach have arisen.

---

[430]Bagley (2015, p. 5). Patents are likely going to be the predominant type of IPRs protection in the field of synthetic biology. For a number of applications, protection via copyright has also been taken into consideration. In the past, the informational content of DNA has been used as a basis to argue that it could be protected by copyright. The analogies between synthetic biology and computer programs have accentuated such ideas, as synthetic biologists could write completely new DNA sequences showing significant originality and personal expression. In spite of this, commentators noted that copyright could be a "poor fit" for synthetic biology, given that the structure of sequences is mostly dictated by the function they ought to perform and, consequently, does not constitute the free expressive choice of its author (Bagley 2015, p. 6). Furthermore, biological subject matter is generally not mentioned amongst the categories of copyrightable subject matter. The structure itself of copyright might also not adapt well to synthetic biology, given that works that monopolise a particular function are excluded from protection. Furthermore, copyright could protect a specific DNA sequence and prevent others from copying it; however, it could not prevent others from designing an identical or similar one (Torrance 2010, p. 201). By contrast, others noted that: "*Copyright may seem desirable to synthetic biologists because it is easy to obtain, for it attaches automatically by operation of law, and because it would enable synthetic biologists to make their creations available under Creative Commons license*" (Samuelson n.d., p. 1). Synthetic DNA sequences were also considered to qualify for copyright protection, if they can be considered: "*Original works of authorship fixed in [a] tangible medium of expression*" (Torrance 2010, p. 639; *U.S. Code - Title 17 - Copyrights* n.d. § 102). In spite of this, in a case where a company tried to register a synthetic biology creation at the Copyright Office in the US, this request was refused by the Office (Samuelson n.d., p. 1). For further discussions on the use of copyright for biotechnologies and synthetic biology (Dietz 2019; Neethu 2018; Lucchi 2018; Murray 2014).

[431]Palombi (2009a, p. 318).

[432]U.S. Supreme Court (2012, p. 21).

## Discovery and Invention

The patentability of synthetic biology inventions might be objected on the same grounds raised for other biotechnology items and, in particular, for genes. Such arguments would emphasise that those items are ever-existing and that their workings and functions are determined by nature. Equally, it could be maintained that they operate according to natural laws which are still not fully understood and do not respond to human will. Therefore, even the creation of artificial genes codifying new traits should be considered a discovery instead of an invention. In synthetic biology, the strength of these arguments is likely to be inversely proportional to the difference between nature and SynBio inventions. The more a product or process is different from nature, the weaker the argument becomes. For example, xenobiology inventions could hardly be considered discoveries if they involve the production of amino acids that are not naturally available.

From a European perspective, if the idea prevails that discoveries constitute the "*mere becoming aware of a pre-existing reality*",[433] then most synthetic biology inventions would be patent eligible. Only with regards to synthetic life inventions could this point reasonably be raised, as it could be argued that they mimic life and its workings.[434] Yet, this argument would likely be unsuccessful. The fact that synthetic life inventions closely mirror nature might raise obstacles for their patentability under the inventiveness and novelty requirements, but it would be complex to argue that such mirroring equals the awareness of a pre-existing reality. On the other hand, definitions of the notion of invention linking this concept to the creation of something that did not previously exist might be more problematic, especially in the field of synthetic life and de-extinction. As inventions in these fields substantially mirror natural ones, they can be said to concern the creation of something that existed in nature beforehand.

Those patentability issues stem from the fact that biotechnological inventions, including SynBio ones, are often inspired or modelled around natural ones.[435] Although synthetic biology is widening the gap between natural and man-made

---

[433]Sterckx and Cockbain (2012, p. 115).

[434]Scholars have held that synthetic genes modelled on existing ones should not be seen as inventions, given that the gene already exists as such and cannot be considered "man-made" (Thornton 2002, p. 82).

[435]Interestingly, over time, patents have been obtained for elements of the periodic table (i.e. Americium and Polonium), which, though generally produced by man, can also be obtained via natural processes. Similarly, patents have been issued for synthetic versions of naturally occurring products (e.g., insulin and adrenalin) and covered not merely the processes to manufacture them, but also the products themselves (Koepsell 2014, p. 46). Authors went as far as to say that, if bionic human arms can be patented, it would be wrong to establish that artificial genes could not because they are identical to natural ones (de Miguel Beriain 2010). Other scholars noted instead that: "*An object that is morphologically identical to another may yet be considered to be different in a legally significant way allowing the patent on one but not the other. All objects thus must have a structural quality, and a genetic quality, and if both are the result of some human intention, and meet the other criteria of patent... then they may be patentable*" (Koepsell 2014, p. 47).

biological systems, the complexity of nature does not allow quick and easy solutions. Therefore, much of what is done in this field is still strongly correlated to nature and it is impossible to say if it will ever be possible to design completely *ex novo* systems. Taking as an example the pharmaceutical field, studies have shown that between 1989 and 1995 over 60% of approved drugs and candidates in the field of cancer and infective diseases were of natural origin. This is possible because "*all natural sources... represent a virtually untapped reservoir of novel drugs awaiting imaginative and progressive organizations*".[436] This shows that even in fields with advanced and well-developed research efforts, nature is still a cornerstone of progress.

Because of this, authors were critical of the idea that humans can effectively design nature, so that it becomes "*the result of an originating design that shaped the final creation*".[437] In support of this theory, it was noted that biotechnological "inventors" are unable to describe their work in written form in a way that can be used to recreate it by third parties.[438] Also, it was noted that, although humans have consciously modified naturally occurring systems for ages, this does not turn them into inventions.[439] Other scholars proposed a test based on the principles of evolutionary biology to determine whether organisms could be protected by patents. This two-prong test required an applicant to show that the organism would have had little chance of developing naturally and that natural selection would have worked against it, had it not been for human intervention and technology.[440] After applying this test to synthetic biology, it was stated that:

> Given the scientific approach of creating organisms from scratch using artificial raw materials or creating minimal genomes using synthetic materials, one can infer that it would be relatively easy for synthetically generated organisms to pass this test. Unless the goal is to mimic an existing organism in every detail, which is not in the current scope of synthetic biology research, this test can be passed. Any attempts to mimic an existing genome in its entirety using synthetic components... would fail this test.[441]

Furthermore, it was maintained that synthetic organisms that survive only under specific lab conditions would pass this test. On the other hand, organisms that can survive in the wild would pass this test if they contain a mechanism that stops them from functioning after a certain time or after a task has been completed.[442]

---

[436]Cragg et al. (1997, p. 58).

[437]Sherman (2015, p. 1224). In the past, authors argued that it is an illusion to consider many creations in the biotechnological field as inventions, especially when transgenic animals having thousands of genes are considered invented if only one or few genes are inserted. Yet, such arguments could be countered by noting that these alterations confer to those animals characteristics and functions not present in nature (Westerlund 2002, p. 31).

[438]Sherman (2015, p. 1224).

[439]Dutfield (2010, p. 540).

[440]Iwasaka (2000, p. 1519).

[441]Bhutkar (2005, p. 23). The problematic nature of inventions using synthetic components to mimic an existing genome may be relevant in the field of synthetic life inventions.

[442]Bhutkar (2005, p. 24).

## Degree of Change

A second set of issues that emerged with regards to the patentability of synthetic biology inventions concerns a question of degree; namely, how much difference is necessary in comparison to the natural product in order for synthetic biology inventions to overcome the discovery and the product of nature exclusions?

The European approach to this issue appears straightforward. The Directive expressly allows the patentability of items that have previously occurred in nature or that are identical to natural ones, thus effectively rendering this question irrelevant. Outside the realm of application of the Directive, the situation might be more complex. Still, items that have been altered, even minimally, compared to their natural version would most likely overcome this exclusion.[443]

The situation is more complex in the US. Following *Chakrabarty* and *Roslin*, the markedly different requirement regained a pivotal role. Hence, projects like JCVI 1.0 and the addition of a new pair of bases to a longer nucleotide sequence could be considered very similar to their natural equivalent. But when is this similarity overcome? Does it depend on the type of modification and on whether new functions and properties were introduced[444]? Is it a problem that synthetic biology *"aims at creating simplified artificial cells and life forms functionally indistinguishable from naturally occurring ones"*[445]? How are sequences, which are completely different from natural ones but produce natural proteins, to be considered[446]? Would minor variations that have an effect on the function of the item pass this test?

Those issues are complex and aggravated by the subjective nature of those criteria and enquiries. Still, it could be argued that markedly different characteristics leading to patentability could be identified if there are significant modifications of either quantitative or qualitative nature to the original product. Quantitatively significant alterations would represent a change compared to natural products and thus avoid the related exclusions. However, if the focus is placed on the informational content of the sequence, pure chemical changes having no impact on the functioning of the system could be problematic. On the other hand, changes that are limited in quantity, but significantly affect the workings of the biological system would overcome this exclusion.[447]

---

[443]Minimal changes would overcome the exclusion in Article 52(2)(a) EPC, but may not be sufficient under Article 54 and 56 EPC.

[444]Lai (2015, p. 1066).

[445]Dutfield (2010, p. 540).

[446]Lai (2015, p. 1065).

[447]This conclusion would be confirmed by the ruling in *Ex parte Latimer*. If the chemical structure of the product is unchanged and if it is impossible to distinguish it from its natural counterpart, the invention would not constitute a new article of manufacture and would not be patentable. Opposing arguments based on the presence of non-functional watermarks might not be sufficient to overcome this exclusion. Likewise, the utility and value of the synthetic product do not change this assessment. By contrast, significant changes to the synthetic version would counter the product of nature exclusion. Lastly, as for *Cochrane*, this decision would not preclude the patentability of methods

Functional differences have also acquired relevance pursuant to the recent USPTO Guidelines on patent eligibility, since those characteristics can be used to pass the markedly different characteristics threshold, so long as they are not innate or occurring independently from the actions of the inventor. Since a change in the function of a biological system is generally connected to a structural change, it would be difficult to imagine one without the other. Because of this, it was suggested that structural differences could be seen as necessary, but not sufficient for patentability and that the functionality would also need to be markedly different.[448] If this approach were adopted, it might have an impact on the patentability of synthetic biology inventions. SynBio projects where both the structure and the functionality are similar, if not identical, to the natural version might be unpatentable.[449] This conclusion seems to be confirmed by the Court's findings in the *Roslin* case, which concerned a similar situation. There the patentability of an item was denied, since it was identical to something already existing in nature and in spite of the fact that it had been the result of a clearly technical process. The difference between the natural and the artificial item laid in the manufacturing process rather than in the characteristics of the product. In *Roslin*, the function, chemical structure and informational content of the product mirrored something occurring in nature and were thus not considered markedly different. Hence, a final product, which has a fundamentally identical natural version and whose only difference compared to it lies in its production method, might not be patentable pursuant to current US law. Interestingly, those concerns seem restricted to US law, as in Europe the Directive expressly states that the presence of a natural equivalent does not affect the patentability requirement.[450]

On the other hand, for inventions with a different structure, it would need to be clarified whether the production of the same substance in increased quantities could be seen as a markedly different functionality. The USPTO believed so, as in its Guidelines it maintained that marked differences could be found when cells express substances they normally would not or when they operate with increased efficiency.

In light of this, simplified cells and life forms might also pass this enquiry. The idea behind this line of work is to maximise the productive output, while minimising the needed input. Therefore, even though they would be used in the production of the same substance, their ability to manufacture it in different conditions or with less input might constitute a new functionality that warrants their patentability. Differences in the function and properties of the final item should also warrant

---

used to manufacture the synthetic product. Yet, even the novelty of the manufacturing process would not be enough to overcome the product of nature exclusion for the item.

[448]Mouta and Tridico (2015).

[449]This conclusion would be supported by earlier US case law (*Kuehmsted*). There, the fact that the two items performed differently, despite their identical chemical formula, was considered sufficient for patentability.

[450]On the other hand, product-by-process claims are permitted only if the product as such fulfils the requirements for patentability. The EPO clarified that a product is not rendered novel merely by the fact that it is produced by means of a new process (EPO 2018, § F.IV.4.12).

patentability. Conversely, the fact that the end product is indistinguishable from the natural one has raised a number of objections, in particular under US law.[451] Scholars have argued that:

> Irrespective of the platform used to produce these proteins or other types of biological materials, those materials are not new – they belong to the natural world.[452]

### Occurrence in Nature

Scholars wondered "*when can facets of nature be ignored for deciding on patentability? Random? No purpose? Rare?*".[453] The answer to this question would depend on the jurisdiction involved. In Europe, the existence, whether current or prior, of an item in nature does not impede *per se* its eligibility if it has been isolated or technically produced.[454]

By contrast, in the US, the approach on this matter has evolved over time. In *General Electric v. De Forest Radio*, the Court focused on whether the substance could occur as such in nature instead of considering whether it indeed occurred.[455] The Federal Circuit in *SmithKline* clarified that human-made compounds are not unpatentable when the compound could occur via a natural physical process.[456] This decision was criticised by authors, who noted that it is problematic when a patent claims also naturally occurring subject matter.[457] More recently, in *Ex parte Allen*, the focus was placed on whether a product is naturally occurring or man-made, instead of whether it could exist in nature.

---

[451] Those issues might emerge in Europe only when the Directive is not applicable. Otherwise, the issue would not arise, since the Directive expressly confirms the eligibility of items that previously occurred in nature.

[452] Palombi (2009b, p. 375).

[453] Dreyfuss (2013, p. 20).

[454] For instance, at the Luxembourg Conference for the drafting of the EPC, a proposal was put forward concerning natural and synthetic products. There: "*After agreeing that patent protection should be given to 'synthetically-produced substances by means of a new process', the Working Party of the Conference nonetheless decided that precise rules should not be laid down as to the protection to be given where the same substance was subsequently discovered to exist naturally*" (Pila 2005a, pp. 765–766). This resembles a theory presented in the past for the patentability of chemical substances, where it was argued that patent protection should be denied on the basis of the theoretical possibility that a naturally occurring equivalent of the synthetic version could be found. On this, see Sect. 4.1.3.1.

[455] U.S. Circuit Court of Appeals - Third Circuit (1928).

[456] Conley (2009, pp. 122–124) and U.S. Court of Appeals for the Federal Circuit (2005).

[457] Conley (2009, pp. 125–126).

### Artificiality

The synthetic origin of SynBio inventions has played a pivotal role in the eligibility debate surrounding them.[458] Still, its actual impact is unclear. Some argued that, although artificiality is relevant from this perspective, it is not the sole criterion needed to determine subject matter patentability.[459] In particular, there is the fear that artificiality will be used to warrant patentability, much like "isolation" has in the past. Specifically, it was argued that:

> Much like the 'isolation' of a naturally occurring biological material which has been used as the point of distinction between a product of nature and a product of man, so it would seem the synthesis of a naturally occurring product in an artificial environment or by some technical means is being used by the inventors as a point of distinction between a product of nature and a product of man.[460]

By contrast, other commentators pointed out that:

> Artificially derived biotechnological inventions, on the other hand, are the proper subject matter deserving of patent protection by virtue of that artificiality. Since this is the type of innovation that the patent monopoly is intended to serve.[461]

Article 3 of the Directive seem to confirm this point, as technically produced items benefit from a favourable eligibility treatment under that norm. This would prove advantageous for SynBio inventions, since they are manufactured via those techniques. However, applicants could not rely on this norm for inventions that are outside the scope of the Directive.

From a US perspective, authors argued that the synthetic origin of an item would not suffice for patentability, as Courts would focus also on the type and relevance of the structural differences between the natural and artificial molecules and their functionality.[462] Indeed, in the US, the synthetic nature of the sequences might have lost its talismanic status, at least in certain circumstances. Before *Myriad*, it had been argued that all synthetic organisms would be considered patentable subject matter, given the wide scope of *Chakrabarty*.[463] A report issued in 2004 by the US Department of Energy claimed that:

---

[458]The notion of synthetic refers to the method employed to produce the sequence, rather than to the novelty of the sequence itself. Synthetic strands could perfectly mirror natural ones and still be considered synthetic. The use of the word "synthetic" does not mean that the sequence will necessarily be a newly devised one with no natural equivalent. Therefore, just as in *Genentech* for recombinant technologies, the notion of synthetic might not describe the product itself, but rather its history (Palombi 2009a, p. 236; UK Court of Appeal (CA) 1989a, § 262).

[459]Palombi (2009a, p. 322).

[460]Palombi (2009b, p. 379). The possibility to patent the synthetic version of a biological item has led to question whether: "*The distinction between 'natural' and 'synthetic' [is] sufficient to safeguard competitive development of broad scientific prospects*" (Dreyfuss 2013, p. 22).

[461]Crowne-Mohammed (2010, p. 469).

[462]Holman (2013, p. 291).

[463]Saunders (2008, p. 84).

> The ability to synthetize genes will potentially lower barriers to patenting DNA sequences and the products that result by facilitating demonstration of utility and raising questions about the barrier of the 'product of nature' doctrine.[464]

The situation further evolved with *Myriad*, where products of nature were held unpatentable, whereas synthetic items were patent protected. This has been read as a possible confirmation of the patentability of synthetic biology inventions. Yet, this conclusion is not settled. The same Court noted that sequences that mirror natural ones are unpatentable, even if they are synthetic.[465] Because of this, commentators argued that the mere creation of synthetic equivalents of natural items does not make them patentable pursuant to the US Supreme Court case law.[466]

Lastly, the artificial nature of synthetic cells and their impact on the final product was critically examined by *Palombi*, who argued that

> The essence of invention is not mimicry but something 'new' and not in the sense of being novel but in the sense of being an 'invention' (BASF; Brogdex; Chakrabarty; Genentech) merely to replicate nature's protein products using natural genetic material, even if that material is synthetic. . ., is not to make something 'new' because the protein will be the same as the natural protein (Kirin Amgen). Therefore while these synthetic cells are artificial and they incorporate into their genetic structure DNA which has been genetically modified from its natural equivalent, do the products which they produce 'display markedly different characteristics not found in nature'? Unless they do, these products are not patentable subject matter. Furthermore, to the extent that the processes employ synthetic genes which are substantially identical to natural genes, the processes will also not be patentable because they too are not new, or are obvious.[467]

### Other Criteria

Alongside synthesis, the level of human labour needed to obtain a SynBio inventions was considered from an eligibility perspective. In Europe, such criterion is not mentioned by norms and decisions; therefore, it is unlikely to be of any relevance. Conversely, in US Courts, the relevance of human labour for patentability is not univocal. On the one hand, in *Myriad*, the Court took into consideration the labour

---

[464]ETC Group (2007a, p. 40). On the other hand, by citing *BASF*, *Palombi* held that the mere fact that a substance was artificial did not render it a new patentable composition of matter. Hence, he concluded that: "*It is the ability of these techniques to harvest the products of the protein-coding regions that are pharmacologically, nutritionally or industrially useful that will be relevant to their patentability*" (Palombi 2009b, p. 375). Similarly, other commentators argued that the source of a product is irrelevant. If a product is equivalent to a natural one, the product of nature exception operates, independently of its method of production (Conley and Makowski 2004, p. 34).

[465]Committee on Science, Technology, and Law of the National Academies of Science, Engineering and Medicine (2013, p. 3). *Myriad* might have eliminated patent protection for some synthetic biology inventions; for example, short cDNA sequences that are indistinguishable from natural DNA. Indeed, claims for cDNA sequences in synthetic biology inventions have already been rejected by the USPTO on this ground (USPTO final rejection Manmohan 2012).

[466]Sarnoff (2012, p. 106).

[467]Palombi (2009a, p. 322).

done by the lab technician to manufacture the product. On the other, in *Roslin*, this aspect did not impact the finding that the organism was not patentable. Hence, labour would not suffice to rule in favour of patentability under US law.

Lastly, from an EU perspective, SynBio inventions will need to display technical character in order to be considered patentable. Whether synthetic biology inventions will pass this test is to be decided on a case-by-case basis. Still, the lower threshold set by the "any hardware" approach should facilitate the passing of this requirement. Nonetheless, patent applications of speculative nature or providing a proof of principle rather than a practical use could be particularly impacted by this requirement.

### Protectable Items

Authors listed a series of synthetic biology inventions and applications that could be patent protected. In particular, *Dreyfuss* considered that new molecules with specific end uses could be patentable. This would include wholly synthetic molecules, minimal genomes, molecules based on natural structures and interactions, molecules that substitute natural structures, derived from living cells and modifying existing molecules. Furthermore, it was argued that synthetic biology building blocks, research tools and assembly techniques would also be patent protected. This conclusion was extended also to design and evaluation techniques realised via computer simulations and models. Lastly, for databases, it was argued that they would not be protected via patents, but rather via *sui generis* European rights.[468]

*McLennan* also examined which items would be patent protected. In her opinion, key processes underpinning synthetic biology would be patentable on the basis of recent US case law. Synthetic DNA constructs and standard biological parts are also likely to meet the US eligibility test, even though their patentability might prove more complex.[469] With regards to methods for creating a synthetic genome by assembling smaller pieces of DNA or standard parts, she argued that, while those techniques are based on natural processes:

[468]Dreyfuss (2013, p. 37) and Committee on Science, Technology, and Law of the National Academies of Science, Engineering and Medicine (2013, p. 4).

[469]McLennan (2017, p. 61). For standard biological parts, she noted that such inventions would often be based on sequences that have their informational origin in natural organisms. Taking this as well as *Myriad* into consideration, she argued that: "*standard biological parts are unlikely to be patentable where they just reproduce a naturally occurring sequence. However, parts are often engineered to have minor changes from the natural sequence... This may be enough to make them patent eligible... In addition, composite parts or biological devices are being produced which combine several basic parts together ... While the genetic sequences of the basic parts might be derived from naturally occurring sequences, the combination of the sequences could result in a stretch of DNA which is not identical to any naturally occurring genetic sequence. This takes synthetic biology even further from naturally occurring phenomena, and it is unlikely that such a device would be regarded as a product of nature*" (McLennan 2017, p. 72).

> These methods go beyond the processes of nature to transform existing matter into something which does not exist in nature... These methods also go beyond adding a routine, conventional, well-understood activity to a process of nature because they transform the existing matter into a new product with different utility... Synthetic biology processes are also likely to meet the machine or transformation test, because they would have transformative steps... It is important to note that if there were claims for synthetic biology processes which did involve natural phenomena, the cases show that these could not become patent eligible merely by adding the use of a computer to perform or implement the process.[470]

Lastly, SynBio breeding methods are also likely to be patent protected, since the rulings in *Tomato* and *Broccoli* (*I* and *II*) were seen as favourable to the synthetic biology industry.[471]

### 4.2.3 Specific Assessment

After having examined some general patentability issues that could emerge in synthetic biology, this section will focus on specific questions that might arise in relation to the main SynBio research areas. Although it is impossible to offer a generally valid conclusion due to the peculiarities of each case, this section will address the norms applicable, the role of the exclusions from patentability and review some patent applications filed and granted in those research areas.[472]

---

[470]McLennan (2017, pp. 67–68). Regarding synthetic DNA, the same author argued that: "*it is likely that synthetic, computer-designed DNA will meet the product of nature exclusion test more easily than earlier biotechnologies, as the genetic sequence will be written by humans. The resulting products will not occur in nature, but will be clearly produced by human intervention*" (McLennan 2017, p. 71).

[471]Minssen and Nordberg (2015). SynBio breeding methods would pass both the quantitative and the qualitative requirements set by the case law. The quantitative requirement prescribes the presence of at least one non-biological feature. If synthetic biology were involved in the method, then this requirement would be met, as SynBio technique represent technical steps requiring human intervention. From a qualitative perspective, such step is required to have a genuine technical effect, that is, to introduce or modify by itself a trait in the genome. While this analysis would be conducted on a case-by-case basis, it is not difficult to imagine that synthetic biology techniques could genuinely and decisively impact the course of a natural breeding method and its result. If this were the case, the breeding method would fall outside the realm of essentially biological ones and could thus be patented. This conclusion is supported by the fact that, for these purposes, it is irrelevant whether the step is new or known or whether it is fundamental or trivial. Hence, the presence of a SynBio step capable of altering the genome would suffice for patentability.

[472]For an overview of the patentability issues raised before the USPTO by a number of applications filed by the Venter Institute (McLennan 2018, pp. 251–299; McLennan 2017).

#### 4.2.3.1 Minimal Genome

Research into the minimal genome is focused on creating the ultimate efficient production organism that operates on the absolute minimum number of genes needed for metabolism and replication in a given environment. The idea behind this approach is that even the smallest known genomes contain a certain degree of redundancy and could be streamlined.[473] Such streamlined genome could be used as a chassis platform, which should result in a "*biological system that is easier to understand and manipulate, as well as one that has more surplus energy available to devote to making or doing something useful*".[474]

Applicable Norms

The patentability of works concerning the minimal genome will be regulated by the EPC alongside the Directive. Inventions in this area undoubtedly concern genetic material, which is capable of reproduction in a biological system.

Subject Matter Patentability

The issue of subject matter patentability was rarely raised with regards to minimal genomes.

Such minimal genome is artificial, as it is synthetically manufactured and does not exist in nature in that form. Yet, the genetic information contained in it has an equivalent in nature, albeit in an extended form. Such similarity raises the question of whether minimal genomes should be considered discoveries or products of nature and thus excluded from patentability.

From a European perspective, minimal genomes will likely be considered inventions.[475] While their information content is based on a natural existing biological system and it is a mere detraction from it, the sequence as such has no natural equivalent. Those inventions represent the result of human work and ingenuity and are obtained via a technical process, which would vouch for their patentability pursuant to Article 3 of the Directive. Lastly, this type of invention could fulfil the technical requirement, as it could be used as a chassis for the production of a variety of substances. Still, whether the specific application solves a technical problem or constitutes a scientific proof of principle will need to be examined on a case-by-case basis. The patentability of minimal genomes is confirmed by the conclusions reached

[473]Schwille (2015, p. 688).

[474]Dutfield (2012, p. 6).

[475]The patentability of some of those inventions could also be established pursuant to the norms protecting microorganism, which will be examined in the following sections. On this, see Sect. 4.2.3.5.

by a number of German academies and foundations, which held that "*it should be possible to protect minimal cells... by copyright (preferably by patents) to provide an economic incentive for investments in the new techniques*".[476]

Conversely, in the US, the patentability of minimal genomes is more controversial. The USPTO has noted that "*minimal genomes, with particular genes omitted, are difficult to claim*".[477] Similarly, *Palombi* considered those inventions to be nothing more than the genome of a natural organism from which specific genes have been deleted. Because of this, he argued, "*in every other respect, the residual genome is identical to the corresponding natural genes found in the naturally occurring genome*".[478] As a result, arguments were presented to show that:

> While this gene set itself may not exist in nature in the same identical way and, to that extend, is artificial as well as having been constructed synthetically, the fact remains that it contains the same, or substantially the same, genetic information as found in the natural genes that correspond to those included in the gene set.[479]

Specifically, although the genome was recognised as an artificial construct, it was held that its characteristics are the same as those displayed in nature, thus leading to problems with the standards set by *American Fruit Growers* and *Chakrabarty*. The function of this genome is identical to the one of its naturally existing equivalent, namely to provide instructions to express proteins. Those proteins are not new, novel or inventive, given that they function as prescribed by nature. Therefore, functions performed "*by an organism with a genome that has been assembled in a laboratory and made using synthetic biology parts do[es] not enhance the patentability of the synthetic organism per se*".[480] Lastly, it was maintained that the lack of enablement would also be relevant under a subject matter patentability perspective, given that in this case it cannot be said that the genome is useful.[481]

Opposing arguments have been presented also by the ETC Group, which held that, "*in practice, the organism is 'being patented for what it is not'*".[482] In criticising the patent applications of Venter on this line of work, the ETC Group noted that:

> The patent application claims any synthetically-constructed organism that lacks at least 55 of 101 genes that they've determined are non-essential. All synthetic biologists developing functionalized microbes are going to have to pay close attention to the claim on a 'non-essential' set of genes. If someone creates another bug that lacks some of the same genes that, Synthia lacks, will the Venter Institute sue them for infringing its patent[483]?

---

[476]Deutsche Akademie der Technikwissenschaften et al. (2009, p. 60).

[477]LeGuyader (2009, p. 15).

[478]Palombi (2009b, p. 378).

[479]Palombi (2009b, p. 385). He cited *American Fruit Growers* in support of his position. On this, see Sect. 4.1.6.4.

[480]Palombi (2009b, p. 385).

[481]Palombi (2009b, p. 386).

[482]World Science (2007).

[483]ETC Group (2007b, p. 2).

Moreover, objections have been raised against the patentability of these inventions due to the lack of understanding connected to their functioning. Even the most recent project of Venter and his team showed the limits of human understanding of biological systems, as they could not assign a function to a large portion of the genes.[484] Because of this, it could be argued that no patents should be granted for something that the patentee has not understood. Equally, *Dutfield* argued against the patentability of whole life forms by stating that the way biochemical elements operate and interact with each other is still poorly understood. Given that this complexity is currently beyond human reach, their patentability might be premature.[485] A somewhat similar line of thought was expressed in the *Morse* case. There, the Justices opposed the breadth of the claims and noted that Morse claimed a process that he had neither described nor invented. Yet, this argument could be rebuffed by looking at the prior practice. In the past, biological substances were covered by patents, even though they were little known or understood. For instance, adrenaline and insulin were patented at the beginning of the twentieth century despite the difficulties that applicants encountered in describing them.[486]

The points made above are confirmed by the analysis of the patent applications filed for the minimal bacterial genome by Venter before both the EPO and the USPTO. In both cases, no patent was granted.[487] In Europe, the application was withdrawn and, in the US, it was abandoned. The application related to:

> A minimal set of protein-coding genes which provides the information required for replication of a free-living organism in a rich bacterial culture medium, wherein… the gene set does not comprise the 101 genes listed.

The claims covered sets of protein-coding genes. Those sets covered the lack of at least 40 of the 101 protein-coding genes that were held non-essential. Interestingly, the applicant sought to claim "*a set of any of claims 1-15, which are recorded on a computer readable medium*". Claims over a genome sequence recorded on a computer-readable medium (i.e. for the digital version of a DNA sequence) are not an isolated case.[488] Companies are increasingly claiming the information content of gene sequences or even entire genome sequences instead of focusing their claims only on tangible molecules. Patent applications have been filed for the sequences of the *Haemophilus influenzae* genome, for the genome of *Mycoplasma genitalium* and

---

[484]Commentators have argued that: "*Anyone who claims that she or he understands how a cell works is either ignorant or ridiculously arrogant… they don't*" (Twilley 2016).

[485]Dutfield (2010, p. 539).

[486]Dutfield (2010, p. 532). The ruling in the *Nissan motor* case before the EPO is in line with this conclusion. There it was noted that the inability of the patent applicant to explain why an effect occurs does not imply the presence of a discovery in lieu of an invention. On this, see Sect. 4.1.3.2.

[487]Glass et al. (2007a, 2008b).

[488]van den Belt (2009, p. 1327).

for that of *Mycoplasma laboratorium*.[489] In all cases, the claims extended to the computer-readable medium on which the sequence was stored. The problematic nature of this type of claims was raised in the opinion of the International Searching Authority. There, the Examiner noted that the subject matter of one of the claims concerned "*genetic information recorded on a computer readable medium. This represents mere representation of information characterized solely by the content of the information, which is excluded from examination*".[490] Commentators noted that the USPTO was not ready to accept such claims. Specifically, it was observed that in the first two examples, where a patent was issued, the reference to the computer-readable medium was limited to the description part rather than to the claims.[491]

The opposition against this type of claims is understandable, as they are seen as a radical departure from current patent law and patent bargain. For example, the mere downloading and printing of the patent documents presented to the Patent Office would constitute an infringement.[492] This argument cannot be dismissed, especially as genetic material is increasingly considered from an informational rather than a chemical perspective. In the Australian prong of the *Myriad* case, the Federal Court raised this point and considered that:

> Much emphasis was placed by the applicants upon the informational character of DNA as a storehouse of genetic information. But the disputed claims are not to information as such. They could never be infringed by someone who merely reproduced a DNA sequence in written or digitised form.[493]

Before the EPO, the application was subjected to a number of objections by the Examination Division. Specifically, objections were put forward concerning its lack

---

[489]A patent on the first has been filed by Human Genome Sciences (HGS), the second refers to an application of the Institute of Genome Research and the third is the one presented by Venter and his team (van den Belt 2009, p. 1327).

[490]Glass et al. (2008b, Document of 16 April 2008, attachment p. 3).

[491]In the case at hand, the reference to the "computer readable medium" is found both in the claims and in the description of the patent application (Glass et al. 2007a). The unsuitability of patent protection for genetic information stored on a computer readable medium was highlighted by *Eisenberg* long before the emergence of synthetic biology. In a 2001 symposium, she noted that: "*I do not think it is an appropriate subject matter for a patent claim... computer-readable medium is the only practical way to perceive and analyse large volumes of DNA sequence information... The claim to this sequence in... computer-readable medium, in effect, gives the patent holder the right to restrict the ability of others to use the information in a computer-readable medium and thus precludes others from perceiving and analysing this sequence information itself. This is distinguishable, I think, fundamentally different, from old-fashioned DNA sequence patents that claimed only molecules. You can read those old-fashioned patents, learn what the DNA sequence is, type or scan it into your computer, search for similarities to other sequences in databases, all without infringing. By contrast, this... would allow... to capture much, if not all, of the informational value of the discovery, not just its tangible value*" ("Molecules vs. information: Should patents protect both?," 2002, p. 200).

[492]van den Belt (2009, p. 1327).

[493]Federal Court of Australia (2013, § 76). However, since the Australian High Court embraced the information approach, it cannot be excluded that this issue will become increasingly relevant, especially when claims are directed to the informational content of the biological material.

of novelty, inventiveness and disclosure.[494] No eligibility concerns under Article 52 EPC were expressed. The application was then abandoned and no divisional application was filed.[495]

The application was challenged also before the USPTO before being abandoned.[496] Both in 2013 and 2015, the Examiner held that some of the claims were directed to non-statutory subject matter. In 2013, following *Myriad*, it was argued that the claims related to naturally occurring nucleic acids, whether isolated or not, thus falling within the laws of nature exception.[497]

The grounds presented in the 2015 rejection were similar. The claims were directed at products of nature, that is, naturally occurring genes. Since the claims did not recite elements that amounted to significantly more than the exception, they were not considered patent eligible. Following the 2014 Interim Guidance on Subject Matter Eligibility, the Examiner noted that:

> According to the Guidance, 'in the case of a nature-based combination, the closest counterpart may be the individual nature-based components that form the combination, i.e., the characteristics of the claimed nature-based combination are compared to the characteristics of the components in their natural state'. Genes in combination are not different in structure or function from the structure or function of naturally occurring genes just because they are combined, which leads to the conclusion that the instant claims are not directed to patent eligible subject matter.[498]

The applicant objected to this view by arguing that the minimal genome is not naturally occurring and is not available as such in nature. Furthermore, it was argued that one of the claims covered a minimal genome comprising a heterologous gene. This position was rejected by the Examiner. Specifically, by applying the 2014 Guidelines on Eligibility, the USPTO established that the claims did not:

> Recite additional elements that amount to significantly more than the judicial exception. The nature-based products claimed are not markedly different from genes found in nature... A claim drawn to one gene or one nucleic acid or a claim drawn to a combination of genes or a combination of nucleic acids is not significantly different than a gene or genes found in nature. Likewise, a subset of the set of all genes in an organism or a combination of genes found in nature even if not found together in nature does not constitute being significantly different.[499]

This is an interesting conclusion, which signals the change of position in the US on the patentability of biotechnologies. Indeed, it could have been argued that, since this minimal combination of genes is not found as such in nature, this would represent a significant difference compared to the natural product. Instead, the

---

[494]Novelty questions were raised also by the presence of US patent 6989265 for a "bacteria with reduced genome" that had already been granted to the Wisconsin Alumni Research Foundation (Palombi 2009b, p. 280).

[495]Fernandez y Branas (2014, p. 191).

[496]EPO (n.d.-a).

[497]USPTO (n.d.) USPTO non-final rejection 15 August 2013.

[498]USPTO (n.d.) USPTO non-final rejection 7 January 2015.

[499]EPO (n.d.-b) USPTO non-final rejection of January 2015.

USPTO focused on the identity of the genes with their natural equivalents, rather than on the presence of that specific set of genes in nature. As the chemical and information characteristics of the genes remained unchanged, the minimal bacterial genome was held unpatentable.[500] By focusing only on single genes, the functions and characteristics of the set are disregarded. Yet, would this approach be acceptable even if the set of genes represented a perfect chassis for production and offered significant improvements (e.g. energy and input savings)? Would it be possible to argue that the specific set of genes acquires an added functionality that is not available in nature and that should be patent protected? It will be interesting to see if this approach to the product of nature exception will be further employed.

The approach recently adopted by the USPTO might seem surprising if one considers that, in 2003, patent applications on bacteria with a reduced genome were filed before the EPO and the USPTO and successively granted.[501] The abstract of the patent concerned:

> A bacterium having a genome that is genetically engineered to be at least 2 to 14% smaller than the genome of its native parent strain. A bacterium with a smaller genome can produce a commercial product more efficiently. The present invention also provides methods for deleting genes and other DNA sequences from a bacterial genome.

The claims concerned both the bacterium, with different degrees of reduction, and a method for manufacturing it.

Despite some differences between the application and the granted patent, the EPO patent covers both product and method claims. The product claims relate to *E. coli* bacteria having a chromosome that is genetically engineered to be 5% to 40% smaller than the chromosome of its native parent strain.[502] On the other hand, the US patents cover product claims only.[503] Specifically, they cover "*an isolated E. coli bacterium having a genome... lacking DNA selected from the group...*" and "*an E. coli bacterium having a genome that is genetically engineered to be at least 5% smaller than the genome of its native parent strain*". In light of the recent US case law and of the rejection of the Venter patent on the minimal bacterial genome, it is doubtful whether those patents would be granted again under the current conditions.

---

[500]A second set of issues is raised by the possibility of having hybrid organisms. In its patent on a minimal bacterial genome, Venter and his team suggested the substitution of the deleted genes with those of other organisms. This organism is yet to be realised. This application was criticised by commentators because of its speculative nature (Palombi 2009b, p. 378). US case law has attached relevance to a new name given to a microorganism. Although this approach was contested, Courts saw it as a sign of the novelty of the item. This point however did not help the patenting of *Synthia*. On this, see Sect. 2.2.2.1.

[501]Blattner et al. (2003, 2004, 2006, 2010). It is also worth noting that recently the EPO and the USPTO granted patents for reduced genome bacteria with improved genetic stability and did not raise subject matter patentability concerns during their analysis (Blattner et al. 2016, 2017). Other applications for minimal bacterial genomes before the USPTO were instead recently abandoned. Also in this case, no specific § 101 concerns were raised (Glass et al. 2015).

[502]Blattner et al. (2010).

[503]Blattner et al. (2006, 2011).

The comparison between those granted patents and the approach recently adopted by the EPO and the USPTO towards the minimal bacterial genome shows the increasing differences in their positions on subject matter eligibility. While no patent was granted by the EPO, the objections of the Examiner related to the novelty, inventiveness and disclosure of the application. No patentability issues were raised. The EPO would likely bypass this enquiry by focusing on Article 3 of the Directive. This would place it outside the scope of the discovery exclusion set by Article 52 EPC. By contrast, the USPTO raised eligibility objections, which were strong enough to hinder the issuing of the patent. Compared to the patents granted just a few years before, it seems that the USPTO has radically changed its approach to subject matter patentability in this sector. Conversely, the EPO has not visibly amended its position. The consequences of such scenario are conflicting decisions on the patentability of minimal genome inventions.[504]

#### 4.2.3.2 Synthetic Life

The work on synthetic life has been most divisive and controversial. Doubts have emerged over whether this approach is feasible and will ever really lead to synthetic life.[505] Similarly, the projects completed until now have been criticised for an incorrect and premature use of the notion of "synthetic life".

##### Applicable Norms

This field will be governed by both the provisions of the EPC and the Directive, given that those inventions would fit the definition of biological material.

Even the fact that a biological system is of wholly synthetic origin should not hinder the applicability of the Directive. Nevertheless, this circumstance might be relevant from a social, moral and policy perspective.

[504]The EPO has generally focused its patent analysis away from the patentability criteria and towards other requirements. Thus, it could be argued that different approaches would not really impact the granting of patents in this field, as an invention that is considered patentable by the EPO might still not fulfil the other stricter patent requirements. While it is true that passing the patentability hurdle does not assure that a patent will be granted, it is however indicative of the approach of a jurisdiction towards certain sectors.

[505]One would need to consider what would happen if truly synthetic life would turn out to be impossible. In this case, doctrines relating to natural substances and their patentability might make a comeback (Schneider 2014, pp. 166–167).

## Subject Matter Patentability

While methods relating to this line of work are likely to be patentable, the patentability of synthetic life inventions requires closer analysis.[506] The assessment will first focus on the patentability of JCVI syn1.0 and then evaluate hypothetical synthetic life inventions, whose genome is not substantially identical to a naturally occurring one. Lastly, the patentability of the Yeast 2.0 project will be examined.

Taking into consideration the patentability of JCVI syn1.0, a number of questions arise. Could this organism be patented or would it fall within the discovery and product of nature exclusion? Would the substantial identity between its genome and that of a naturally occurring organism affect its patentability? Would the technical origin of this invention be of relevance?

As previously mentioned, the genome of JCVI syn1.0 is substantially identical to that of the naturally occurring *Mycoplasma mycoides* (except for the insertion of watermarks and the removal of the infectious section). The role of those watermarks is not functional, but was needed to differentiate the original from the synthesised version of the genome, as otherwise the two would be indistinguishable. Since the two genomes would be impossible to separate and given that the sequence was dictated by nature, it could be argued that JCVI syn1.0 represents a discovery instead of a patentable invention.

Furthermore, if inventions were seen as creations that did not previously exist in nature, the prior existence of the genome sequence would be problematic. Equally, as noted by other authors, genes that have not been substantially modified or have been artificially reproduced lay in a grey area between discovery and invention, as it is complex to establish whether the cause of their creation can be ascribed to nature or to humans.[507] Indeed, it was noted that, while artificiality is important, it is not the only criterion to be used in determining whether there is an invention. The core of this concept is something new and that can be considered an invention, rather than a mere imitation of something already existing.[508] Many natural phenomena are replicated by mankind (e.g. fire), but this does not change the fact that those elements are natural. Hence, the question is whether manufacturing via human intention is relevant for patentability purposes. The answer is yes, at least as far as the patentability of the process is concerned and as long as it regards intentional processes used to manufacture those elements. However, those considerations cannot extend to the products of such processes. The idea behind this is that, while they owe their origin to different processes, they are still the same physical phenomenon.[509] Lastly, it might be suggested that this achievement is a product of skill rather than a product of invention.

---

[506]The patentability of JCVI syn1.0 may also be established pursuant to the norms protecting microorganism, which will be examined in the following sections.

[507]Resnik (2002, p. 149).

[508]Nose (2010, p. 20).

[509]Koepsell (2014, pp. 47–48).

Yet, those arguments could be opposed by noting that JCVI syn1.0 exists in such form only because of human intervention. The synthesis of the genome, the inclusion of watermarks and the insertion in an emptied recipient cell are tasks that could only be performed by humans with the aid of technical means. Therefore, the near-identity of the genome would not exclude the patentability of JCVI syn1.0. Equally, although it could be argued that there was no human design in JCVI syn1.0, this might be reductive, given that this test requires human design and not an inventive human design. While the inventiveness of the design will be assessed under that specific requirement, the fact alone that JCVI syn1.0 exists *per se* in that form because of human design may overcome the exclusion of Article 52 EPC. This conclusion would be supported by theories that consider an item an invention if it can be obtained as such solely via the intervention of human intelligence and if one essential step to obtain the product requires human effort.[510]

Further objections could be raised under the technical character requirement. If the synthetic and the naturally existing organism were identical, which type of technical problem would be solved by the former? In *Relaxin*, the EPO rejected arguments holding that genes perform the same function in nature and when isolated, despite the fact that artificial Relaxin provided no utility above and beyond its natural counterpart.[511] Because of this, it could be argued that JCVI syn1.0 does not perform the same function of *Mycoplasma mycoides* and therefore it has its own utility. The patentee would also need to show the actual solving of a technical problem and not merely allege this possibility. As stated in the *Multimeric Receptors* case, the industrial exploitation of the invention is needed to prove that it is more than a mere research finding.[512] Hence, if this is not apparent, objections to its patentability could be moved by arguing that this was a mere research endeavour instead of an invention.

Yet, the most decisive argument under European law is offered by Article 3 of the Directive, which establishes the patentability of biological materials manufactured via technical processes, even if they previously occurred in nature. The substantial identity of the sequences could thus be overcome by focusing on their manufacturing process.[513]

In light of the above and despite the objections that could be raised against the patentability of JCVI syn1.0, this would most likely be considered an invention pursuant to European law, given that its existence in such form is dependent on human intervention and a technical process.

By contrast, the patentability of this invention pursuant to US law is more controversial. Following the *Myriad* and *Roslin* decisions, Courts have been focusing on the presence of marked differences between the natural product and the

---

[510]Westerlund (2002, p. 51).

[511]Sterckx and Cockbain (2012, p. 242) and Technical Board of Appeal of the EPO (TBA) (2002c).

[512]Lai (2015, pp. 1050–1051).

[513]Still, the identity of the sequence might play a role in the assessment of other patentability requirements pursuant to the EPC and the Directive.

invention. Differences in the chemical composition and informational content were taken into consideration, albeit in a somewhat inconsistent manner. In the case of JCVI syn1.0, it could be argued that the lab technician created something new, as it was the case for cDNA in *Myriad*. However, in that case, the Justices noted that cDNA was distinct from the DNA from which it was derived and hence it was not a product of nature. By contrast, here it is complex to establish whether JCVI syn1.0 constitutes something different from the naturally existing *Mycoplasma mycoides*, as the chemical and informational content of the two organisms are substantially identical. Indeed, even though the genome was placed into the empty cell of another bacterium, the synthetic genome took over the running of the bacterium, which ended up behaving like a *Mycoplasma mycoides* one. Similarly, the watermarks do not introduce a new functionality in the bacterium; instead, they are needed to distinguish the two genomes, which shows just how similar the two sequences really are. Therefore, if naturally occurring is interpreted to mean that a product has the same effect it always had, does not gain additional utility via human manipulation and performs in its own natural way, then JCVI syn1.0 could be falling within the product of nature exclusion. However, a different conclusion could be reached if the non-infectious nature of JCVI syn1.0 were considered a markedly different feature.

Even the synthetic nature of the invention might not, *per se*, solve questions on the patentability of JCVI syn1.0. Following *Myriad*, synthetic strands should be patentable. Yet, short strands that are identical to natural ones would not. This might raise issues on the patentability of synthetically made organisms that are very similar to natural ones, such as JCVI syn1.0. After *Roslin*, those concerns have grown stronger, since the Court required the presence of a marked difference even for a product that had clearly been manufactured via a technical process. There, the Court concentrated on the genetic identity of the natural and synthetic item to determine the unpatentability of the latter. The markedly different characteristics prescribed by *Chakrabarty* were lacking there and this could not be overcome by the labour that the patentee had invested in creating the clone. Similarly, pursuant to *Cochrane*, a synthetic substance that is identical in properties and composition to a natural one would not be patentable merely because it has been produced artificially for the first time. Although the process would be patentable, the end product could not be considered a new composition of matter. If this reasoning were extended to JCVI syn1.0, the production method could be covered by a patent, while JCVI syn1.0 itself would not. Additionally, *American Fruit Growers* would require a new or distinctive form, quality or property. Lastly, even the attribution of a new name to the bacterium (JCVI syn1.0) will not, *per se*, favour its patentability.

Further objections have been raised by noting that this invention, despite being synthetic, is still aiming at effects that already exist in nature.[514] Recent Court decisions held that the presence of a function similar to that found in nature would lead to patentability only if the item has a unique DNA structure, different from

[514]Nose (2010, p. 20).

anything found in nature.[515] If this reasoning were applied to JCVI syn1.0, its patentability would be doubtful, given that neither its structure nor its function is truly different from anything found in nature.

By contrast, the USPTO Guidelines could hint to another direction. Marked differences that would allow patentability were found in substances containing one structural modification that did not occur naturally and even if it did not affect functionality. If the rationale behind it is to prevent the pre-emption of the natural item, this could play in favour of the patentability of near-identical synthetic biology inventions.

Fewer patentability questions would arise for synthetic life inventions presenting changes compared to naturally occurring ones. Synthetic strands that have been significantly altered would be considered patentable under European law and, most likely, also under US law. Applying the reasoning of *Myriad*, one could argue that the lab technical produced something new and that the sequence does not mirror a naturally existing one. Hence, those synthetic life inventions would be patentable.

This reasoning could be applied also to the Yeast 2.0 project. The changes made to the genome (i.e. unnecessary sequences removed and new landing sites) as well as the inclusion of a diversity generator would ensure that the resulting item is different enough from naturally existing yeast to warrant its patentability under US law. On the other hand, its patentability under European law could have been established even in the absence of those changes, as explained above.

A number of patents have been filed to cover this line of research.

In 2006 and 2007, applications for "synthetic genomes" were filed before the EPO and the USPTO.[516] The abstract of the applications referred to:

> Methods... for constructing a synthetic genome, comprising generating and assembling nucleic acid cassettes comprising portions of the genome, wherein at least one of the nucleic acid cassettes is constructed from nucleic acid components that have been chemically synthesized, or from copies of the chemically synthesized nucleic acid components. In one embodiment, the entire synthetic genome is constructed from nucleic acid components that have been chemically synthesized, or from copies of the chemically synthesized nucleic acid components. Rational methods may be used to design the synthetic genome... Synthetic genomes or synthetic cells may be used for a variety of purposes, including the generation of synthetic fuels, such as hydrogen or ethanol.[517]

The patentees claimed not only the method for constructing the synthetic genome, but also the synthetic genome itself as well as a synthetic cell comprising a synthetic genome. The application encompassed bacterial genomes, minimal genomes, eukaryotic cell organelles, genomes that are substantially identical to naturally occurring ones, non-naturally existing genomes and synthetic genomes. The method claims would cover any method of constructing the synthetic genome using nucleic acid cassettes. Similarly, the product-by-process claims would extend the patent

[515] U.S. Court of Appeals for the Federal Circuit (2014a).

[516] Venter and Smith (2008) and Venter et al. (2007).

[517] Venter and Smith (2008).

monopoly to things that have been constructed synthetically using such method. This would include cellular components that have been chemically synthesised or made from copies of chemically synthesised components. Authors criticised this application by holding that it covers a wide range of genome construction and by pointing out its speculative nature.[518]

Other concerns emerged because the genome that was used for it (i.e. *Mycoplasma genitalium*) was not invented by Venter and his team. In particular, it was noted that the genome is substantially identical to the one existing in nature, apart from having been tweaked in order not to infect humans. Indeed, the claims refer to the full-scale version of the *Mycoplasma genitalium* genome. Furthermore, the empty cell in which the genome is inserted is a naturally occurring one, whose genome has been removed. Therefore, all "*the essential components are either naturally occurring or substantially identical thereto*".[519] Such identity has been seen by authors as problematic under *American Fruit Growers* and *Chakrabarty*. In particular, it was stated that:

> The fact that it is incapable of infecting humans is not the kind of difference which would distinguish it in 'form, quality, or property' (Brogdex), nor the kind of functionality that would be 'markedly different characteristics to any found in nature' (Chakrabarty). Consequently the patent specification merely provides particulars of a method that produces something which frankly is not patentable subject matter. That it is synthetic does not make it an invention (BASF).[520]

This position has been shared by other scholars, who noted that:

> Tweaking the genome so as to render it harmless to humans may not be enough to make the difference between the synthetic and natural versions distinguishable either in 'form, quality or property' nor the kind of functionality that would be 'markedly different characteristics to any found in nature'.[521]

By contrast, the recent USPTO subject matter eligibility examples on life sciences would point to the contrary. In assessing a sample claim for an inactivated virus, the Office compared the virus to its natural counterpart and observed that the:

> Like the Chakrabarty bacterium... the inactivated virus has markedly different characteristics, due to the non-natural chemical modification of its nucleic acids and the resultant

---

[518]Palombi (2009a, pp. 319–320) and Palombi (2009b, p. 379). In particular, *Palombi* argued that this type of claim is unacceptable pursuant to *BASF*, as it would cover an entire technological field. If the application were to be accepted as such, it would lead to wide monopolies over both methods and products. Furthermore, the application covered any method of constructing a synthetic genome using nucleic acid cassettes. According to the OECD, this represents an example of a fundamental patent, covering the basic starting point of a technology and which could impede follow-on research in that field (OECD 2014, p. 96). Similar concerns were expressed by *Sulston*, who held that the very broad scope of the application filed by Venter could grant him a monopoly over a wide number of techniques (van den Belt 2014, p. 32). This application has been considered shaky in terms of both the enablement and the novelty requirements ("Patenting the parts," 2007).

[519]Palombi (2009b, p. 387).

[520]Palombi (2009a, p. 321).

[521]Agovic (2014, p. 111).

> change in the virus's ability to... cause disease. While in other fact patterns, a functional change may be enough by itself to confer eligibility, for this claim the functional change is a result of the structural change and thus is inseparable from it. Because the inactivated virus has markedly different characteristics from what exists in nature, it is not a 'product of nature' exception. Thus, the claim is... eligible subject matter.[522]

A third set of issues raised by this patent concerns its products. The synthetic bacterium could constitute a platform for the production of energy sources. However, such synthetically produced hydrogen and ethanol are no different from natural ones or from those produced via other methods. This, coupled with the lack of information on how to manufacture such energy sources, was seen as problematic.[523]

The patent was granted in Europe in 2013 in amended form.[524] While claims for the construction of a synthetic double-stranded genome were admitted, the patent no longer contained product claims for the synthetic genome and the synthetic cell comprising the synthetic genome.

By contrast, no US patent was awarded so far.[525] Over time, some of the claims have been challenged by the USPTO for lack of description, enablement, obviousness, with only limited and indirect references to subject matter patentability.[526] It would thus seem that the limited changes made to the genome were enough to render it markedly different.

Similar concerns were raised for another patent for the "installation of genomes or partial genomes into cells or cell-like systems", which was filed both before the USPTO and the EPO in 2006.[527] The abstract of the application refers to:

> A method... for introducing a genome into a cell or cell-like system. The introduced genome may occur in nature, be manmade with or without automation, or may be a hybrid of naturally occurring and manmade materials... The genome is installed into a naturally occurring cell or into a manmade cell-like system. A cell-like system or synthetic cell resulting from the practice of the provided method may be designed and used to yield gene-expression products, such as desired proteins... the provided method makes possible a broader field of experimentation and bioengineering than has been available using prior art methods.

This application refers to insulin peptides that could be collected from the synthetic cell. This was seen as a sign that the patentee intended to manufacture insulin via this system. However, such prospect was challenged by authors who opposed the idea of a new round of patents over human insulin. Despite the different

---

[522]USPTO (2016, p. 4).

[523]Palombi (2009a, p. 321).

[524]Venter et al. (2013).

[525]For the current status of the proceedings (EPO n.d.-c).

[526]EPO (n.d.-d). In a non-final rejection from 9 December 2008, the USPTO argued that some method claims were directed to non-statutory subject matter, as they claimed a process of nature (EPO n.d.-d, Document of 9 December 2009, pp. 6–7). This rejection was withdrawn following an amendment of the claims.

[527]Glass et al. (2007b, 2008a).

production methods, it was maintained that such insulin was structurally and functionally identical to the naturally occurring product.

Moreover, this application has been criticised for its formulation and type of claims. In addition to product claims, it contains a number of method claims. Even though method claims are often less problematic from a patentable subject matter perspective, the ones presented by Venter and his team might differ. The reason for this is that they seem to be product claims rather than method ones. This conclusion is supported by the fact that the application does not detail the steps of the method.[528]

The patent was granted in 2014 at the EPO and in 2016 at the USPTO, albeit with a number of changes.[529] The original patent application concerned methods for making synthetic cells as well as the synthetic cells themselves, whereas the granted patents contain no product claims, but rather include claims to methods and production processes.[530]

#### 4.2.3.3 Xenobiology

Even before the advent of synthetic biology, companies had been preparing modified forms of DNA bases to be used for genetic testing. Over time, those skills improved and nowadays scientists have created semi-synthetic organisms having an expanded genetic alphabet that contains nucleotides not found in nature.[531] The development of new proteins and amino acids has also been at the heart of this discipline. For example, companies are commercialising synthetic proteins that are different from naturally occurring ones both chemically and functionally.[532]

Applicable Norms

The provisions of the EPC will be applicable to xenobiology inventions. Conversely, the application of the Directive is not straightforward.

Xenobiology inventions pertain to three main research areas: the creation of a new backbone for nucleic acids (e.g. TNA), the expansion of the genetic alphabet via the introduction of new nucleic bases and non-canonical amino acids.

The application of the Directive to nucleic acids having a different backbone is uncertain. Those constructs could be equated to a necklace, where the beads remain

---

[528]Palombi (2009b, p. 386). The application was criticised by the EPO Examiner for lacking the essential technical features required for an invention (Glass et al. 2008a; Document of 9 July 2012, p. 2).

[529]Glass et al. (2014, 2016).

[530]A patent for methods of genome installation in a recipient host cell was applied for in 2008 both before the EPO and the USPTO. However, such patent was never granted. The rejection of the USPTO was not based on eligibility grounds (Glass et al. 2010).

[531]Bera (2015, p. 195).

[532]Holman (2015, p. 428).

the same, but whose cord is replaced with one made of another material. The idea behind this line of research is to create systems that do not interfere with other biological systems, thus implementing the engineering concept of orthogonality. Such systems would not be under the control of other biological systems and would be invisible to them.

In general, it could be argued that nucleic acids with a different backbone are biological material, since they contain genetic information in the form of the usual DNA bases (i.e. A, C, G, T). Yet, this view could be opposed by arguing that those systems are invisible to other biological systems. Indeed, it might be problematic to include within the realm of biological systems elements that cannot be detected and read by other biological entities and whose purpose is to be invisible to them. Furthermore, problems could emerge regarding the ability of those inventions to reproduce or to be reproduced. Based on the current scientific status, those products are not capable of reproduction.[533] While this obstacle may be overcome in the next years, for now it might not be possible to consider those inventions as biological material pursuant to Article 2 of the Directive.

Similar concerns arise for inventions containing new nucleotide bases, as they might not fit within the definition of biological material contained in the Directive. On the one hand, such bases could be seen as biological material, given they carry genetic information, albeit non-canonical one. Since the norm does not differentiate between canonical and non-canonical genetic information, it could be assumed that both types of materials would fall within the realm of this definition. However, it could also be argued that the addition of bases that have no natural equivalent goes beyond the intended meaning of genetic information and biological material. More vocal opposition on this point could be raised once the new bases are not limited to one within the entire DNA sequence, but become more numerous. This could raise the question of whether those systems are biological and lead to reconsider how such concept adapts to this field.

As for the second prong of the definition of biological material (i.e. material reproducible in a biological system), this assessment will need to be carried out on a case-by-case basis. Recent studies have shown that those bases can replicate, albeit at a limited rate and for a limited time. Yet, this is not the case for all novel bases, as they are generally not capable of reproduction. For this reason, the latter could fall outside the scope of the definition of biological material and hence of the Directive.

Similar issues may exist for non-natural amino acids. On the one hand, they could fall within the scope of the Directive due to their use of genetic material. On the other hand, they might become so different from natural ones that they may no longer fit the definition of biological material. If this were the case, the Directive would not

[533]Specifically, they are not capable of reproduction in vivo via polymerase. The term polymerase refers to an enzyme that synthesises in vivo long chains of nucleic acids in order to assemble DNA and RNA on the basis of a template. This process, in all its variations, constitutes the general way in which nucleic acids are assembled in nature.

govern those inventions. This conclusion would be reinforced in case they could not reproduce in a biological system.

In light of the above, the application of the Directive to xenobiology inventions would be justified in the majority of cases, with the sole exception of inventions not capable of reproducing or being reproduced. The norms do not distinguish between canonical and non-canonical biological materials; hence, the latter could be included within the definition of biological material.

The application of the Directive to xenobiology inventions would have a number of advantages. First, it would allow uniformity within the European patent panorama. Second, it would avoid complications intrinsically related to tracing a line between xenobiology inventions that can be considered biological materials and those that cannot. Arguments to the contrary usually brought against this solution (e.g. inability to be detected by other biological systems and growing gap with natural biological systems) would not on their own justify the exclusion of xenobiology from the scope of application of the Directive on the basis of the formulation of the norms and the current case law.

However, for all those products, it must be noted that questions over the applicability of the Directive to them might intertwine with considerations on whether they can be considered biological from a policy, social and moral perspective. While such policy questions have not emerged for now, it cannot be excluded that they will play a role in the shaping of this discipline.

#### Subject Matter Patentability

Xenobiology inventions embody the kind of change that synthetic biology hopes to bring about in biological systems; a change that goes beyond what is available in nature and that is based on human inventiveness and ingenuity. Substantial man-made alterations would overcome both the discovery and the product of nature exclusion, thus hinting to the patentability of such inventions. For this reason, scholars have argued that:

> There is little doubt that objections to the patenting of isolated DNA sequences, according to which they are mere discoveries or products of nature, will not hold if claimed sequences included bases other than A, C, G and T, or if the sequences are made of other nucleic acids than DNA.[534]

Yet, opposing arguments will need to be assessed before confirming the patentability of such inventions.

Inventions with a new backbone would likely be patentable both in Europe and in the US. From a European perspective, those creations would fall outside the discovery exclusion and thus be considered inventions, as long as they display technical character. Additionally, if the Directive were applicable, their patentability could be

[534]Dutfield (2012, p. 201).

established under Article 3 of the Directive, since they are obtained via a technical process.

Those inventions would be considered patentable also pursuant to US law, as they cannot be considered a product of nature.

Nonetheless, it could be argued that xenobiology products could already be available in the universe, albeit unknown to man.[535] This argument channels the opinion that man cannot overcome the laws of nature and, consequently, all his creations are either already existing or could possibly exist in nature. Similar positions have been expressed at the turn of the last century regarding the patentability of chemical substances.[536] Still, those arguments are not solid. All human inventions are based on and utilise laws of nature. Their use cannot be avoided and this is recognised by the patent system, which only excludes the patentability of discoveries and scientific theories *per se*. Hence, this argument should not hinder the patentability of xenobiology inventions under both the EPC and the most recent US case law.[537]

Inventions containing new nucleotide bases and unnatural amino acids would likely be patentable under European law. Those inventions are not available as such in nature and therefore would fall outside the realm of the discovery exclusion set by Article 52 EPC and be patentable. This conclusion would be reached, even if the Directive were to apply, given that xenobiology inventions are obtained via technical processes.

Likewise, those xenobiology inventions would most likely be patentable under US law. Following *Myriad* and *Roslin*, the patentability test focuses on whether an invention is markedly different from naturally occurring items. Considering that xenobiology aims at the development of new amino acids and nucleotide bases, this should be enough to pass the markedly different test.

However, questions may arise if only one or few new bases were inserted in the product. There, it could be argued that the difference between the natural and the resulting product is not sufficient to warrant patentability. Indeed, the level of change currently introduced by xenobiology is minimal. Works generally see the addition of one or few unnatural bases to the natural existing strand. Analogously, changes to amino acids are also limited in number. Despite the apparent similarities, the impact of those limited changes is different for new bases and amino acids and this might affect their patentability. Minor changes in the amino acid sequence could have a

[535]It needed not be forgotten that humans have a limited knowledge of nature's inner workings, as confirmed by the fact that only about 1% of bacteria are culturable and could be examined so far.

[536]On this, see Sect. 4.1.3.1.

[537]From a European law perspective, this concern would not arise if the Directive were applicable. Indeed, the Directive expressly establishes the patentability of items, even if they previously occurred in nature. Furthermore, xenobiology inventions have not been detected in nature so far. Their existence is thus only attributable to human ingenuity and abilities. Lastly, even if natural xenobiology items were to exist, for this exclusion to operate, they would need to have the same structure and function both in nature and in their synthetic version. Hence, this line of argument would not be sufficient to reject their patentability.

drastic impact on the function of the resulting protein. Hence, changes that impact the functioning of the biological system should be considered as patentable, even if the chemical alteration was minimal. By contrast, the addition of unnatural DNA bases might not have an impact on the functioning of the system. If an increased number of bases could be added, questions on their impact and relevance might decrease.

In spite of this, arguments linking the patentability of inventions containing new bases with their number seem unfounded. The concept behind the addition of one or thousands of new bases is no different. While a high number of additions substantially changes the biological system, the addition of even one new base still represents a change compared to the naturally occurring product. Therefore, those creations would escape the discovery and product of nature exclusion even if just one base had been added. This conclusion seems consistent with prior US case law, if one considers that the mere removal of introns from a DNA sequence was considered a marked difference by the *Myriad* Court. This assessment would be confirmed by the Guideline examples published by the USPTO, where the substitution of one base was considered sufficient for patentability, even if it was silent.

Nevertheless, other scholars have been more sceptical of the patentability of xenobiology inventions. *Palombi* reported that in 2005 an application for a modified human growth hormone had been filed. The modification included one or more unnatural amino acids. Although the applicant held that the unnatural amino acids enhanced the production of the polypeptide, he argued that those remained substantially the same and performed essentially the same function. The question would then be whether, pursuant to *American Fruit Growers*, the resulting protein could be considered different in *"form, quality or property"*.[538]

Despite arguments to the contrary, both the EPO and the USPTO have granted patents in this area. Amongst those is the 2005 application cited above by *Palombi.*[539] The resulting European patent covered a polypeptide linked to an unnatural amino acid.[540] The same company, Ambrx, obtained patent protection in the US for a method of making polypeptides of less than 100 amino acids in length comprising two or more unnatural amino acids.[541] Interestingly, the application was

---

[538]Palombi (2009a, pp. 323–324). He noted that in the International Preliminary Report on Patentability of May 2007, objections were moved towards this patent as the term "non-naturally encoded amino acid" was considered unclear, as it could mean either an artificial amino acid or refer to the substitution of a naturally occurring one. The novelty of this invention was also challenged. Authors considered if such modifications would affect the product (i.e. the protein) or if they would merely improve its production. One could argue that if the proteins expressed are identical or substantially identical to natural ones, despite having been expressed by non-naturally occurring genes, they could not be considered new, given that they would be a synthetic replica of something existing in nature. They would thus fail the *Chakrabarty* test and could be excluded from subject matter patentability. Similarly, they could incur into novelty and obviousness objections (Palombi 2009b, pp. 391–393).

[539]Cho et al. (2012a).

[540]Related inventions were patented in the US as well (Cho et al. 2012b).

[541]Cho et al. (2009).

not limited to the method, but comprised also the polypeptides themselves. Another patent granted in the US claimed processes relating to non-standard nucleotides, whose bases are joined by hydrogen bonding patterns that are different from standard ones. This way, the number of nucleotides would be extended beyond the four found in standard DNA and RNA.[542]

Lastly, a withdrawn European patent application filed by Venter and his team claimed a method for expressing a protein containing one or more non-standard amino acids as well as a method to engineer microorganisms to survive on non-standard amino acids.[543] The US version of this patent was also abandoned. In this case, the USPTO did not cite eligibility amongst the possible grounds for rejection.[544] By contrast, the approach adopted by the EPO is interesting from an Article 52 EPC perspective. In assessing the clarity of the claims under Article 84 EPC, the EPO Examiner held that:

> The terminology 'synthetic genome' is a product-by-process definition which cannot be considered to impact any technical features on a 'genome' per se. Applicant indeed uses a natural genome... in all of his examples and admits that one does not need to make the genome synthetically... Applicant keeps 'going on' about the synthetic genome aspect, yet this is of no relevance at all. The ED can not and will not acknowledge that this terminology has any technical meaning regarding a product. One could 'design' the E.coli K-12 genome again. It would then have been designed but be identical to that in nature. In the sense of a product both would be identical. Changing the same codon in one or the other would also lead to identical products. Argument that by designing one could achieve different features is only hypothetically interesting until these actual features are reflected in the claims. Applicant argues that a synthetic genome is easily recognized... This is however clearly incorrect since the cited prior art documents fulfil these features yet do not comprise synthetic genomes according to applicants interpretations.[545]

#### 4.2.3.4 Protocells

Research in the field of protocells aims at constructing cell systems from the bottom-up employing exclusively non-living elements as raw materials. If successful, this work would "*result in the very transition that supposedly occurred for the first time*

[542]Benner and Yang (2013). An earlier application, filed in 2002 and granted before the EPO and the USPTO, concerned the in vivo incorporation of unnatural amino acids (Schultz et al. 2006, 2015; Anderson et al. 2004; Schultz et al. 2003). This is connected to the prior wave of work in this field that started before synthetic biology emerged as a discipline and that later evolved into it. The abstract of the patent concerned methods and compositions for in vivo incorporation of unnatural amino acids as well as compositions including proteins with unnatural amino acids. The initial claims covered, amongst others, both products (e.g. cells and proteins comprising the unnatural amino acids) and methods for producing them. The final version of the claims was however much narrower.

[543]Krishnakumar et al. (2010a).

[544]Krishnakumar et al. (2010b) and EPO (n.d.-e).

[545]Krishnakumar et al. (2010a, Document of 5 February 2013, pp. 2–3).

≈ *3.5 billion years ago: the transition from non-living to living*".[546] Indeed, it is postulated that protocells must have existed at some point in time during the evolution of the universe. For this reason, authors argued that "*this enterprise is not a genuinely biological one, because biology is only what results from it*".[547]

Before addressing the patentability issues connected to protocells, it is worth to briefly consider the concept of cell itself. Cells are defined through the presence of three features: a container, a metabolism and an information content. Specifically, containers are chemical boundary systems that enclose the cell (e.g. a membrane). The notion of metabolism concerns the self-reproducing chemical motor of the cell, whereas the information content refers to the chemical information stored in the cell via DNA and RNA.[548] The day protocells will acquire these three characteristics, they will become regular cells and their patentability will be regulated accordingly.

### Applicable Norms

Protocells represent a futuristic research area in the field of synthetic biology, as the development of a wholly functional protocell is still far to come. Yet, commentators have already noted that this field might not be governed by the provisions of the Directive and the related EPC Rules.[549]

Currently, protocells are a container in which genetic information is found. This would satisfy the first prong of the definition of biological material. However, protocells are not self-sustaining systems and they are not presently capable of reproduction. For this reason, they would fail the second prong of the definition found in Article 2 of the Directive.

Nonetheless, studies have shown cases in which protocells were randomly able to divide. Scientist kept on adding lipids to the cell to the point that the container stretched and broke into two parts. In those cases, it would be theoretically possible to argue that those protocells are capable of reproduction. This way, they would fall within the realm of application of the Directive. Nevertheless, this conclusion does not seem defendable. Even though the norms do not explicitly distinguish between purposeful and random reproduction, the latter should not be equated with the former. Patent law protects technical teachings that can be repeated by others skilled in the art. However, in this case, the addition of lipids to a membrane does not guarantee when, where and how the cell would divide. Every time that the protocell divides, the resulting cells could take a different shape and content.

Furthermore, it would be problematic to argue that protocells reproduce within a biological system. Indeed, in the current technological status, protocells are non-living entities and lay in a grey area between chemistry and biology. In fact,

---

[546]Schwille (2015, p. 688).

[547]Schwille (2015, p. 689).

[548]Schwille (2015, p. 688).

[549]Rutz (2010a, p. 13).

even the regulatory framework applicable to them is borrowed from the chemical sector instead of the biological one.

As long as the situation remains as such (i.e. protocells remain non-living systems incapable of purposeful reproduction), they would not fall within the realm of the Directive. Hence, their patentability would be judged according to the EPC, as it is the case for chemical inventions. However, should they become capable of reproduction within a biological system, the norms of the Directive would apply.

This outcome is valid for two reasons. First, the concept of biological material traces back to the notion of life and living processes. Therefore, even if an extensive interpretation of this concept were considered, this definition would not cover materials that are currently not alive. Second, the inability of protocells to reproduce clearly fails the second prong of the definition of biological material. As a result, an extension of the applicability of the Directive to protocells would currently be unwarranted.

### Subject Matter Patentability

Protocells are likely to be patentable subject matter pursuant to European law. This conclusion was foreseen also by a group of German academies and foundations, which stated that: "*it should be possible to protect… protocells by copyright (preferably by patents)*".[550]

Those inventions fall outside the patentability exclusion set in Article 52 EPC for discoveries as such, as their existence is the sole result of human intervention. Nevertheless, it could be argued that protocells have already existed at some point in time and therefore they are not inventions. Yet, this argument is not solid. The existence of protocells in a distant past is only a scientific postulation. Scientists could only guess the characteristics and conditions that lead to their development, but do not possess specific information about them. It would be unreasonable to consider them discoveries or natural products, when scientists are creating them without blueprints and without knowing whether they were really available, and if so in what form, in the past.

The patentability of protocells might also be objected under the technical requirement. The solving of a technical problem via technical means might be difficult to prove at the early stages of development of this field, as the first creations are likely to constitute a proof of principle of the feasibility of this approach. However, with time, things may change. If protocells can be used to advance the practical goals of synthetic biology, they would likely meet the technical requirement test.

Lastly, protocells would be patentable if the Directive were to apply. If protocells were considered biological material, their production via a technical process would render them patentable pursuant to Article 3 of the Directive. In this case, even the postulated prior existence of protocells would not affect this conclusion.

---

[550] Deutsche Akademie der Technikwissenschaften et al. (2009, p. 60).

Similar conclusions would be reached in the US, considering that protocells would likely fall outside the scope of the product of nature exclusion.

#### 4.2.3.5 Microorganisms

High hopes have been placed on the possibility that synthetic biology microorganisms will be able to operate in the future as true bio-factories for the production of a variety of substances. Much like in factories, the aim here is to ameliorate their efficiency and their use for human purposes. In general, those techniques could lead to increased production quantities, substances displaying increased purity levels as well as the manufacturing of new substances.

Companies are increasingly developing synthetic biology microorganisms aimed at producing high-value molecules, such as fine chemicals or chemicals that can be used as energy sources. Those results are achieved by inserting metabolic pathways in the organisms.[551]

Despite overlaps with other areas of synthetic biology research, patents for microbiological inventions need to be examined separately given the particular treatment granted to microorganisms in general.

Applicable Norms

The Directive as well as the EPC will be applicable to microbiological processes and products, as both the Directive and the EPC Rules specifically address the patentability of microbiological processes and materials. Given the extensive interpretation of the term microorganism promoted by the EPO, this coverage would extend to

[551]For example, the company Codexis patented technologies to engineer cells and organisms having enhanced characteristics as well as technologies for engineering synthetic metabolic pathways into living organisms (Holman 2015, pp. 431–432). Guild Biosciences is commercialising synthetic organisms that display new and useful functions. For examples, the company has engineered viruses and single cell organisms that operate as biosensors. Those could then be used in the environment to detect and neutralise harmful agents (Holman 2015, pp. 434–435). Several approaches have been employed to devise efficient bio-factories (e.g. minimal genomes, xenobiology). One of those consists in the reengineering of proteins with enhanced functionalities. Recently, the company Genencor has modified the amino acid sequence of a naturally occurring protein to manufacture enzymes with improved functionalities (e.g. improve catalytic activity and stability, altered temperature dependence). Even though the tools employed to alter those sequences are often connected to conventional biotechnology techniques, they fulfil the aspirations of synthetic biology. In one case, Genencor implemented one amino acid substitution in the protein sequence, which resulted in an increased catalytic efficiency (Holman 2015, p. 428). Catalytic efficiency can be defined as an increase in the reaction rate of a chemical reaction which allows it to become more efficient, thus making sure that more products are generated at a faster rate (LibreTexts.org 2015). Patents for methodologies used in engineering microbial genomes have been granted also to another company, Amyris. The latter has patented engineered cells capable of manufacturing high-value chemicals (e.g. isoprenoids) (Holman 2015, pp. 432–434).

unicellular organism with dimensions not visible to the naked eye and that can be manipulated and propagated in a laboratory.

### Subject Matter Patentability

Microbiological processes and organisms are generally patentable, given the provisions found in TRIPS, in the EPC and in the Directive. Therefore, microorganisms and microbiological processes obtained via synthetic biology techniques will likely be patent eligible both in Europe and in the US.

This conclusion is validated by the fact that patent protection is currently available for microorganisms developed and altered via biotechnology techniques. Drawing on similarities with the patentability of those engineered bacteria, authors maintained that:

> Future processes to produce or modify chemical compounds using 'synthetic' microorganisms as well as the synthetic microorganisms as such would in principle be regarded as patentable inventions.[552]

A confirmation of the patentable nature of microbiological products obtained via synthetic biology techniques derives from the clearly technical nature of the processes involved, which would fall within the definition of Article 3 of the Directive. This would be the case even if the items previously occurred in nature.

Similar conclusions could be drawn for SynBio techniques that extensively and substantially alter microbiological items and processes. In those circumstances, the level of human intervention and its impact on the molecular and functional features would point towards the presence of inventions instead of discoveries. The broad scope of the definition of microbiological products and processes as well as the declaration in the 1988 Proposal for the Biotech Directive that this term should be construed extensively so that "*future developments in biotechnology in respect of animate matter... can benefit from the principles of the Directive*" would lead to the extension of the norms on microbiological processes also to those of synthetic origin.[553]

Following this line of thought, commentators have argued that protocells and cells with a synthetic minimal genome might be considered microorganisms as well, although the Boards of Appeal would need to provide further interpretation on this point. If this were the case, Article 53(b) EPC and Rule 27(c) EPC would be applicable and would establish the patentability of the microbiological process and its products.[554]

Regarding the patentability of whole organisms, it was believed that the concerns expressed in the past would be relevant also for synthetic biology. It was maintained

[552]Fernandez y Branas (2014, p. 190).

[553]*Proposal for a Council Directive on the legal protection of biotechnological inventions (COM/88/496FINAL - SYN 159)* (1988) Explanatory Memorandum.

[554]Fernandez y Branas (2014, p. 190).

that product-by-process claims would be easier to justify. Equally, product claims regarding modified cellular mechanisms should be acceptable. Nonetheless, issues might emerge for simplified organisms. For those, it was argued that:

> One should continue to be cautious in accepting the patentability of these metabolically enhanced organisms. Indeed, for synthetic biology, the real proof of concept requires the complete construction of a life form in the laboratory: SynBio scientists haven't quite proven that a cell is a kind of biochemical machine, and religious biologists... hang on tightly to this uncertainty. Proof will come when the first discrete, self-maintaining, stable organic creature... is created from scratch in the lab.[555]

Other authors have also expressed doubts over the patentability of those organisms and their products. For instance, it has been maintained that:

> Even though synthetic microorganisms are different from naturally occurring microorganisms, it is the DNA encoding the valuable proteins that is substantially identical to the corresponding, naturally occurring DNA. This is important because the resulting protein otherwise will not have the same bioequivalence in the human body. Irrelevant of the platform used to manufacture these proteins or other types of biological materials, those materials are not new as they are part of the natural world.[556]

Yet, these arguments do not seem robust. The EPC and the Directive contain norms expressly establishing the patentability of microbiological products and processes. Given the clear formulation of these provisions, it would be complex to claim their unpatentability. Furthermore, the norms do not require the presence of a new function for patent eligibility.[557]

The situation might be different in the US. After *Roslin*, a final product that is fundamentally identical to a natural one and whose only difference compared to it lies in its production method might not be patentable, unless it displays markedly different characteristics.

A number of patents have been granted in this field. One of the best-known examples concerns the production of Artemisin. The manufacturing of a semi-synthetic version of Artemisin has been claimed in patent applications, some of which have been withdrawn.[558] Nonetheless, the USPTO issued a patent for Artemisinic epoxide and methods for producing it to Jay Keasling and his team at UC Berkeley.[559] This patent covers "*artemisinic epoxide, and methods of*

---

[555]Dutfield (2012, p. 201).

[556]Nose (2010, p. 17). On a similar note, questions have been raised with regard to the final product manufactured by those improved biological systems. The issue revolved around whether those modifications changed the final product or merely improved its production. In discussing proteins, it was held that this distinction is pivotal as: "*Even if these genes are substantially different from natural genes and are enhanced, if the proteins which these synthetic cells express are identical to, or are substantially identical to, natural proteins then the proteins themselves are not new. The idea that 'anything under the sun made by man' is patentable subject matter suggests that artificiality is the key to invention, but there is... much more to invention than that*" (Palombi 2009a, p. 322).

[557]Increased production and purity could be seen as new functionalities.

[558]Dietrich et al. (2007).

[559]Dietrich et al. (2012).

*synthetizing artemisinic epoxide in a genetically modified host cell*" as well as "*methods for producing artemisin*".[560] The patentability of such work was confirmed by authors, who noted that:

> There are no particular problems associated with the patentability of processes for the production of drugs using advanced techniques of synthetic biology, the production of artemisinin in yeast being a typical example.[561]

Other patent applications covering microorganisms capable of enhanced production have been filed both before the EPO and the USPTO.[562] One of the examined applications concerned "*yeast strains capable of increased prenyl phosphate production… enabling increased terpenoid molecule production*". The claims covered a method for producing isoprenoid compounds in yeast as well as the yeast host cell. Equally, another application related to "*engineered microalgae that exhibit enhanced lipid production during exponential growth. Such engineered microalgae are useful, for example, for the production of biofuels*".[563] In this case, the claims cover the engineered microalgae as well as a method to produce lipids from it.

Although these applications have been filed only recently, in light of the considerations expressed above, it is likely that both would be considered patent eligible by the EPO. As for the US, so far no § 101 issues were raised by the USPTO for either patent.[564] For this reason, their eligibility seems established also in this jurisdiction.

#### 4.2.3.6 De-extinction

In the past decades, efforts to regenerate now extinct organisms have gone from being a sci-fi idea to an actual research field. Although the feasibility and timing of those projects are still highly uncertain, the use of synthetic biology methodologies may render them possible.

In the mid 1990s, companies claimed to have revived thousands of prehistoric microorganisms that had been preserved in chunks of amber. This led to the granting of patents on the isolated viable culture of microorganisms obtained from resin.[565] Recently, George Church tried to modify the genes of an Asian elephant to render

---

[560]Dietrich et al. (2012). The genetically modified host cell could be a yeast or a bacterial cell. The production of Artemisin via synthetic biology techniques is taking advantage of different research areas. For instance, the use of DNA with a different backbone for its production is being investigated. The goal would be to obtain greater control over the microorganism and reduce any background noise that could influence the output.

[561]Fernandez y Branas (2014, p. 195).

[562]Allen and Dupont (2014), Jørgen Hansen (2013a, b) and Dietrich et al. (2008).

[563]Allen and Dupont (2014).

[564]EPO (n.d.-f, n.d.-g).

[565]Rohrbaugh (1997, p. 373).

them more mammoth-like.[566] In general, attempts to recreate extinct animals are far more likely to succeed for organisms that have been extinct for a short time.[567]

Given the current status of research, recreated animals could only be developed in the medium to long-term. Therefore, their patentability will be evaluated in abstract and on the basis of the current legal panorama in order to establish the patent issues that might be raised by them.

#### Applicable Norms

Although those types of inventions will not be available in the near future, it can be hypothesised that they would be governed by both the EPC and the Directive. Indeed, those inventions would concern biological materials that operate within biological systems and are capable of reproduction. The fact that such biological systems would correspond to previously existing creatures would not *per se* be relevant from this perspective.

#### Subject Matter Patentability

While methods used to bring extinct organisms back to life would most likely be patentable, the patentability of the organisms themselves is far more controversial.[568]

Moral issues aside,[569] the recreation of once extinct organisms raises the question of whether they should be seen as discoveries or products of nature. The concepts of discovery and invention have been interpreted in a variety of ways. Some definitions tie the notion of invention to the creation of something that did not previously exist. Such understanding would be problematic here, as de-extinction concerns the recreation of something that existed in nature beforehand.

Still, those creations would probably not fall within the discovery exclusion under European law if they employ cloning and genetic engineering techniques.[570] The techniques employed ensure that the recreated organism is not a perfect copy of the previously existing one, but represents, at best, an altered version of it. Hence,

---

[566] Bera (2015, p. 207).

[567] Carlin et al. (2014, p. 4).

[568] It has been argued that, while the methods employed in this field are patent eligible, the patentability of the products obtained via them is uncertain (Swedlow 2015, p. 186). The patentability of such products might be opposed also on novelty and obviousness grounds (Hagglund 2007, p. 385).

[569] On this, see Sect. 5.2.3.6.

[570] The eligibility of inventions obtained via breeding methods would be problematic, given the exclusion from patentability of essentially biological processes. On this, see Sect. 4.1.5.2. The application of the norms on the patentability of animal and vegetable varieties has not been examined in this context.

the resulting organism cannot be considered a discovery, but would rather be a genetically modified version of the organism manufactured via technical means. The use of a technical process for its creation would corroborate its patentability pursuant to the Directive. Importantly, this conclusion applies even if the biological material previously occurred in nature.

By contrast, issues could derive from the application of the technical requirement. While the passing of this test is to be established on a case-by-case basis, those recreated organisms would need to address a technical problem in order to be patentable, which might be difficult for those types of inventions. Indeed, they could be viewed as lacking practical application and be rather an instrument to extend human knowledge instead of human practical capabilities.

In America, a number of scholarly works have analysed the patentability of those inventions, which might blur the line between the product of nature exclusion and man-made patentable subject matter. Organisms that are the result of human ingenuity rather than a natural phenomenon would be considered patentable subject matter.[571] Pursuant to *Allen,* the level of human ingenuity needed is fairly low. Also, the fact that the recreated organism is governed by laws of nature would not impact this conclusion.[572]

Whether an invention can be considered a natural product or the work of human ingenuity will depend on the method employed to manufacture it. In theory, cloning might be seen as insufficient under *Chakrabarty*, as the organism does not present markedly different characteristics compared to those found in nature.[573] Furthermore, the recent *Roslin* case seems to open new scenarios, as cloned animals would generally not be considered patent eligible.[574] Yet, those arguments could be opposed by noting that the organism is not an exact copy of the extinct one and that in *Myriad* the mere removal of introns was considered sufficient for patentable subject matter.[575] Also, *Roslin* referred to organisms existing in nature and not to extinct ones.

By contrast, facsimiles produced via genetic engineering techniques are likely to be considered markedly different, since they are a man-made combination of different genomes that does not exist in nature. Conversely, species made via artificial selection may not be patentable, if they employ conventional breeding techniques.[576] The result would be different if they were to employ SynBio breeding techniques.

The arguments presented above could be influenced by whether Courts and patent offices will focus on the surface similarities between the species or whether they will

---

[571]Rohrbaugh (1997, p. 386).

[572]Rohrbaugh (1997, p. 386).

[573]Carlin et al. (2014, p. 51).

[574]The method employed to create the animal is covered by a patent granted to the Roslin Institute (Swedlow 2015, p. 192).

[575]Carlin et al. (2014, p. 52).

[576]Carlin et al. (2014, p. 51).

examine the genetic similarities and distinctions more closely. A focus on the intent to reproduce a once naturally occurring organism could be seen as problematic for the patentability of those inventions, although it could also be argued that such similarities never reach the level of a complete equivalence between the current and the extinct species.[577] Patentees should demonstrate that the recreated organisms have structural, functional and other characteristics that differentiate them from their extinct predecessors.[578] Minor differences might already be enough to prove that the two organisms are dissimilar. On the basis of *Myriad*, the use of synthetic DNA might also support the patentability of such organisms.[579]

The patentability of recreated organisms will also depend on whether nature is intended as all organisms that ever lived in it or whether it is restricted to organisms currently living in it. It could be argued that the prior extinction of those organisms places them automatically outside of the product of nature category.[580] Products of nature are living things found in the world, thus suggesting that a naturally occurring living organism is to be present in nature at the time of the invention.[581] As by definition extinct organisms no longer exist in nature, recreated organisms should not be seen as products of nature. Even the presence of their complete DNA in fossils should not affect this conclusion.[582] Firstly, the remains on which the organism is based upon have completely different properties from the final living organism.[583] Fossil remains, albeit containing the complete DNA of an organism, do not correspond to living creatures, as the features and qualities of the latter are clearly distinct from the former. Also, the patent would encompass only the living organism and not the remains themselves. Secondly, the existence of the cloned organism is dependent on a human activity. Indeed, even in the presence of a complete genome, the organism could not have been recreated by nature alone. Hence, while the recreated organism is derived from nature, the organism itself is not a product of nature.[584] For this reason, it was argued that:

> If an extinct animal itself exists in nature in the form of remains, the fact that human intervention is necessary to transform it into a living form that is not present in nature,

---

[577] Carlin et al. (2014, p. 51).

[578] Swedlow (2015, p. 186).

[579] Swedlow (2015, p. 193).

[580] Swedlow (2015, p. 193).

[581] This interpretation seems to be confirmed by *Chakrabarty*. On this, see Sect. 4.1.6.4.

[582] It would be possible to argue that the remains used to recreate the once extinct animals were found in nature and derive from a naturally occurring organism. Even the process of regenerating them could be seen as not inventive enough and as a mere reverse engineering of nature's handiwork. *Funk* could be cited in support of this conclusion, by holding that the resulting organism has no properties beyond its natural ones (Hagglund 2007, p. 406).

[583] Artificially preserved biological materials would also be patentable, if their existence or form depends mostly on a human inventive activity. This includes methods for their preservation. Instead, naturally preserved biological materials or materials that would have existed in nature in a similar form would be considered products of nature (Rohrbaugh 1997, pp. 386–387).

[584] Hagglund (2007, pp. 406–407).

which is new and different as compared to the remains, renders the living form patentable.[585]

In support of this conclusion, a number of US decisions were cited. Following *American Fruit Growers*, it was argued that the recreated organism possesses a new or distinctive form, quality or property compared to those currently present in nature. In that decision, the Court did not indicate that the scope of the product of nature doctrine is broader than what currently exists in nature. *Allen* reinforced this conclusion, as it did not demand that an organism has never existed in nature in order to be patentable. Specifically, in *Allen*, the patentability of the oysters was confirmed even though rare events in nature could generate them. The point was not whether a rare event would create the organism, but rather whether it would exist in nature in the ordinary course of things. If applied to the case at hand, the patentability of such organisms would be expected, as they do not occur because of rare natural events, but only due to human input. Lastly, it could be argued that the recreated organism passes the *Chakrabarty* test, where it requires a difference with the characteristics found in nature, as the interpretation of nature presented above seems to exclude extinct organisms.[586]

The situation might be different for near-extinct organisms. They are a copy of an organism and do not present characteristics (e.g. form, quality and other properties) that are not available in the original organism. This identity with a naturally existing organism places them within the realm of the product of nature doctrine. Only once all naturally existing organisms of that kind have died out, it could be argued that the organism is no longer a product of nature. This finding does not contrast with *Allen*, as in that case there was no actual knowledge of the current existence of a polyploid oyster. Conversely, in the case of near-extinct animals, the organism is surely known to still exist. In light of the above, the product of nature doctrine would exclude the patentability of clones of near-extinct animals.[587] This would impact the goals and protection offered to initiatives like the Frozen Ark Project.[588]

Although research in this field is still far from realising its goals, scholars have already considered the implications of patents on human chimeras. In 1997, two scholars sought to patent a method to create a human chimera (i.e. a hybrid between a human and an animal).[589] Within 2 years, the USPTO had rejected the application by holding that Congress did not intend US Patent Law to cover human beings.[590] In 2002, the same scholars filed again an application claiming that the invention was not directed at a human being or embryo, but concerned instead a man-made chimeric animal obtained from a chimeric embryo. They argued that, even if it had

[585]Hagglund (2007, p. 410).

[586]Hagglund (2007, pp. 407–409).

[587]Hagglund (2007, p. 432).

[588]On this, see Sect. 2.2.3.1.

[589]Patent application filed before the USPTO by Stuart Newman and Jeremy Rifkin (Application 08/993,563 (Newman-Rifkin I)) (Rabin 2006, p. 517).

[590]Pursuant to European law, the patentability of this type of applications would be challenged under the morality clause. On this, see Sect. 5.2.3.6.

covered humans, the norms do no exclude the patentability of the claims merely because they concern human beings. In spite of this argument, the application was rejected again on eligibility grounds, showing that the USPTO does not consider human beings as patentable.[591] This finding would likely apply to claims for the recreation of our ancestor, the Neanderthal.[592]

In conclusion, if those inventions were to be assessed on the basis of the current legal panorama, they would be considered patentable subject matter both before the EPO and, in most cases, also before the USPTO. Still, they would likely face morality and public order objections, especially during the initial embryonic phase.[593]

#### 4.2.3.7 Data Processing and Encoding

The analogy between biology and data technology has become one of the pillars of synthetic biology. Nucleotide sequences can be seen as chemical structures or as codes carrying data and information. Although the debate on which is the correct approach is still ongoing, recent decisions and scholarly works have shifted the interest towards the informational content of those molecules. Cells and organisms have been seen as "*highly sophisticated information-processing systems*", whereas genes have been considered "*repositories of information written in a surprisingly similar manner to the one that computer scientists have devised for the storage and transmission of other information*".[594]

Recent applications in the field of synthetic biology have done just the same. Biostorage, biosensors and biocomputers are being devised to take advantage of the information carrying capacity of nucleotides and of the possibility of inserting those systems in environments where regular sensors and computers would not be able to operate.

Applicable Norms

The application of the Directive to inventions concerning data processing and encoding is not straightforward.

On the one hand, those inventions clearly incorporate genetic information, intended as a sequence of DNA bases. However, such sequences do not have a biological content, but rather a purely informational one. They represent a code that

[591]Rimmer (2008, pp. 98–103). This 2002 patent application is known as "Newman-Rifkin II" (Rabin 2006, p. 517).

[592]At the EPO, the question would instead revolve around the morality of this invention. On this, see Sect. 5.2.3.6.

[593]Casabona (2014, p. 184).

[594]Dutfield (2012, p. 185).

can be read by humans and that stores information for human needs, instead of being a sequence that is directed at biological processes and systems. Indeed, the information contained therein is either useless or unreadable for biological systems. Yet, from a normative perspective, such systems contain *strictu sensu* biological information. On the other hand, it is more controversial whether the information-storing sequences can also be reproduced in biological systems. If this were the case, the provisions of the Directive would be applicable to them. Otherwise, the claim will not be governed by the Directive and would need to be assessed pursuant to the EPC.[595]

In any event, it could be doubted that the Directive and the EPC are the best way to protect those inventions. In general terms, both normative instruments might not be optimal to protect products whose value lays in their non-biological informational content. From this perspective, the nucleotide sequences represent a code, much like a binary code. Hence, it could be argued that they should be protected via other forms of IPRs (e.g. copyright).

#### Subject Matter Patentability

Methods and apparatuses used in this sector will most likely be patentable. However, the patentability of products in this field is more complex.

Those inventions might overcome both the discovery and the product of nature exclusions set by Article 52 EPC and § 101 USC 35, since they present a structure that does not exist in nature and that was achieved via human manipulation and ingenuity.

However, the understanding of nucleotide strands as either chemicals or information carriers has an impact on biological data creations. If one were to focus on the chemical structure of the sequence, differences in the strands would warrant patentability, even if they concern a non-coding area. Conversely, if an information perspective were adopted, two diverging approaches could be taken. On the one hand, it could be argued that the biologically relevant information remains unchanged and that the invention concerns a product of nature. On the other hand, the dissimilarity in the informational content altogether (i.e. including the non-coding sections directed at humans) could be taken into consideration. This would differentiate the item from naturally occurring ones. Yet, both solutions are not entirely convincing. The biological function remains the same in both the natural product and in the version including the encoded text, as it is dictated by the unchanged sections. Therefore, the weight of the claim is to be carried by the non-coding regions that have no biological function. Still, the addition of the data storing sequences effectively changed the chemical structure and overall

[595]Storing systems where the information is expressed in DNA outside and independently of any biological systems would not be governed by the Directive either, as long as their genetic information is not reproducible in a biological system.

informational content of the strands. Given this situation, it is complex to predict which approach will be chosen by patent offices and Courts.

From a European perspective, it is likely that this issue will be overlooked and that their patentability will be affirmed by focusing on the technical process employed for the production of these strands pursuant to Article 3 of the Directive. Equally, the fact that they did not previously exist as such in nature would avoid the discovery exclusion.

As for the technical requirement, the passing of this test will need to be examined ad hoc. In general, it is foreseeable that inventions addressing realistic technical problems would fulfil it (e.g. detecting substances in the body via biosensors). This could extend to biocomputers, despite the doubts that usually concern their subject matter eligibility, if they can be shown to address and solve a technical problem. The passing of the technical requirement for systems in which a non-biologically relevant text is encoded might be more complex and would need to be assessed on a case-by-case basis to show the specific technical problem solved each time.

In the US, *Myriad* would be applicable in this case. cDNA was held patentable even though the changes made did not affect the biological functionality of the system (i.e. removing non-coding introns). Coupling this with the fact that the technician definitively made something new and changed the chemical structure of the sequence, it could be argued that those inventions would overcome the product of nature exclusion.

Recently, a number of patent applications in this sector have been filed. A 2010 application was granted in 2016 by the EPO for encoding text into nucleic acid. The abstract explained that the invention related to:

> Methods and apparatus… for encoding human readable text conveying a non-genetic message into nucleic acid sequences with a substantially reduced probability of biological impact and decoding such text from nucleic acid sequences… Synthetic nucleic acid sequences comprising such human readable text, and recombinant or synthetic cells comprising such sequences are provided, as well as methods of identifying cells, organisms, or samples containing such sequences.

The original claims covered synthetic nucleic acid sequences corresponding to human readable symbols. Such sequences were not genetically viable and could comprise a watermark, which could even be a trademark, a copyright notice or normal data. Apparatuses for transforming and converting those sequences were claimed, as were computer-readable media for use in an encoding and decoding machine. The granted patent comprised methods for generating sequences corresponding to human readable symbols and for creating non-human organisms comprising a watermark, non-human recombinant or synthetic organisms or cells with a synthetic nucleic acid sequence and the needed apparatuses. Interestingly, in the course of the examination procedure before the EPO, the applicant was asked to amend claims in order to address the objection of the Examiner that some of them were problematic under Article 52(2)(d). To counteract the critique that some claims constituted a mere presentation of information, the applicant included in the claims a step for making nucleic acid.[596]

A patent application for encoding text into nucleic acid was also presented recently before the USPTO.[597] Interestingly, the Office raised a number of § 101 concerns during its examination. The claims were considered to be non-statutory subject matter, as they amounted to nothing more than the abstract idea of a procedure for organising information. Specifically, it was held that "*the claims appear to tie up the abstract idea itself, rather than confining the use of the abstract idea to a particular technological application*".[598]

Applications for recombinase-based logic and memory systems in living cells were presented before the USPTO and before the EPO.[599] The claims related to systems operating in cells containing logic functions (e.g., AND, OR, NOT A, NOT B et sim.) as well as a cell comprising those systems. A method for altering gene expression was also claimed. During the examination procedure before the USPTO, claims were rejected on patentability grounds based on the 2014 Interim Guidance on Patent Subject Matter Eligibility.[600] In July 2016, the Examiner affirmed that:

> The markedly different analysis compared the nature-based product limitation to its naturally occurring counterpart in its natural state... When the... claims are viewed as a whole there are no markedly different characteristics to the nature-based products of a bacterial genome. The cell that encompasses the plurality of nucleic acid is also not markedly different.[601]

From this, it can be deduced that the USPTO did not contest the eligibility of those inventions in general, but rather the presence of markedly different changes in the case at hand. Interestingly, the functionality claimed by the applicant was also not considered markedly different.

Lastly, an application was presented for a digital to biological converter.[602] The application, which is currently pending before the EPO and has expired before the USPTO, concerned:

> A system for receiving biological sequence information and activating the synthesis of a biological entity. The system has a receiving unit for receiving a signal encoding biological sequence information transmitted from a transmitting unit. The transmitting unit can be present at a remote location from the receiving unit. The system also has an assembly unit connected to the receiving unit, and the assembly unit assembles the biological entity

---

[596] Hutchison et al. (2012, Document of 31 March 2015, p. 2). Interestingly, in the granted patent, the reference to the computer-readable medium was found only in the description and not in the claims.

[597] Hutchison et al. (2011, 2012, 2016a, b) and EPO (n.d.-g).

[598] Hutchison et al. (2016a, Document of 12 October 2018, pp. 4–5).

[599] Lu and Siuti (2014, 2015).

[600] EPO (n.d.-h).

[601] EPO (n.d.-h).

[602] Venter et al. (2014, 2015). Similarly, an application claiming a genomically-encoded memory in live cells has also been filed (Lu and Farzadfard 2016). It claims synthetic-biology platforms for in vivo genome editing, which enable the use of live cell genomes as tape recorders for long-term storing of event histories and analog memories. Another patent was applied for and granted by the USPTO for in vivo gene sensors (Collins and Lu 2013, 2014).

> according to the biological sequence information. Thus, according to the present invention biological sequence information can be digitally transmitted to a remote location and the information converted into a biological entity, for example a protein useful as a vaccine, immediately upon being received by the receiving unit and without further human intervention after preparing the system for receipt of the information.

In rejecting some of the claims, the USPTO did not cite subject matter eligibility as a problematic feature.[603]

#### 4.2.3.8 Algorithms and Simulation Tools

The synthetic biology goal to devise innovative biological systems that go above and beyond what is found in nature is currently limited by a lack of biological designer skills. Biological systems are highly complex and minimal variations can impact their viability, function and safety. Therefore, current biological designs follow natural blueprints quite closely. In fact, "*we know how to build but not yet with clarity what to build*".[604] With the evolution of synthetic biology, it is foreseen that increasingly new systems will be devised. Their viability and functionality will first be modelled and tested in a virtual environment, before they are concretely realised.

This sector is penned to be one of the main propellers of the evolution of synthetic biology. The more accurate those modelling and simulations tools become, the more SynBio will be able to achieve its goals. Independently of the low number of patents granted so far in this field, it is useful to examine the normative and patentability scenarios posed by them, in particular since patent activity for algorithms for interactions between genes, promoters and transcription factors is likely to be high.[605]

##### Applicable Norms

Algorithms and simulation tools used in synthetic biology relate only indirectly to biotechnological inventions. Indeed, they represent biological materials and microbiological processes solely in abstract terms and not in real ones. For this reason, the Directive would not apply to them and their patentability will be established only pursuant to the EPC.

---

[603]EPO (n.d.-i).

[604]Bera (2015, p. 195).

[605]Edwards (2010). The author noted that it is possible that applicants would attempt to obtain broad claims, covering entire families of transcriptional logic. Yet, the limited number of patents in this field could be seen as indicative of the reluctance to protect those methods and products via patents (Edwards 2010).

### Subject Matter Patentability

The exclusion from patentability of mathematical methods is based on their abstract and purely intellectual nature. Simulation tools could be viewed as abstract, since they offer a visual representation of synthetic biology inventions. Given that visual representations cannot be equated with the actual inventions, this might constitute an obstacle to patentability. Likewise, the use of mathematical models and algorithms to determine the behaviour and characteristics of biological systems might raise patentability concerns, given the theoretical nature of such work and the distance between this phase and the manufacturing one.

The *Infineon*, *Halliburton* and *Logikverifikation* rulings shaped the criteria for subject matter eligibility in this field. Despite a more restrictive prior case law, both the EPO and national Courts have recognised the changed and increasingly pivotal role of abstract modelling and simulation steps in current manufacturing techniques. Despite the lack of direct interactions with the outside world, those tools constitute an essential part of the manufacturing process by improving the development of the final product. In the case of synthetic biology, it would be fair to say that, without those tools, there would be no development of new biological systems, as scientists would not be able to abandon nature's blueprint and would need expensive and time-consuming trial and error experiments in the lab to try to achieve the desired functionality. But even then, the construction of new biological systems would be fundamentally unachievable via those techniques, as the goals of synthetic biology require the rethinking and streamlining of biological systems as well as realistic predictions on their viability, functionality and safety. Those aspects confer a practical application to SynBio simulation tools, especially in times when product development and manufacturing are increasingly separated. If SynBio simulation methods were seen as a step within a complex technical production method, they could not be denied technical character merely because no tangible biological systems are incorporated at that stage of production. Furthermore, their aim to improve the effectiveness of the final product would constitute the solving of a technical problem emerging during a technical process. Lastly, the fact that the simulation tools are restricted to specific applications directed at a technical field that is not excluded from patentability (i.e. biotechnology) is advantageous, since simulations in industrial fields connected to highly technical processes are more likely to be patentable.

The EPO Guidelines go in a similar direction, as they confirm that specific applications of computer-implemented simulation methods can have technical character. Those methods are essential in the manufacturing process and cannot be denied technical effect because they do not incorporate a physical end product. Only an undefined technical purpose could lead to a different conclusion.[606]

Therefore, if SynBio simulation tools are employed within a technical industrial process with the clear goal of improving the design and functionality of the end

[606]EPO (2018, § G.II.3.3).

product, this would overcome the exclusion set in Article 52(2) EPC. Yet, it is important for those tools to have an apparent technical application and purpose, as otherwise their unpatentable nature could be argued. This is especially the case for earlier abstract modelling stages. Equally, higher levels of abstraction could be prejudicial under § 101 USC 35.

In the US, the USPTO rejected applications in this field if they did not prove to be concrete, tangible and useful. For instance, one of the first SynBio patent applications in this field was rejected by the USPTO. It concerned a method to infer cellular networks to recognise and model the way in which cells operate. This was achieved by assessing biological interaction networks that have been generated by a computer and that are, at least to some extent, based on real biological information. In rejecting this application, the Examiner noted that inferring cellular networks was not something concrete, tangible and useful. He based his conclusion on the case law available at the time.[607]

In recent times, the patentability of synthetic biology algorithms has been considered problematic, since *Bilski* confirmed the patent ineligibility of abstract ideas. Furthermore, the need to disclose the algorithm might dissuade patents on them. Indeed, patent protection might be of limited use once the algorithm has been disclosed, given that competitors can obtain it from the patent documents and apply it without the patentee finding out.[608]

From a patent perspective, a now withdrawn application expressly claimed methods comprising a number of bioinformatics algorithms to predict and identify specific DNA and amino acid sequences as well as bioinformatics models. Product claims for cells with a heterologous genetic circuit were also presented. Interestingly, the application claimed also a "*computer readable storage medium encoded with instructions, executable by a processor, for designing a host cell comprising a heterologous genetic circuit*".[609]

Similar claims are found in a recent application for structure based predictive modelling presented before the EPO and the USPTO.[610] The claims concern a computer program comprising computer-readable storage media with instructions to implement a method for conducting directed evolution. The claims extend to a computer system, in addition to covering the method of conducting directed evolution itself. The method begins with the receipt and filtering of data and results in an improved sequence activity model. Genetic algorithms are employed in some of the steps. In its opinion on the application, the International Searching Authority noted that one of the claims related to non-technical subject matter, as it concerned the modelling of the properties of molecules. Specifically, it held that:

> Only the purposive use of information modelling in the context of a solution to a technical problem may contribute to the technical character of an invention. The Examiner is in the

[607]Mohan-Ram and Waxman (2008, p. 2) and U.S. Court of Appeals for the Federal Circuit (2007).

[608]Holman (2015, p. 444).

[609]Lou et al. (2012).

[610]Sarmiento et al. (2015).

present case of the opinion that the broad formulations 'training an improved sequence activity model' does not constitute adequately defined technical purpose for the method of claim 1 because said model is not used in the solution of a technical problem.[611]

The USPTO, in a non-final rejection from June 2016, raised the issue of subject matter patentability. Some of the claims were seen as methods:

> Directed to processing information and converting one form of numerical representation into another. In other words, the method steps simply address the concept of gathering and combining data by reciting steps of organizing information through material relationships. The gathering and combining merely employs mathematical relationships to manipulate existing information to generate additional information. All of these concepts relate to organization and processing information which can be performed mentally and is an idea of itself. Therefore, the method steps are directed to an abstract idea which is a judicial exception... none of the hardware offers a meaningful limitation beyond generally linking the system to a particular technological environment, that is, implementation via computers. The additional elements in the claims other than the abstract idea per se amount to no more than a recitation of generic computer structure... the instruction to apply the abstract idea using some unspecified, generic computer is not enough to transform the abstract idea into a patent-eligible invention.[612]

In a subsequent document, the Examiner pointed out that the method simply concerns the gathering and combining of data through mathematical steps, which could also be done mentally. Therefore, the application would fall within the abstract idea exception to patentability.[613] In another occasion, it was also noted that Courts have made it clear that "*to transform an unpatentable judicial exception into a patent-eligible application, one must do more than simply state the judicial exception while adding the words 'apply it*'".[614] For this reason, the performance of the claimed invention on a generic computer structure that performs generic computer functions was not considered enough to transform the abstract idea into an eligible invention.[615]

The designing goals of SynBio are also exemplified in a 2015 patent application for systems and methods for synthetic biology design and host cell simulation.[616]

---

[611]Sarmiento et al. (2015, Document of 29 March 2016, attachment pp. 2–3).

[612]EPO (n.d.-i, Document of 20 June 2016 pp. 2–5).

[613]Sarmiento et al. (2015, Document of 18 November 2016 p. 3).

[614]Sarmiento et al. (2015, Document of 14 December 2017 p. 4).

[615]Sarmiento et al. (2015, Document of 14 December 2017 pp. 5–6). On this, see also Sarmiento et al. (2015, Document of 19 June 2018). In Europe, the same patent was granted. During the review, the Examiner mentioned Art. 52(2) EPC since a claim had to be amended to explain that the method was computer-implemented (Sarmiento et al. 2015, Document of 23 May 2018 p. 3).

[616]Jayaraman et al. (2015). The design system covers a model conversion component that receives genetic circuit data indicative of user-specified genetic circuit design, identifies the constituent parts of the genetic circuit design and their connections, obtains mathematical models corresponding to them and, finally, provides a composite model generating genetic circuit output data based on the input data. The cell simulation component receives the genetic circuit output data from the composite model and generates host cell output data that represents the physiological state of the host cell. Such physiological state concerns, amongst others, the growth rate and composition of the cell, as well as its metabolism.

Before the EPO, this application did not raise eligibility issues. The USPTO noted instead that the claims recited an abstract idea. In particular, the Examiner held that the claims "*do not integrate the abstract idea into a practical application; . . . do not confine the use of the abstract idea to a particular technology; . . . do not solve a problem rooted in. . . the use of a particular technology*".[617] Hence, the application was not eligible for patent protection.

In conclusion, the patentability of those inventions will generally depend on their solving a specific technical problem within a technical manufacturing process and on their level of abstraction.

#### 4.2.3.9 Completely *ex Novo*

The creation of completely new biological systems fulfilling human needs is one of the main goals of synthetic biology. The idea is that men will be able to devise and produce biological systems following engineering principles and without being bound by nature's blueprint. Given its complexity, this goal could be achieved, if ever, only in years from now. Indeed, for the moment, even the feasibility of such objective is impossible to determine.

In spite of this, it is important to theoretically examine the patentability of completely new biological systems, as those inventions represent one of the main objectives of synthetic biology. Another reason for this analysis is the relevance of those inventions in the legal literature. Authors often base their assessment of the patentability of synthetic biology on those types of developments, disregarding the fact that truly novel biological systems are a futuristic aspiration rather than a reality.

##### Applicable Norms

Those types of synthetic biology inventions will be governed by the EPC. Doubts could however be raised concerning the application of the Directive. Indeed, it would be possible to argue that those systems are non-biological, since they have been created *ex novo*, have been devised by men following engineering principles and are not based on natural models.

Although it is debatable whether such systems fit the societal and policy notion of biological, from a legal perspective the matter is more straightforward. The Directive speaks solely of biological materials, microbiological processes and biotechnological inventions, without addressing either the origin or the novelty of the material. The Directive applies to inventions that contain genetic information reproducible within a biological system. Since those systems are likely to employ nucleotides, they would fulfil the first prong of this definition. Similarly, if this information can be reproduced in a biological system, they would squarely fall within the definition of

[617] Jayaraman et al. (2015, Document of 10 May 2019, p. 6).

biological material set by the Directive and thus be governed by it. Conversely, different conclusions may be reached if those systems were not capable of reproduction. However, as the creation of those systems is still far ahead in the future, it can only be postulated that such situations will have to be assessed on a case-by-case basis.

Even the existence of wholly synthetic systems would not hinder, *per se*, the applicability of the Directive. As previously noted, the norms do not distinguish between the origin of the biological materials (i.e. natural or synthetic) and thus this aspect is unlikely to play a role.[618] Equally, the fact alone that those systems do not have any naturally existing ancestor and are completely synthetic in origin may not, as such, impede the application of the Directive. Nevertheless, it cannot be excluded that policy reasons might demand a change of approach in the future.

### Subject Matter Patentability

The patentability of completely new and human devised biological systems is generally accepted, as they would require "*considerable human involvement based on methodological technical teachings*".[619] Systems that are significantly different from any natural equivalent and are the result of human ingenuity overcome the main critiques moved to biotechnological inventions: namely that those patents claim discoveries or products of nature. The mimicking of natural systems, the limited understanding of nature's inner workings and the ability to introduce only a limited number of alterations have led commentators to argue that "*the creativity of life comes from within. No amount of human tinkering can change that*".[620] From this perspective, humans are not seen as inventors of biological systems; hence, they should not be rewarded with a patent, as they have merely taken advantage of nature and the possibilities intrinsically offered by it.[621]

The creation of completely new biological system would overcome those critiques. Even in the absence of completely novel systems, it is obvious that the wider the difference between a biological system and what exists in nature is, the weaker the arguments on the discovery and product of nature exclusions become. Several authors have shared this conclusion.[622] *Ludlow* believed that:

[618]However, such considerations may have an impact on policy and societal issues.

[619]Schwartz and Minssen (2015, p. 202).

[620]Dutfield (2010, p. 539).

[621]This is one of the reasons why patentability of biotechnological inventions has encountered strong opposition from a legal, policy and social perspective.

[622]It has been argued that synthetic products and processes that are the result of an innovative and original biological process are inventions, since they were obtained via human creativity and innovative ideas (Falcone 2012, p. 5).

> When science is capable of creating its own genes rather than taking them from, or modelling synthetic constructs on, nature, the resulting organism or product will clearly not be only a discovery.[623]

Likewise, *Resnik* argued that substantially modified genes would constitute inventions, given that, when the primary cause for creation of an item are humans, an invention exists.[624]

Similarly, the OECD argued that:

> Synthetic DNA sequences are... more easily patentable than DNA sequences derived from natural sources. One of the main criticisms of the patent system in biotechnology was whether patents should be granted to products of nature, because products of nature are 'discoveries', which are not patentable. In synthetic biology, however, DNA sequences, systems, cells and organisms are designed by humans. Human-made DNA sequences can therefore receive patent protection without touching on the issue of discovery from nature.[625]

Similar positions were expressed by *Torrance*, who held that:

> Genes constructed using synthetic biological techniques will have their origins in human imagination and will, thus, not be products of nature... synthetic genes would remain patentable subject due to their non-natural origins.[626]

In light of the above statements, it is important to differentiate between the artificial origin of the DNA sequence and the originality of the sequence itself. The term synthetic DNA describes the method used to produce a nucleotide sequence.[627] As in the case of JCVI syn1.0, the resulting synthetic strand or organism could mirror a naturally occurring one. Therefore, the fact that a sequence was produced in vitro and without relying on pre-existing strands for its manufacturing does not render it *per se* a completely new item with no natural equivalent. A synthetic sequence that mirrors a naturally occurring one would be patentable in Europe, since it is obtained via a technical process. On the other hand, the product of nature exclusion might prove an obstacle to their patentability in the US and in Australia, as the technical origin of the item is *per se* no guarantee of patentability.

By contrast, *ex novo* biological systems based on human designs are likely to be considered inventions on both sides of the Atlantic. Those new models, which are obtained via in vitro gene synthesis, would be seen as the product of considerable human ingenuity and creativity. This would affect their patentability in the US, as *Myriad* would have little impact on inventions that are substantially different from naturally existing ones. This approach would be saluted by those who held that the

---

[623]Ludlow (1999, p. 303).

[624]Resnik (2002, pp. 142–149).

[625]OECD (2014, p. 96).

[626]Torrance (2010, p. 640).

[627]The concept of DNA synthesis could be intended to encompass DNA replication, PCR techniques as well as gene synthesis. In the context of synthetic biology, references to DNA synthesis usually concern the latter technique, which is also known under the name of artificial gene synthesis.

*Myriad* saga might impact the development of synthetic biology. Indeed, from a policy perspective, the ruling of the Court would seem to favour patentability in the field of synthetic biology. In particular, it was noted that:

> Allowing such sequences while denying the patentability of isolated ACGT sequences may be good policy if the objective is to encourage synthetic biology… accepting genuinely new sequences devised by scientists as patentable inventions, would likely encourage research to shift towards the development and application of truly original synthetic genes.[628]

## 4.2.4 Conclusions

Despite the absence of case law on the patentability of synthetic biology inventions and the limited number of patents granted, a few elements have emerged from the analysis conducted above.

The EPO has not directly raised objections against the subject matter eligibility of synthetic biology (apart from some of its employees observing that the Directive may not apply to a number of those inventions). Yet, the areas not covered by the Directive are the ones where eligibility questions are less likely to arise, given that those inventions could rarely be seen as discoveries. Nonetheless, the inapplicability of the Directive could have additional legal effects within Europe. Gone the framework of the Directive and the ECJ, States might develop national approaches on patentability issues and thus harm the original intent of the Directive to harmonise the European legal panorama on biotechnologies and to remove different national approaches that could impede the internal market.[629]

From an eligibility perspective, it is most likely that SynBio inventions will be held patentable in Europe. Article 3 of the Directive on the patentability of biological materials produced via technical means would be fundamental here, given that synthetic biology inventions could only be achieved through those means. On this, the formulation of the Directive is clear and leaves no room for interpretations that would hinder the patentability of synthetic biology inventions. In light of this and given that the EPO does not adopt a strict approach to eligibility, synthetic biology inventions would generally be considered patent eligible by the EPO. Yet, this conclusion is balanced by the consideration that the passing of this requirement does not necessarily lead to the granting of a patent. The EPO has often focused its analysis on other patentability criteria, with the consequence that it is common for an application to pass this requirement, only to be later rejected on other grounds. The analysis made above seems to confirm this.

The situation may be diametrically different in the US, due to recent cases and the new Eligibility Guidelines of the USPTO. The patentability hurdle is likely to be higher in the US compared to Europe. However, the actual impact of such

[628] Dutfield (2012, p. 201).

[629] Recitals 3, 5, 6, 7 of the Directive.

divergence on the granting of a patent may be limited. Indeed, the above analysis showed that most granted patents were issued before both offices, and so were most rejected applications. The difference laid in the grounds of those rejections. Still, since the change of position of US authorities is quite recent, it is possible that the full repercussions of this approach will become apparent only in the future.[630]

The European panorama is unlike to change in the near future, as a different approach could appear only once the language of the Directive is reformulated. Since it is not likely that the EU would address such politically and socially charged matter unless it was really necessary, this approach might remain unvaried for years. As a result, the conclusions expressed above on the patentability of synthetic biology inventions in Europe are more settled than their US counterparts.

Under such conditions, industry players would likely prefer the European approach due both to its favourable assessment of technologies produced via technical processes and the clear-cut nature of its patentability requirement. The approach followed by the EPO has the advantage of being predictable and of raising limited patentability concerns. This would definitely be welcomed by industry players and investors alike. On the other hand, the US approach is more problematic given its unpredictability and its currently stricter standards. Industry players generally avoid raising eligibility questions, since decisions on this issue can have far-reaching consequences for the patentability of entire classes of inventions. Subject matter patentability is an "in or out" criteria that does not allow much fine-tuning of the scope and reach of a patent, as instead provided by the novelty and inventiveness requirements.

From the perspective of civil society organisations and NGOs, the US approach appears to be more favourable, at least at first. The newly found interest in this criterion of US Courts and of the USPTO gives those actors more room to oppose synthetic biology patents on eligibility grounds. However, this advantage may be short lived. Technical progress will increasingly bring this discipline outside the scope of the product of nature doctrine. If this were the case, civil society organisations and NGOs might shift their focus to European authorities, where they could rely on the morality clause to try to block the patentability of entire classes of inventions.

## References

Adcock M, Llewelyn M (2000) TRIPs and the patentability of micro-organisms. Bio-Sci Law Rev 4:91–101

[630]If applicants were not able to protect their inventions in the US, they could do so in Europe, as the Directive explicitly considers patentable a number of inventions that would fail the US patentability test (Bagley 2015, p. 6). Equally, the EPO policy of "*in dubio pro patente*" could facilitate the patentability of SynBio inventions in Europe (Schneider 2014, p. 157). Differences in the approach to patentability could affect trade and investment patterns in this sector (Saunders 2008, p. 89).

Advocate General Jacobs (2001) Opinion of Advocate General in Kingdom of the Netherlands v European Parliament and Council of the European Union (C 377/98)
Aerts RJ (2019) A switch on a switch on a switch: the status of harmonisation of biotech patent law in Europe. Eur Intell Prop Rev 41:541–545
Agovic A (2014) Patenting SynBio in Anglo-America and Europe: chaos or opportunity. In: de Miguel Beriain I, Casabona CMR (eds) Synbio and human health: a challenge to the current IP framework? Springer, Dordrecht, pp 101–121
Allen A, Dupont CL (2014) Engineered microalgae with enhanced lipid production. US20140162330 A1
Amendments suggested by the European Parliament to the Draft Directive on the legal protection of biotechnological inventions ((1992) OJ C 125/112), 1992
Amendments suggested by the European Parliament to the Draft Directive on the legal protection of biotechnological inventions ((1992) OJ C 305/160), 1992
Anderson JC, Chin JW, Liu DR, Magliery TJ, Meggers EL, Mehl RA, Pastrnak M, Santoro SW, Schultz P, Wang L, Zhang Z (2004) In vivo incorporation of unnatural amino acids. EP1490483 A2.
Andrews LB (2014) The progress clause: an empirical analysis based on the constitutional foundation of patent law. N C J Law Technol 15:537–561
Aplin TF, Davis J (2009) Intellectual property law: text, cases, and materials, 1st edn. Oxford University Press, Oxford
Aplin TF, Davis J (2017) Intellectual property law: text, cases, and materials, 3rd edn. Oxford University Press, Oxford
Arcuri S (2011) They call that natural? An analysis of the term "naturally occurring" and the application of genes to the patent act. Valparaiso Univ Law Rev 40:743–787
Australian Government (2015) Examination practice following the High Court decision in D'Arcy v Myriad Genetics Inc [WWW Document]. www.ipaustralia.gov.au. URL https://www.ipaustralia.gov.au/sites/default/files/examination_practice_following_the_high_court_decision_in_darcy_v_myriad_genetics_inc.pdf. Accessed 2 Mar 2020
Australian Government (2016) Commissioner's proposed revised examination practice for consultation 16 October 2015 to 6 November 2015 [WWW Document]. www.ipaustralia.gov.au. URL https://www.ipaustralia.gov.au/sites/default/files/20151208_proposed_practice_note_consultation.pdf?acsf_files_redirect. Accessed 2 Mar 2020
Baccelli M, Hiratsuka M (2009) EU-Japan - Patent protection of software simulators: Practice at the European Patent Office and open questions [WWW Document]. https://aippi.org. URL https://aippi.org/enews/2009/edition08/software-simulators.html. Accessed 27 Feb 2020
Bagley MA (2015) Digital DNA: the Nagoya protocol, intellectual property treaties, and synthetic biology. Synthetic Biology Project. https://doi.org/10.2139/ssrn.2725986
Baldock C, Cook TM, Karet I, Rollins T, Wood J (2000) Report Q 150: patentability requirements and scope of protection of Expressed Sequence Tags (ESTs), Single Nucleotide Polymorphisms (SNPs) and entire genomes. Eur Intellect Prop Rev 22:39–42
Beauchamp C (2013) Patenting nature: a problem of history. Stanford Technol Law Rev 16:257–312
Benner SA, Yang Z (2013) Polymerase incorporation of non-standard nucleotides. US8614072 B2
Bentham A, Reinhard C, Wright S (2019) EPO plant patent-eligibility update - Political pressure brings further uncertainty for applicants [WWW Document]. www.jakemp.com. URL https://www.jakemp.com/en/knowledge-centre/briefings/epo-plant-patent-eligibility-update-political-pressure-brings-further-uncertainty-for-applicants. Accessed 2 Mar 2020
Bently L, Sherman B, Gangjee D, Johnson P (2018) Intellectual property law, 5th edn. Oxford University Press, Oxford
Bentz WG (2000) Patent protection for bioinformatics tools [WWW Document]. http://pubs.acs.org. URL https://pubsapp.acs.org/subscribe/journals/ci/30/i10/html/10legal.html. Accessed 2 Mar 2020

Bera RK (2015) Synthetic biology and intellectual property rights. In: Ekinci D (ed) Biotechnology. IntechOpen, pp 195–232

Bergia S (2013) Art. 45 - Invenzioni. In: Sironi GE, Vanzetti A (eds) Codice della proprietà industriale. Giuffrè, Milano, pp 630–643

Betten J (2000) Patent law: Germany: "Viterbi algorithm". Int Rev Intellect Prop Compet Law 31:442–446

Bhutkar A (2005) Synthetic biology: navigating the challenges ahead. J Biolaw Bus 8:19–29

Blattner F, Posfai G, Herring C, Plunkett G, Glasner J (2003) Bacteria with reduced genome. US20030138937 A1

Blattner F, Csorgo B, Posfai G (2016) Reduced genome bacteria with improved genetic stability. US9340791 B2

Blattner F, Csorgo B, Posfai G (2017) Reduced genome bacteria with improved genetic stability. EP2768943 B1

Blattner F, Glasner JD, Herring CD, Plunkett G, Posfai G, Twose TM (2004) Bacteria with reduced genome. EP1483367 A2

Blattner F, Posfai G, Herring CD, Plunkett G, Glasner JD (2006) Bacteria with reduced genome. US6989265 B2

Blattner F, Glasner JD, Herring CD, Plunkett G, Posfai G, Twose TM (2010) Bacteria with reduced genome. EP1483367 B1

Blattner F, Posfai G, Herring CD, Plunkett G, Glasner JD (2011) Bacteria with reduced genome. US8043842 B2

Boguniewicz A (2014) Discovering the undiscoverable: patent eligibility of DNA and the future of biotechnical patent claims post-Myriad. Wash J Law Technol Arts 10:35–50

Bostyn SJR (1999) The patentability of genetic information carriers. Intellect Prop Q:1–36

Bostyn SJR (2004) Patenting DNA sequences (polynucleotides) and scope of protection in the European Union: an evaluation. Office for Official Publications of the European Communities, Luxembourg

Brabin C (2014) Intellectual property law in the realm of biology - striking the right balance. Eur Intellect Prop Rev 36:687–692

Brief for the United States as amicus curiae in support of neither party - in the US Court of Appeals for the Federal Circuit in Association for Molecular Pathology, et al. v. United States Patent and Trademark Office (Myriad) (2010)

Brief of James D. Watson, Ph.D. as amicus curiae in support of neither party to the Supreme Court of the United States in the Association for Molecular Pathology, et al. v. Myriad Genetics, Inc., et al. (Myriad) (2013)

Bundesgerichtshof (Federal Supreme Court) 27.03.1969 File: X ZB 15/67 Rote Taube (Red Dove) (1970) IIC - International Review of Intellectual Property and Competition Law 136–142

Burk DL (2013) Are human genes patentable? Int Rev Intellect Prop and Compet Law 44:747–749

Calvert J, Joly P-B (2011) How did the gene become a chemical compound? The ontology of the gene and the patenting of DNA. Soc Sci Inf 50:157–177

Carlin NF, Wurman I, Zakim T (2014) How to permit your mammoth: some legal implications of "de-extinction". Stanford Environ Law J 33:3–57

Casabona CMR (2014) Patentability, synthetic biology and human genome. In: de Miguel Beriain I, Casabona CMR (eds) Synbio and human health: a challenge to the current IP framework? Springer, Dordrecht, pp 175–186

Cho HS, Daniel TO, Hays A-M, Wilson TE, Litzinger DC, Mariani R, Kimmel BE, Keefe WM. (2009) Biosynthetic polypeptides utilizing non-naturally encoded amino acids. US7638299 B2

Cho HS, Daniel TO, Dimarchi R, Hays A-M, Wilson TE, Sim B-C, Litzinger D (2012a) Modified human growth hormone polypeptides and their uses. EP1740608 B1

Cho HS, Daniel TO, Hays A-M, Wilson TE (2012b) Modified human interferon polypeptides and their uses. US8119603 B2

Collins JJ, Lu TK-T (2013) In vivo gene sensors. US20130053432 A1

Collins JJ, Lu TK-T (2014) In vivo gene sensors. US8795958 B2
Committee on Science, Technology, and Law of the National Academies of Science, Engineering and Medicine (2013) Ownership and sharing - Setting the patent framework for innovation in synthetic biology. Committee on Science, Technology, and Law of the National Academies of Science, Engineering and Medicine
Common position (EC) No 4/94 adopted by the Council with a view to adopting European Parliament and Council Directive 94/. . ./EC of . . . on the legal protection of biotechnological inventions (94/C 101/01), 1994. , OJ C 101, 9.4.1994, pp 65–75
Conference proceedings: new perspectives in science education: 5th Conference, Florence, Italy, 17-18 March 2016, 2016. Libreriauniversitaria.it, Padova
Conley JM (2009) Gene patents and the product of nature doctrine. Chicago-Kent Law Rev 84:109–132
Conley JM, Makowski R (2004) Rethinking the product of nature doctrine as a barrier to biotechnology patents in the United States - and perhaps Europe as well. Inf Commun Technol Law 13:3–40
Constitution of the United States of America (1789)
Contreras JL, Cuticchia AJ (eds) (2013) Bioinformatics law: legal issues for computational biology in the post-genome era. American Bar Association, Chicago
Convention on the unification of certain points of substantive law on patents for invention (ETS No. 047), 1963
Cook TM (2000) The human genome project: crucial questions the biotechnology Directive does not answer. Eur Lawyer 1:56–59
Cornish WR, Llewelyn D, Aplin TF (2019) Intellectual property: patents, copyright, trademarks and allied rights, 9th edn. Sweet & Maxwell, London
Council for Trade-Related Aspects of Intellectual Property Rights (2006) Review of the provisions of Article 27.3(B) - Summary of issues raised and points made - Note by the Secretariat
Council of the European Union (2017) Council conclusions on the Commission Notice on certain articles of Directive 98/44/EC of the European Parliament and of the Council on the legal protection of biotechnological inventions (2017/C 65/02)
Court of Appeal of England and Wales (EWCA) (2006) Aerotel Ltd. v Telco Holdings Ltd [2006] EWCA Civ 1371
Covone MG (2015) Presentation: patents in the field of biotechnology. Palermo
Cragg GM, Newman DJ, Snader KM (1997) Natural products in drug discovery and development. J Nat Prod 60:52–60
Crowne-Mohammed EA (2010) Review of the as such exclusions to patentable subject matter in the United Kingdom: lessons for Canadian and American courts. Albaum Law J Sci Technol 20:457–486.
D'Antonio V (2004) Invenzioni biotecnologiche e modelli giuridici: Europa e Stati Uniti. Jovene, Napoli
Danish Council of Ethics (2004) Patenting human genes and stem cells a report. The Danish Council of Ethics, Copenhagen
de Miguel Beriain I (2010) Patenting human life, a question of morals? [WWW Document]. www.sybhel.org. URL http://www.sybhel.org/?p=419. Accessed 29 Feb 2020
Demers L (2013) Product of nature doctrine: Myriad's effect beyond nucleic acids. https://doi.org/10.2139/ssrn.2279754
Deutsche Akademie der Technikwissenschaften, Deutsche Akademie der Naturforscher Leopoldina, Deutsche Forschungsgemeinschaft (ed) (2009) Synthetische Biologie: Stellungnahme. Wiley-VCH-Verlag, Weinheim
Dietrich JA, Yoshikuni Y, Keasling JD, Chang MC-Y (2007) Artemisinic epoxide and methods for producing same. WO2008073498 A2
Dietrich JA, Yoshikuni Y, Keasling JD, Chang MC-Y (2008) Artemisinic epoxide and methods for producing same. US20080187983 A1

Dietrich JA, Yoshikuni Y, Keasling JD, Chang MC-Y (2012) Artemisinic epoxide and methods for producing same. US8101399 B2
Dietz W (2019) Protecting our genetic code: can copyright succeed where patents have failed? Southwest Law Rev 48:437–456
Directive 96/9/EC of the European Parliament and of the Council of 11 March 1996 on the legal protection of databases, 1996., OJ L 77, 27.3.1996, pp 20–28
Drahos P (1999) Biotechnology patents, markets and morality. Eur Intellect Prop Rev 21:441–449
Dreyfuss R (2013) Presentation: the patent landscape: recent decisions from the US Supreme Court. Presented at the Ownership and Sharing: Setting the Patent Framework for Innovation in Synthetic Biology workshop, London
Dutfield G (2010) Who invents life: intelligent designers, blind watchmakers, or genetic engineers? J Intellect Prop Law Pract 5:531–540
Dutfield G (2012) "The genetic code is 3.6 billion years old: it's time for a rewrite": Questioning the metaphors and analogies of synthetic biology and life science patenting. In: Lever A (ed) New frontiers in the philosophy of intellectual property. Cambridge University Press, Cambridge, pp 172–202
Dutfield G, Suthersanen U (2008) Global intellectual property law, 1st edn. Edward Elgar, Cheltenham
Economic and Social Committee (1996) Opinion of the Economic and Social Committee on the "Proposal for a European Parliament and Council Directive on the legal protection of biotechnological inventions" 96/C 295/03
Edwards C (2010) Will IP law choke synthetic biology work? [WWW Document]. www.thebiomachine.com. URL http://blog.thebiomachine.com/2010/06/james-boyle-synthetic-biology-patent-fears.html. Accessed 12 July 2016
Eisenberg RS (2000) Re-examining the role of patents in appropriating the value of DNA sequences. Emory Law J 49:783–800
Eisenberg RS (2006) The story of Diamond v. Chakrabarty: technological change and the subject matter boundaries of the patent system. In: Ginzburg JC, Dreyfuss RC (eds) Intellectual property stories. Foundation Press, New York, pp 327–357
Eisenberg RS (2013) Prometheus rebound: diagnostics, nature and mathematical algorithms. Yale Law J Online 122:341–349
Enlarged Board of Appeal of the EPO (EBA) (2010a) G 0002/07 (Broccoli/Plant bioscience)
Enlarged Board of Appeal of the EPO (EBA) (2010b) G 0001/08 (Tomatoes/State of Israel)
Enlarged Board of Appeal of the EPO (EBA) (2015a) G 0002/12 (Tomato II)
Enlarged Board of Appeal of the EPO (EBA) (2015b) G 0002/13 (Broccoli II)
EPO (2013) Patents for software? European patent law and practice [WWW Document]. https://ciencias.ulisboa.pt. URL https://ciencias.ulisboa.pt/sites/default/files/fcul/inovacao/PI-Pack-INPI-E-Patents-for-Software-EPO.pdf. Accessed 3 Mar 2020
EPO (2016) Case law of the Boards of Appeal (8th edition) [WWW Document]. www.epo.org. URL https://www.epo.org/law-practice/case-law-appeals/case-law.html. Accessed 3 July 2016
EPO (2018) Guidelines for examination [WWW Document]. www.epo.org. URL https://www.epo.org/law-practice/legal-texts/guidelines.html. Accessed 3 Sept 2019
EPO (2019) EPO Contracting States discuss next steps regarding the patentability of plants obtained by essentially biological processes [WWW Document]. www.epo.org. URL https://www.epo.org/news-issues/news/2019/20190329.html. Accessed 4 Mar 2020
EPO (n.d.-a) EPO Global Dossier - Patent US54636406 [WWW Document]. www.epo.org. URL https://register.epo.org/ipfwretrieve?apn=US.54636406.A&lng=en. Accessed 3 Mar 2020a
EPO (n.d.-b) EPO Global Dossier - Patent US63535506 [WWW Document]. www.epo.org. URL https://register.epo.org/ipfwretrieve?apn=US.63535506.A&lng=en. Accessed 3 Mar 2020b
EPO (n.d.-c) EPO Global Dossier - Patent US68188708 [WWW Document]. www.epo.org. URL https://register.epo.org/ipfwretrieve?apn=US.68188708.A&lng=en. Accessed 3 Mar 2020c

EPO (n.d.-d) EPO Global Dossier - Patent US201113699198 [WWW Document]. www.epo.org. URL https://register.epo.org/ipfwretrieve?apn=US.201113699198.A&lng=en. Accessed 3 Mar 2020d

EPO (n.d.-e) EPO Global Dossier - Patent US201113993859 [WWW Document]. www.epo.org. URL https://register.epo.org/ipfwretrieve?apn=US.201113993859.A&lng=en. Accessed 3 Mar 2020e

EPO (n.d.-f) EPO Global Dossier - Patent US201615052781 [WWW Document]. www.epo.org. URL https://register.epo.org/ipfwretrieve?apn=US.201615052781.A&lng=en. Accessed 3 Mar 2020f

EPO (n.d.-g) EPO Global Dossier - Patent US201314105994 [WWW Document]. www.epo.org. URL https://register.epo.org/ipfwretrieve?apn=US.201314105994.A&lng=en. Accessed 3 Mar 2020g

EPO (n.d.-h) EPO Global Dossier - Patent US201313969215 [WWW Document]. www.epo.org. URL https://register.epo.org/ipfwretrieve?apn=US.201313969215.A&lng=en. Accessed 27 Feb 2020

EPO (n.d.-i) EPO Global Dossier - Patent US201414498881 [WWW Document]. www.epo.org. URL https://register.epo.org/ipfwretrieve?apn=US.201414498881.A&lng=en. Accessed 3 Mar 2020i)

EPO, USPTO, JPO (1988) Joint statement of the EPO, USPTO and JPO. Biotechnol Law Rev 7:159

ETC Group (2007a) Extreme genetic engineering: an introduction to synthetic biology. ETC Group

ETC Group (2007b) Patenting Pandora's bug

European Commission (1992) Amended proposal for a Council Directive on the legal protection of biotechnological inventions (COM (92) 589 final - SYN 159)

European Commission (2000) Legal protection of biotechnological inventions - Frequently asked questions on scope and objectives of the EU Directive (98/44) (No. MEMO/00/39)

European Commission (2002) Report from the Commission to the European Parliament and the Council - Development and implications of patent law in the field of biotechnology and genetic engineering (No. COM/2002/0545 Final). Brussels

European Commission (2016) Commission notice on certain articles of Directive 98/44/EC of the European Parliament and of the Council on the legal protection of biotechnological inventions (2016/C 411/03)

European Commission's Directorate-General for Health & Consumers (2010) Synthetic biology from science to governance. European Commission's Directorate-General for Health & Consumers, Brussels

European Court of Justice (ECJ) (2001) (C 377/98) Kingdom of the Netherlands v. European Parliament and Council of the European Union

European Court of Justice (ECJ) (2010) (C 428/08) Monsanto Technology LLC v Cefetra BV and Others

European Court of Justice (ECJ) (2011) (C 34/12) Oliver Brüstle v. Greenpeace e.V

European Court of Justice (ECJ) (2018) (C 528/16) Confédération paysanne and others v Premier ministre and ministre de l'agriculture, de l'agroalimentaire et de la forêt

European Patent Office (1999) Official Journal EPO 8-9/1999 - Notice dated 1 July 1999 concerning the amendment of the Implementing Regulations to the European Patent Convention - The new provisions relating to biotechnological inventions

European Patent Office: European Patent Convention, Arts. 52(1)(2)(3), 54, 56, 57 - "Auction Method/Hitachi," (2005) International Review of Intellectual Property and Competition Law 36, 235–241

Falcone A (2012) Biologia di sintesi, bio-bricks e brevetti biotech fra libertà della ricerca scientifica e tutela della salute umana: Alcune considerazioni sui principi. Forum di quaderni costituzionali 9:1–27

Falcone A (2014) Synthetic biology, biotechnology patents and the protection of human health. A consideration of the principles at stake. In: de Miguel Beriain I, Casabona CMR (eds) Synbio and human health. Springer, Dordrecht, pp 55–76
Federal Court of Australia (2013) Cancer Voices Australia v Myriad Genetics Inc, [2013] FCA 65
Fernandez y Branas FJ (2014) Patentability of synthetic biology under the European Patent Convention (EPC). In: de Miguel Beriain I, Romeo Casabona CM (eds) Synbio and human health. Springer, Dordrecht, pp 187–199
Fernolend R (2015) Germany: Patentability of mathematical methods [WWW Document]. www.managingip.com. URL https://www.managingip.com/Article/3482774/Germany-Patentability-of-mathematical-methods.html. Accessed 2 Mar 2020
Finkel E (2016) Patenting human genes was always a bad idea, now it's history [WWW Document]. https://cosmosmagazine.com. URL https://cosmosmagazine.com/society/patenting-human-genes-was-always-bad-idea-now-its-history. Accessed 2 Mar 2020
Fox N, Corbett W (2014) UK and EPO approaches to excluded subject-matter and inventive step: are Aerotel and Pozzoli heading for the rocks? Eur Intellect Prop Rev 36:569–576
Friedrich C (2007) Presentation: Patenting of gene sequences. The view of the EPO. Presented at the Current challenges in intellectual property rights and biotechnology workshop
Fritz SC, Grünbeck EK, Hijazi A (2015) Key to the European Patent Convention, 1st edn. E. Grünbeck, Munich
Gerber S, Campbell P (2019) Patent eligibility remains uncertain, even after recent efforts to bring clarity [WWW Document]. www.law.com. URL https://www.law.com/2019/01/11/patent-eligibility-remains-uncertain-even-after-recent-efforts-to-bring-clarity/?slreturn=20200202161045. Accessed 2 Mar 2020
German Federal Patent Court (BPatG) (1977) Naturstoffe, 16 W (Pat) 64/75
German Federal Patent Court (BPatG) (1996) Viterbi-Algorithmus, 20 W (Pat) 12/94
German Federal Supreme Court (BGH) (1969) Rote Taube, X ZB 15/67.
German Federal Supreme Court (BGH) (1972) Imidazoline, X ZB 2/71
German Federal Supreme Court (BGH) (1975) Bäckerhefe, X ZB 4/74
German Federal Supreme Court (BGH) (1995) Klinische Versuche, X ZR 99/92
German Federal Supreme Court (BGH) (1999) Logikverifikation, X ZB 11/98
German Federal Supreme Court (BGH) (2001) Suche fehlerhafter Zeichenketten, X ZB 16/00
German Federal Supreme Court (BGH) (2015) Flugzeugzustand, X ZB 1/15
German patent law 1877, 1877
Glass JI, Smith H, Hutchison C, Alperovich N, Assad-Garcia N (2007a) Minimal bacterial genome. US20070122826 A1
Glass JI, Young L, Lartigue C, Assad-Garcia N, Smith H, Hutchison C, Venter J (2007b) Installation of genomes or partial genomes into cells or cell-like systems. US20070269862 A1
Glass JI, Young L, Lartigue C, Assad-Garcia N, Smith HO, Hutchison C, Venter JC (2008a) Installation of genomes or partial genomes into cells or cell-like systems. EP1963515 A2
Glass JI, Alperovich N, Clyde A Hutchinson C, Lartigue C, Merryman C, Smith HO, Vashee S, Venter JC (2010) Methods of genome installation in a recipient host cell. EP2147099 A1
Glass JI, Young L, Lartigue C, Assad-Garcia N, Smith HO, Hutchison C, Venter JC (2014) Installation of genomes or partial genomes into cells or cell-like systems. EP1963515 B1
Glass JI, Smith HO, Hutchison CA, Alperovich NY, Assad-Garcia N (2015) Minimal bacterial genome. US2015344837 A1
Glass JI, Young L, Lartigue C, Assad-Garcia N, Smith HO, Hutchison CA, Venter JC (2016) Installation of genomes or partial genomes into cells or cell-like systems. US9434974 B2
Glass JT, Smith HO, Clyde A Hutchison C, Alperovich NY, Assad-Garcia N (2008b) Minimal bacterial genome. EP1951874 A1
Hacon R, Pagenberg J (2008) Concise European patent law, 2nd edn. Kluwer Law International, Alphen aan den Rijn
Hagglund R (2007) Patentability of cloned extinct animals. George Mason Law Rev 15:381–446
Hansen J (2013a) Method of producing isoprenoid compounds in yeast. US20130137138 A1

Hansen J (2013b) Method of producing isoprenoid compounds in yeast. EP2571991 A1

Heath C (2009) The scope of DNA patents in the light of the recent Monsanto decisions. Int Rev Intellect Prop Compet Law 40:940–957

High Court of Australia (2015) D'Arcy v. Myriad Genetics Inc. & Anor, [2015] HCA 35

Holman CM (2013) Editorial: In Myriad the Supreme Court has, once again, increased the uncertainty of U.S. patent law. Biotechnol Law Rep 32:289–293

Holman CM (2015) Developments in synthetic biology are altering the IP imperatives of biotechnology. Vanderbilt J Entertain Technol Law 17:385–462

Hultquist SJ, Harrison R, Yang Y (2002) Patenting bioinformatic inventions: emerging trends in Europe. Nat Biotechnol 20:517–518

Hutchison CA, Montague MG, Smith HO (2011) Encoding text into nucleic acid sequences. WO2011053868 A1

Hutchison CA, Montague MG, Smith HO (2012) Encoding text into nucleic acid sequences. EP2494060 A1

Hutchison CA, Montague MG, Smith HO (2016a) Encoding text into nucleic acid. US2016168579 A1

Hutchison CA, Montague MG, Smith HO (2016b) Encoding text into nucleic acid sequences. EP2494060 B1

Huttermann A, Storz U (2009) A comparison between biotech and software related patents. Eur Intellect Prop Rev 31:589–592

Huys I, Van Overwalle G, Matthijs G (2011) Gene and genetic diagnostic method patent claims: a comparison under current European and US patent law. Eur J Hum Genet 19:1104–1107

Intellectual Property Office (IPO) (2016) Examination guidelines for patent applications relating to biotechnological inventions in the Intellectual Property Office

Intellectual Property Office (IPO) (2017) Manual of patent practice

Iwasaka RMT (2000) From Chakrabarty to chimeras: the growing need for evolutionary biology in patent law. Yale Law J 109:1505–1534

Jayaraman P, Poh CL, Wang HJ (2015) Systems and methods for synthetic biology design and host cell simulation. WO2015199614 A1

Kane EM (2004) Splitting the gene: DNA patents and the genetic code. Tenn Law Rev 71:707–767

Kane EM (2010) Patenting genes and genetic methods: what's at stake? J Bus Technol Law 6:1–33

King PE, Roberts RM, Moshirnia AV (2010) The confluence of European activism and American minimalism: patentable subject matter after Bilski. Santa Clara Comput High Technol Law J 27:247

Koepsell D (ed) (2011) Innovation and nanotechnology: converging technologies and the end of intellectual property. Bloomsbury Academic, London

Koepsell D (2014) Synthetic biology and IP: how do definitions of "products of nature" affect their implications for health? In: de Miguel Beriain I, Casabona CMR (eds) Synbio and Human Health. Springer, Dordrecht, pp 45–53

Kohler J (1900) Handbuch des deutschen Patentrechts in rechtsvergleichender Darstellung. J. Bensheimer, Mannheim

Konski AF (2020) Adjusting to Alice: USPTO's view of its examination guideline [WWW Document]. www.natlawreview.com. URL https://www.natlawreview.com/article/adjusting-to-alice-uspto-s-view-its-examination-guidelines. Accessed 4 May 2020

Krishnakumar R, Glass JI, Merryman CE (2010a) System and method for producing synthetic microorganisms capable of translating proteins containing non-standard amino acids. EP2217700 A1

Krishnakumar R, Glass JI, Merryman CE (2010b) System and method for producing synthetic microorganisms capable of translating proteins containing non-standard amino acids. US20100297698 A1

Kurzchalia T (1995) Specific incorporation of unnatural amino acid into protein in vivo DE19500932 A1

Lai JC (2015) Myriad genetics and the BRCA patents in Europe: the implications of the U.S. Supreme Court decision. UC Irvine Law Rev 5:1041–1075

Lai JC (2016) D'Arcy v. Myriad genetics: a demand for the "made" or "non-information" and clear subject matter? Int Rev Intellect Prop Compet Law 47:537–568

Landers AL (2015) Patentable subject matter as a policy driver. Houston Law Rev 53:505–547

LeGuyader J (2009) Presentation: patents and synthetic biology. Presented at the Patenting Synthetic Biology - A Transatlantic Perspective workshop, Washington DC

Lesser W (2011) Nature or nurture: is there a case basis for a judicially created product of nature exclusion? Are genes somehow different? J Marshall Rev Intellect Prop Law 11:318–366

LibreTexts.org (2015) Catalytic efficiency of enzymes [WWW Document]. www.libretexts.org. URL. https://chem.libretexts.org/Bookshelves/Biological_Chemistry/Supplemental_Modules_(Biological_Chemistry)/Enzymes/Enzymatic_Kinetics/Catalytic_Efficiency_of_Enzymes. Accessed 2 Mar 2020

Lou C, Moon TS, Rhodius V, Stanton B, Tamsir A, Temme K, Voigt C (2012) Synthetic biology tools. WO2012170436 A1

Lu T, Farzadfard F (2016) Genomically-encoded memory in live cells. WO2016025719 A1

Lu TK-T, Siuti P (2014) Recombinase-based logic and memory systems. US20140315310 A1

Lu TK-T, Siuti P (2015) Recombinase-based logic and memory systems. EP2932443 A1

Lucchi N (2018) Genetic copyright: an alternative method for protecting and using essential public knowledge assets? Eur Intellect Prop Rev 40:766–774

Ludlow K (1999) Genetically modified organisms and their products as patentable subject-matter in Australia. Eur Intellect Prop Rev 21:298–312

Macchia G (2012) Patentability requirements of biotech inventions at the European Patent Office: ethical issues. In: Bin R, Lorenzon S, Lucchi N (eds) Biotech innovations and fundamental rights. Springer, Milan, pp 37–43

Manmohan P (2012) Oligonucleotides and process for detection of swine flu virus. US 13/509,535

Marks J, Carron V (2018) Courts are trending toward broader patent eligibility [WWW Document]. www.fooddive.com. URL https://www.fooddive.com/news/courts-are-trending-toward-broader-patent-eligibility/542742/. Accessed 2 Mar 2020

Marterer N (1987) The patentability of micro-organisms per se. IIC 18:666–681

Martinez-Lopez A (2020) Decision T1063/18 of the EPO Technical Board re-sows the seed of uncertainty for the patentability of plant products derived from essentially biological processes. Eur Intellect Prop Rev 42:19–27

McBride MS (2002) Bioinformatics and intellectual property protection. Berkeley Technol Law J 17:1331–1364

McGarrigle P, Norviel V (2008) Laws of nature and the business of biotechnology. Santa Clara High Technol Law J 24:275

McLennan A (2017) Patent law and the emerging science of synthetic biology: an examination of principle and practice. Biotechnol Law Rep 36:59–116

McLennan A (2018) Regulation of synthetic biology: biobricks, biopunks and bioentrepreneurs. Edward Elgar, Cheltenham

Merges RP, Duffy JF (2017) Patent law and policy: cases and materials, 7th edn. Carolina Academic Press, Durham

Merriam Webster's Collegiate Dictionary (2014) Biological, 11th edn. Merriam-Webster, Springfield

Minssen T (2015) The impact of Broccoli II and Tomato II on European patents in conventional breeding, GMO's and synthetic biology: a grand finale of a juicy patents tale? [WWW Document]. http://blogs.harvard.edu. URL https://blog.petrieflom.law.harvard.edu/2015/06/17/the-impact-of-broccoli-ii-tomatoes-ii-on-european-patents-in-conventional-breeding-gmos-and-synthetic-biology-the-grand-finale-of-a-juicy-patents-tale/. Accessed 2 Mar 2020

Minssen T, Nordberg A (2015) The impact of Broccoli II & Tomatoes II on European patents in conventional breeding, GMO's and synthetic biology: the grand finale of a juicy European patents tale? Biotechnol Law Rep 34:81–98

Mohan-Ram VS, Waxman MJ (2008) Synthetic biology patent applications expected to present new challenges. Life Sci Law Ind 2:1–3
Molecules vs. information: should patents protect both?, 2002. Boston Univ J Sci Technol Law 8:190–217
Mouta CA, Tridico AC (2015) Structure or function: Which will ultimately have the most influence on patent eligibility of biotech inventions? [WWW Document]. www.finnegan.com. URL https://www.finnegan.com/en/insights/structure-or-function-which-will-ultimately-have-the-most.html. Accessed 2 Mar 2020
Murray MD (2014) Post-Myriad genetics copyright of synthetic biology and living media. Oklahoma J Law Technol 10:1–56
Myszczuk AP, De Meirelles JML (2014) Patents and living matter: the construction of a patent system attractive to biotechnology. In: de Miguel Beriain I, Casabona CMR (eds) Synbio and human health. Springer, Dordrecht, pp 77–86
Neethu R (2018) Rekindling the debate on genetic copyright in Europe in the era of biobanks and synthetic biology. Eur Intellect Prop Rev 40:172–177
Noonan K (2014) Docs @ BIO: Bloomberg BNA issues report on PTO's patent eligibility guidance [WWW Document]. www.patentdocs.org. URL https://www.patentdocs.org/2014/07/docs-bio-bna-issues-report-on-ptos-patent-eligibility-guidance.html. Accessed 2 Mar 2020
Nose A (2010) Thesis: the impact of patenting on synthetic biology inventions. University of Tilburg, Tilburg
Nuffield Council on Bioethics (2002) The ethics of patenting DNA: a discussion paper. Nuffield Council on Bioethics, London
Obranovich T (2016) Biotechnology and patentability: navigating unchartered waters. Managing Intellect Prop 256:8–11
Odell-West A (2011) "Gene"-uinely patentable? The distinction in biotechnology between discovery and invention in US and EU patent law. Intellect Prop Q:304–321
OECD (2014) Emerging policy issues in synthetic biology. OECD, Paris
Oksanen K, Virtanen P, Soisalon-Soininen E, Kemppinen J (2011) Arguments in considering the similarity of algorithms in patenting. Scripted 8:138–153
Orford J (2014) Association for molecular pathology v. Myriad Genetics Inc.: are genes information or molecules? Sydney Law Rev 36:557
Palombi L (2003) Patentable subject matter, TRIPS and the European Biotechnology Directive: Australia and patenting human genes. Univ N S W Law J 26:782–792
Palombi L (2005) The impact of TRIPS on the validity of the European Biotechnology Directive. J Int Biotechnol Law 2:62–70
Palombi L (2009a) Gene cartels: biotech patents in the age of free trade. Edward Elgar, Cheltenham
Palombi L (2009b) Beyond recombinant technology: synthetic biology and patentable subject matter. J World Intellect Prop 12:371–401
Palombi L (2016) Association for Molecular Pathology v Myriad Genetics (US) and D'Arcy v Myriad Genetics (AU): are gene patents in Europe a threatened species? Eur Intellect Prop Rev 38:231–236
Patenting the Parts (2007) Nat Biotechnol 25:822–822
Phukan PD (2014) Patenting proteins after Myriad. Federal Circuit Bar J 23:619–646
Pila J (2005a) Article 52(2) of the convention on the grant of European patents: what did the framers intend? A study of the travaux preparatoires. Int Rev Intellect Prop Compet Law 36:755–787
Pila J (2005b) Dispute over the meaning of 'invention' in Article 52(2) EPC: the patentability of computer-implemented inventions in Europe. Int Rev Ind Prop Copyright Law 36:173–191
Pila J (2010) On the European requirement for an invention. Int Rev Intellect Prop Compet Law 41:906–926
Pila J (2011) The future of the requirement for an invention: Inherent patentability as a pre- and post-patent determinant. In: Arezzo E, Ghidini G (eds) Biotechnology and software patent law: a comparative review of new developments. Edward Elgar, Cheltenham, pp 55–90

Pila J (2014a) Patent eligibility and scope revisited in light of Schütz v. Werit, European law, and copyright jurisprudence. In: Dreyfuss RC, Ginsburg JC (eds) Intellectual property at the edge: the contested contours of IP. Cambridge University Press, Cambridge, pp 382–401

Pila J (2014b) Isolated human genes: the patent equivalent of a non-copyrightable sound recording. Law Q Rev 180–185

Pizzoferrato A (2002) Brevetto per invenzione e biotecnologie, trattato di diritto commerciale e di diritto pubblico dell'economia. CEDAM, Padova

Porter G (2009) The drafting history of the European Biotechnology Directive. In: Plomer A, Torremans P (eds) Embryonic stem cell patents: European law and ethics. Oxford University Press, Oxford, pp 3–26

Price WNI (2015) Big data, patents, and the future of medicine. Cardozo Law Rev 37:1401–1453

Proposal for a Council Directive on the legal protection of biotechnological inventions (COM/88/496FINAL - SYN 159), 1988

Purevdorj D (2011) Thesis: patenting of bioinformatics inventions related to drug discovery. Munich Intellectual Property Law Center, Munich

Rabin S (2006) The human use of humanoid beings: chimeras and patent law. Nat Biotechnol 24:517–519

Radick G (2002) Discovering and patenting human genes. In: Bainham A, Sclater SD, Richards M (eds) Body lore and laws: essays on law and the human body. Hart Publishing, London, pp 63–78

Rai AK (2000) Addressing the patent gold rush: the role of deference to PTO patent denials

Reeve N (2015) None of this is real: the patentability of computer simulations in Europe [WWW Document]. www.reddie.co.uk. URL https://www.reddie.co.uk/2019/03/14/can-you-patent-a-computer-implemented-simulation-in-europe/. Accessed 2 Mar 2020

Resnik DB (2002) Discoveries, inventions, and gene patents. In: Magnus D, Caplan AL, McGee G (eds) Who owns life? Prometheus Books, Amherst, pp 135–159

Rimmer M (2003) Beyond blue gene: intellectual property and bioinformatics. Int Rev Intellect Prop Compet Law 34:31–49

Rimmer M (2008) Intellectual property and biotechnology: biological inventions. Edward Elgar, Cheltenham

Roberts T (2010) Discovery and invention: where is the boundary between them and should we care? Bio-Sci Law Rev 11:35–38

Robienski J, Simon J (2014) Synthetic biology: challenges and legal questions. In: de Miguel Beriain I, Casabona CMR (eds) Synbio and human health. Springer, Dordrecht, pp 123–143

Rogers DL (2013) After Prometheus, are human genes patentable subject matter? Duke Law Technol Rev 11:434–508

Rohrbaugh ML (1997) The patenting of extinct organisms: revival of lost arts. AIPLA Q J 25:371–417

Rose D (2020) Adjusting to Alice: USPTO report analyzes patent examination outcomes after Alice Corp. v. CLS Bank International [WWW Document]. www.natlawreview.com. URL https://www.natlawreview.com/article/adjusting-to-alice-uspto-report-analyzes-patent-examination-outcomes-after-alice. Accessed 4 Mar 2020

Rummler F (2005) Computer program inventions before the German courts: a review. Int Rev Intellect Prop Compet Law 36:225–234

Rutz B (2009) Synthetic biology and patents. A European perspective. EMBO Rep 10:S14–S17

Rutz B (2010a) Presentation: patent issues in SynBio applications. Presented at the Synthetic biology: From science to governance workshop, Brussels

Rutz B (2010b) Patentability of synthetic biology inventions in Europe. Biotechnol J 5:11–13

Sampson T (2002) Rewriting the genetic code: the impact of novel nucleotides on biotechnology patents. Eur Intellect Prop Rev 24:409–414

Samuelson P (n.d.) Are gardens, synthetic DNA, yoga sequences, and fashion copyrightable? [WWW Document]. http://law.scu.edu. URL http://law.scu.edu/wp-content/uploads/Samuelson-Gardens-DNA-Yoga-Fashion-Abstract.pdf. Accessed 2 Mar 2020

Sarmiento RJ, Baskerville DS, Zhang X (2015) Structure based predictive modeling. US20150134315 A1

Sarnoff JD (2012) The current status of patent eligibility of medical and biotechnological inventions in the United States. In: Rimmer M, McLennan A (eds) Intellectual property and emerging technologies: the new biology. Edward Elgar, Cheltenham, pp 84–116

Saunders M (2008) Creating life from scratch: the patentability of synthetic organisms. Tulane J Technol Intellect Prop 11:75–89

Schertenleib D (2003) The patentability and protection of DNA-based inventions in the EPO and the European Union. Eur Intellect Prop Rev 25:125–138

Schilling SH (2011) DNA as patentable subject matter and a narrow framework for addressing the perceived problems caused by gene patents. Duke Law J 61:731–773

Schneider I (2014) Exclusions and exceptions to patent eligibility revisited: examining the political functions of the "discovery" and "ordre public" clauses in the European Patent Convention and the arenas of negotiation. In: de Miguel Beriain I, Casabona CMR (eds) Synbio and human health. Springer, Dordrecht, pp 145–173

Schultz P, Wang L, Anderson J, Chin J, Liu D, Magliery T, Meggers E, Mehl R, Pastrnak M, Santoro S, Zhang Z (2003) In vivo incorporation of unnatural amino acids. US20030082575 A1

Schultz P, Wang L, Anderson JC, Chin JW, Liu DR, Magliery TJ, Meggers EL, Mehl RA, Pastrnak M, Santoro SW, Zhang Z (2006) In vivo incorporation of unnatural amino acids. US7045337 B2

Schultz P, Wang L, Anderson JC, Chin JW, Liu DR, Magliery TJ, Meggers EL, Mehl RA, Pastrnak M, Santoro SW, Zhang Z (2015) In vivo incorporation of unnatural amino acids. EP1490483 B1

Schwartz RM, Minssen T (2015) Life after Myriad: the uncertain future of patenting biomedical innovation & personalized medicine in an international context. Intellect Prop Q 189–241

Schwille P (2015) Jump-starting life? Fundamental aspects of synthetic biology. J Cell Biol 210:687–690

Sena G (1999) Directive on biotechnological inventions: patentability of discoveries. Int Rev Intellect Prop Compet Law 30:731–738

Shemtov N (2009) The characteristics of technical character and the ongoing saga in the EPO and the English courts. J Intellect Prop Law Pract 4:506–514

Sherman B (2015) What does it mean to invent nature? UC Irvine Law Rev 5:1193

Sieman JS (2011) Majority holds § 101 does not exclude claims directed to a specific, tangible application; Judge Moore finds majority's analysis incomplete [WWW Document]. www.finnegan.com. URL https://www.finnegan.com/en/tools/classen-immunotherapies-inc-v-biogen-idec/analysis.html. Accessed 2 Mar 2020

Singer M, Stauder D (2003) European Patent Convention: a commentary, 3rd edn. Sweet & Maxwell, Köln

Smalley LW (2013) Will nanotechnology products be impacted by the Federal Courts' product of nature exception to subject-matter eligibility under 35 U.S.C. 101? J Marshall Rev Intellect Prop Law 13:397–443

Sommer T (2013) Can law make life (too) simple?: From gene patents to the patenting of environmentally sound technologies, 1st edn. DJØF Publishing, Copenhagen

Spanish Bioethics Committee, Portuguese National Ethics Council for the Life Sciences (2011) Synthetic biology - A joint report by the Spanish Bioethics Committee and the Portuguese National Ethics Council for the Life Sciences. Spanish Bioethics Committee - Portuguese National Ethics Council for the Life Sciences, Lisbon - Barcelona

Stauder D, Luginbuehl S, Singer R, Singer M (eds) (2013) Europäisches Patentübereinkommen: Kommentar, 6th edn. Heymann, Köln

Stauder D, Luginbuehl S, Almer KJ, Blumer F, Singer R, Singer M (2016) Europäisches Patentübereinkommen: Kommentar, 7th edn. Heymanns, Köln

Sterckx S, Cockbain J (2012) Exclusions from patentability: how far has the European Patent Office eroded boundaries? Cambridge University Press, Cambridge

Swedlow MR (2015) The woolly-mammoth in the room: the patentability of animals brought back from extinction through cloning and genetic engineering. Washington J Law Technol Arts 11:183–196

SYBHEL (2014) Final report summary - SYBHEL (Synthetic Biology for Human Health: Ethical and Legal Issues) (Executive Summary). European Commission

Technical Board of Appeal of the EPO (TBA) (1986) T 0208/84 (Computer-related invention/Vicom)

Technical Board of Appeal of the EPO (TBA) (1988) T 0320/87 (Hybrid plants/Lubrizol)

Technical Board of Appeal of the EPO (TBA) (1989) T 0038/86 (Text processing)

Technical Board of Appeal of the EPO (TBA) (1994a) T 0059/93 (Method for interactive rotation/IBM)

Technical Board of Appeal of the EPO (TBA) (1994b) T 0453/91 (Method for physical VLSI-chip design/IBM)

Technical Board of Appeal of the EPO (TBA) (1995a) T 877/92 (Removal of carcinogenic hydrocarbons/Exxon)

Technical Board of Appeal of the EPO (TBA) (1995b) T 0356/93 (Plant cells)

Technical Board of Appeal of the EPO (TBA) (1997) T 869/95 (Nissan Motor)

Technical Board of Appeal of the EPO (TBA) (1998) T 1173/97 (Computer program product/IBM)

Technical Board of Appeal of the EPO (TBA) (1999a) T 919/99 (Human stem cell/Biocyte)

Technical Board of Appeal of the EPO (TBA) (1999b) T 0935/97 (Computer program product II/IBM)

Technical Board of Appeal of the EPO (TBA) (2000a) T 931/95 (Controlling pension benefits system/PBS Partnership)

Technical Board of Appeal of the EPO (TBA) (2000b) T 1194/97 (Data structure product/Philips)

Technical Board of Appeal of the EPO (TBA) (2002a) T 0641/00 (Two identities/Comvik)

Technical Board of Appeal of the EPO (TBA) (2002b) T 0049/99 (Information modelling/International computers)

Technical Board of Appeal of the EPO (TBA) (2002c) T 0272/95 (Relaxin/Howard Florey Institute)

Technical Board of Appeal of the EPO (TBA) (2002d) T 0338/00 (Multimeric Receptors/Salk Institute)

Technical Board of Appeal of the EPO (TBA) (2004) T 0258/03 (Auction method/Hitachi)

Technical Board of Appeal of the EPO (TBA) (2005) T 0914/02 (Method for determining nuclear core loading arrangements/GE)

Technical Board of Appeal of the EPO (TBA) (2006a) T 0424/03 (Clipboard formats I/Microsoft)

Technical Board of Appeal of the EPO (TBA) (2006b) T 1227/05 (Circuit simulation I/Infineon Technologies)

Technical Board of Appeal of the EPO (TBA) (2007a) T 1213/05 (Breast and ovarian cancer/University of Utah)

Technical Board of Appeal of the EPO (TBA) (2007b) T 0471/05 (Optical system/Philips)

Technical Board of Appeal of the EPO (TBA) (2008a) T 0666/05 (Mutation/University of Utah)

Technical Board of Appeal of the EPO (TBA) (2008b) T 0080/05 (Method of diagnosis/University of Utah).

Technical Board of Appeal of the EPO (TBA) (2008c) T 0331/06 (Order processing/Advanced transaction systems)

Technical Board of Appeal of the EPO (TBA) (2010) T 0784/06 (Genotype determination/Beckman)

Technical Board of Appeal of the EPO (TBA) (2012) T 0531/09 (Checkpoint simulation/Accenture)

Technical Board of Appeal of the EPO (TBA) (2018) T 1063/18 (Extreme dark green, blocky peppers/Syngenta)

Teschemacher R (1982) Patentability of microorganisms per se. Int Rev Intellect Prop Compet Law 13:27–41

The US genetic information nondiscrimination Act of 2008 (2008) 122 Stat. 881

Thornton J (2002) Congratulations to the "inventors" of our genes. Newcastle Law Rev 6:77
Toole AA, Pairolero NA (2020) Adjusting to Alice - USPTO patent examination outcomes after Alice Corp. v. CLS Bank International. USPTO
Torrance AW (2010) Synthesizing law for synthetic biology. Minnesota J Law Sci Technol 11:629–665
Torrance AW, Kahl LJ (2013) Bringing standards to life: synthetic biology standards and intellectual property. Santa Clara High Technol Law J 30:199–230
Trevisan L, Cuonzo G (2013) Proprietà industriale, intellettuale e IT, 1st edn. IPSOA, Milano
Tritton G, Davis R, St. Quintin T (2018) Tritton on intellectual property in Europe, 5th edn. Sweet & Maxwell, London
Twilley N (2016) What's the point of streamlining nature? [WWW Document]. www.newyorker.com. URL https://www.newyorker.com/tech/annals-of-technology/whats-the-point-of-streamlining-nature. Accessed 29 Feb 2020
U.S. Boards of Patent Interferences and Appeals (1987) Ex parte Allen
U.S. Boards of Patent Interferences and Appeals (2006) Ex parte Bilski
U.S. Boards of Patent Interferences and Appeals (2009) Ex parte Srinivas Gutta and Kaushal Kurapati
U.S. Circuit Court of Appeals - Third Circuit (1928) General Electric Co. v. De Forest Radio Co., 28 F.2d 641 (3d Cir. 1928)
U.S. Circuit Court, S.D. New York (1911) Parke-Davis & Co. v. H.K. Mulford Co., 189 F. 95 (S.D.N.Y. 1911)
U.S. Code - Title 17 - Copyrights, n.d. , 17 U.S.C (2006)
U.S. Commissioner of Patents (1889) Ex parte Latimer
U.S. Court of Appeals for the Federal Circuit (1991) Amgen, Inc. v. Chugai Pharmaceutical Co., Ltd., and Genetics Institute, Inc., 927 F.2d 1200 (Fed. Cir. 1991)
U.S. Court of Appeals for the Federal Circuit (2005) SmithKline Beecham Corporation and Beecham Group, P.L.C. v. Apotex Corp., Apotex Inc., and TorPharm, Inc., 403 F.3d 1331 (Fed. Cir. 2005)
U.S. Court of Appeals for the Federal Circuit (2007) In re Nuijten, 500 F.3d 1346 (Fed. Cir. 2007)
U.S. Court of Appeals for the Federal Circuit (2011) The Association for Molecular Pathology, et al. v. United States Patent and Trademark Office, et al. (Myriad), 689 F.3d 1303 (Fed. Cir. 2012)
U.S. Court of Appeals for the Federal Circuit (2014a) In re BRCA1– and BRCA2–based hereditary cancer test patent litigation. University of Utah Research Foundation, the Trustees of the University of Pennsylvania, HSC Research and Development Limited Partnership, Endorecherche, Inc., and Myriad Genetics, Inc., v. Ambry Genetics Corporation, 774 F.3d 755 (Fed. Cir. 2014)
U.S. Court of Appeals for the Federal Circuit (2014b) In re Roslin Institute (Edinburgh), 750 F. 3d 1333 (Fed. Cir. 2014)
U.S. Court of Appeals for the Seventh Circuit (1910) Kuehmsted v. Farbenfabriken of Elberfeld Co., 179 F. 701
U.S. Court of Customs and Patent Appeals (CCPA) (1931) In re Marden, 47 F.2d 958 (C.C.P.A. 1931)
U.S. Court of Customs and Patent Appeals (CCPA) (1938) In re Merz, 97 F.2d 599 (C.C.P.A. 1938)
U.S. Court of Customs and Patent Appeals (CCPA) (1970) Application of Bergstrom, 427 F.2d 1394 (C.C.P.A. 1970)
U.S. Court of Customs and Patent Appeals (CCPA) (1977) In re Bergy, 563 F.2d 1031 (C.C.P.A. 1977)
U.S. Department of Energy (2004) Synthetic genomes: technologies and impact [WWW Document]. https://science.osti.gov. URL https://science.osti.gov/-/media/ber/berac/pdf/Syn_bio.pdf. Accessed 3 Mar 2020
U.S. District Court District of Maryland (2014) Classen Immunotherapies, Inc. v. Biogen Idec

U.S. District Court Southern District of New York (2010) Association for Molecular Pathology, et al. v. United States Patent and Trademark Office, et al. (Myriad)
U.S. Patent Act - United States Code Title 35 - Patents (1952)
U.S. Supreme Court (1852) Le Roy v. Tatham, 55 U.S. 156 (1852)
U.S. Supreme Court (1853) O'Reilly v. Morse, 56 U.S. 62 (1853)
U.S. Supreme Court (1874) American Wood-Paper Co. v. Fibre Disintegrating Co., 90 U.S. 566 (1874)
U.S. Supreme Court (1884) Cochrane v. Badische Anilin & Soda Fabrik (BASF), 111 U.S. 293 (1884)
U.S. Supreme Court (1931) American Fruit Growers, Inc. v. Brogdex Co., 283 U.S. 1 (1931)
U.S. Supreme Court (1948) Funk Brothers Seed Co. v. Kalo Inoculant Co., 333 U.S. 127 (1948)
U.S. Supreme Court (1950) Graver Tank & Mfg. Co. v. Linde Air Products Co., 339 U.S. 605 (1950)
U.S. Supreme Court (1966) Graham v. John Deere Co., 383 U.S. 1 (1966)
U.S. Supreme Court (1972) Gottschalk v. Benson, 409 U.S. 63 (1972)
U.S. Supreme Court (1978) Parker v. Flook, 437 U.S. 584 (1978)
U.S. Supreme Court (1980) Diamond v. Chakrabarty, 447 U.S. 303 (1980)
U.S. Supreme Court (1981) Diamond v. Diehr, 450 U.S. 175 (1981)
U.S. Supreme Court (2006) Laboratory Corporation of America Holdings, DBA Labcorp, v. Metabolite Laboratories, Inc. et al., 548 U.S. 124 (2006)
U.S. Supreme Court (2012) Mayo Collaborative Services, DBA Mayo Medical Laboratories, et al. v. Prometheus Laboratories, Inc., 566 U.S. 66 (2012)
U.S. Supreme Court (2013) Association for Molecular Pathology, et al. v. Myriad Genetics, Inc., et al. (Myriad), 569 U.S. 576 (2013)
U.S. Supreme Court (2014) Alice Corporation Pty. Ltd. v. CLS Bank International, 573 U.S. 208 (2014)
UK Court of Appeal (CA) (1989a) Genentech's patent, [1989] RPC 147
UK Court of Appeal (CA) (1989b) Merrill Lynch's application, [1989] RPC 561
UK Court of Appeal (CA) (2008) Symbian Ltd v Comptroller General [2008] EWCA Civ 1066
UK High Court of Justice (Patents Court) (2011) Re Halliburton Energy Services Inc, [2011] EWHC 2508 (Pat)
UK House of Lords (HL) (1996) Biogen Plc v Medeva Plc, [1997] RPC 1
UK House of Lords (HL) (2004) Kirin-Amgen Inc and others v Hoechst Marion Roussel Limited and others etc, [2005] RPC 169
UK Patent Act - The Patents Act 1949 (Chapter 87), 1949
UK Patents Court (2005) CFPH LLC, patent applications by [2005] EWHC 1589 (Pat)
Ulen N (2019) Is DNA really a natural product: it's time to separate fact from legal fiction: an examination of DNA patentability as a biological algorithm in the post-Myriad era. Chicago-Kent Law Rev 94:205–226
USPTO (2000) Utility examination guidelines
USPTO (2001) Utility examination guidelines.
USPTO (2014a) 2014 Interim guidance on patent subject matter eligibility
USPTO (2014b) 2014 Interim guidance on subject matter eligibility - Nature-based product examples
USPTO (2016) Subject matter eligibility examples: life sciences
USPTO (2018) Changes in examination procedure pertaining to subject matter eligibility, recent subject matter eligibility decisions (Berkheimer v. HP, Inc.)
USPTO (2019) 2019 Revised patent subject matter eligibility guidance
USPTO (n.d.) EPO Global Dossier - Patent US54636406 [WWW Document]. www.epo.org. URL https://register.epo.org/ipfwretrieve?apn=US.54636406.A&lng=en. Accessed 4 Mar 2020
van den Belt H (2009) Philosophy of biotechnology. In: Meijers A (ed) Philosophy of technology and engineering sciences. North-Holland, Amsterdam, pp 1301–1340

van den Belt H (2014) Synthetic biology and global health in the age of intellectual property. In: de Miguel Beriain I, Casabona CMR (eds) Synbio and human health. Springer, Dordrecht, pp 19–43

Van Overwalle G (2007) Reshaping bio-patents: measures to restore trust in the patent system. In: Somsen H (ed) The regulatory challenge of biotechnology - human genetics, food and patents. Edward Elgar, Cheltenham, pp 238–256

Van Overwalle G (2008) Biotechnology and patents: global standards, European approaches and national accents. In: Wüger D, Cottier T (eds) Genetic engineering and the world trade system. Cambridge University Press, Cambridge, pp 77–108

Venter JC, Smith HO (2008) Synthetic genomes. EP1968994 A2

Venter JC, Smith HO, Iii CAH, Gibson DG (2013) Synthetic genomes. EP1968994 B1

Venter JC, Smith H, Hutchison C (2007) Synthetic genomes. US20070264688 A1

Venter JC, Gibson D, Gill JE (2014) Digital to biological converter. US20140274808 A1

Venter JC, Gibson DG, Gill JE (2015) Digital to biological converter. EP2885408 A1

Vishnubhakat S, Rai AK (2015) When biopharma meets software: bioinformatics at the patent office. Harv J Law Technol 29:205–241

Vogt T (2010) Presentation: patentability in biotechnology. Presented at the 4th Annual Forum for SMEs - Information on European Bioinformatics Resources Workshop, Munich

Warren-Jones A (2001) Patenting rDNA: human and animal biotechnology in the United Kingdom and Europe. Lawtext Pub, Witney

Watson BT (2014) Carbons into bytes: patented chemical compound protection in the virtual world. Duke Law Technol Rev 12:25–40

Wegner HC (1976) Patenting nature's secrets: microorganisms. Int Rev Intellect Prop Compet Law 7:235–246

Westerlund L (2002) Biotech patents: equivalence and exclusions under European and U.S. patent law. Kluwer Law International, The Hague

Wilkof N (2017) More on broccoli, tomatoes, and the patentability of a plant or animal obtained by means of an essentially biological process [WWW Document]. http://ipkitten.blogspot.com. URL http://ipkitten.blogspot.com/2017/07/more-on-broccoli-tomatoes-and.html. Accessed 2 Mar 2020

Willgoos C, Gont I (2015) Federal Circuit rules Myriad's "primers" and methods are not patent-eligible subject matter. Intellect Prop Technol Law J 27:3–6

World Intellectual Property Organization (WIPO) (1978) Geneva treaty on the international recording of scientific discoveries, (IPLT Text 1-003)

World Science (2007) First patent claimed on man-made life form, and challenged [WWW Document]. www.world-science.net. URL http://www.world-science.net/othernews/070607_mycoplasma.htm. Accessed 3 July 2016

Zimmer F-J, Quest B (2011) When is a process for the production of plants "essentially biological"? Decisions of the Enlarged Board of Appeal (G1/08 and G2/07). Biotechnol Law Rep 30:459–464

Zimmer F-J, Zeman SM, Hammer J, Goldbach K, Allekotte B, Goldbach K (2015) Protecting and enforcing life science inventions in Europe under EPC and EU law: from antibodies to zebrafish, 2nd edn. C.H. Beck, München

# Chapter 5
# The Morality of Synthetic Biology Inventions

## 5.1 Exceptions to Patentability for Inventions Contrary to *Ordre Public* or Morality

The morality clause has been invoked in several controversies concerning biotechnological inventions. Yet, it has seldom been successful. With the advent of synthetic biology, questions were raised on the impact that this clause will have on this new technology.

Differently from the exclusions in Article 52(2) EPC, scholars and organisations have often envisioned the possibility that SynBio inventions might raise morality issues. However, the concerns listed under the morality heading do not always have a legally relevant connotation. The patent system denies protection to inventions that would be contrary to *ordre public* and morality. Hence, concerns not falling within this realm will not be relevant from a patent perspective. As previously seen in Chap. 2, many of the critiques moved against synthetic biology might not necessarily fall within this category, as they rather signal a discomfort with this technology from a safety, personal ethic and regulatory perspective. Whether those concerns will be enough to hinder patentability will be examined next.

Such analysis is necessary, as this issue is still virtually untapped. Until now, very few scholars have addressed the legal morality issues concerning synthetic biology (as opposed to non-legally relevant ones). Likewise, the morality clause has not been applied so far to synthetic biology inventions either before Boards or Courts.

To carry out this evaluation, norms and decisions on the morality clause in Europe will be examined, in so far as those may have a bearing on synthetic biology. Subsequently, those findings will be applied to synthetic biology to identify if and how could morality issues arise. In this assessment, reference will be made to patents and patent applications presented before the EPO as well as to the findings outlined in Chap. 2 on the ethics, safety and acceptance of this technology.



I. de Lisa, *The Patentability of Synthetic Biology Inventions*,
https://doi.org/10.1007/978-3-030-51206-4_5

### 5.1.1 Introduction

Exceptions to the patentability of inventions infringing against *ordre public* and morality have been a constant presence in patent law, despite the idea that the patent system should be value-neutral.[1] Of the 73 national patent legislations examined in 2010 by the WTO, all but six contained morality provisions.[2] The most notable exception were the United States, where no explicit morality clause was found.[3] In spite of its widespread presence, the exception has only rarely been used before patent offices and Courts.[4]

[1]For both pros and cons opinions, Schneider (2014, p. 152), Bonadio (2012, p. 435), Derclaye (2009, p. 254) and Moufang (1998, p. 77). The positivist and natural schools of law have had a say in this debate. Positivists tend to argue that law should be separated from morality and based solely on textual logic and reason. By contrast, naturalists believe that laws reflect the moral principles of society and that, therefore, the two cannot be separated. In light of this, positivists are likely to argue in favour of a reduced impact of the morality clause, whereas naturalists would believe that immoral inventions should not be given legal character (Liddell 2012, p. 155; Bently et al. 2010, p. 43). *Bently* and *Sherman* noted that: "*One of the defining characteristics of patent law over the last century has been, not only its highly technical and specialized nature, but also its startling and marked isolation from matters cultural, political and ethical*" (Bently and Sherman 1995, p. 275). The Patent Statute of 1623 would grant patents only if they were: "*Not contrary to law or mischievous to the State*" (Derclaye 2009, p. 258; UK Government 1623). Other authors considered the 1844 French Patent Act as the first patent law that dealt with morality (Ugurlu 2014, p. 22). The 1883 Paris Convention for the Protection of Intellectual Property also contained a morality clause, albeit for copyright (Plomer 2009, p. 176). For an overview of laws in other jurisdictions, Moufang (1998, p. 75).

[2]Bently et al. (2010).

[3]Moral concerns are not incorporated in the US patent system. The USPTO has shown unwillingness to be a moral policy-maker and has been reluctant to use the nineteenth century moral utility doctrine (Ottolia 2011, pp. 334–335). This doctrine was coined in the *Lovell v. Lewis* case in 1817. It held that an immoral invention was not useful, since something is useful only when it serves a beneficial purpose (Liddell 2012, p. 145; U.S. Circuit Court (Massachusetts) 1817). For an overview of the US approach, Bagley (2003).

[4]Authors have argued that effective decisions over whether something is moral or not will be made by the purchasing power of the market (Milius and Townend 2008, p. 15). Schatz argued that: "*A patent is only of any value if the inventor can actually use his invention in the market-place. But an invention whose exploitation would be contrary to ordre public or morality is already untradeable for that reason alone. So its creator has no sensible reason to try to patent it. This explains why, in practice, Art. 53(a) EPC and the corresponding national provisions are hardly ever used*" (Schatz 1998, p. 12). During the first 18 years of existence of the EPO, this objection had never been used and the situation was not much different at national level, where this clause was invoked in only a handful of older cases concerning human sexuality. By contrast, items that could threaten human life and the environment were not opposed via this proviso (e.g. poisons, dangerous chemicals, pesticides) (Schatz 2000, p. 217). Matters changed with the advent of biotechnology, as the ambivalence of this research field rendered ethical monitoring a necessity (Van Overwalle 2011, pp. 449–450). This field sparked controversies about whether humans were "playing God", interfering with nature and the common genetic heritage of mankind. Equally, concerns were expressed over the disruption of ecosystems (Mills 2010, pp. 14–18).

Morality provisions have been inserted in patent law by a number of international institutions. Given this variety, the interpretation of those clauses can vary. Amongst the organisations that have included morality clauses are the Council of Europe, the EPO, the WTO and the European Union.[5] Authors noted that, over time, the scope of the clause has been expanded to cover the environment and the protection of human, animal and plant life and health. Equally, human cloning, modifications of human germ lines, the use of embryos and alterations of the genetic identity of animals were also addressed by this clause.[6]

The rationale of its inclusion is not strictly related to patent law, but rather to the power and discretion of national governments. For example, the addition of the morality clause in the 1963 Strasburg Convention was justified by stating that States should be able to "*exercise power, provided by national laws, to deny a patent when its granting would be unacceptable under profiles of public order or morality*".[7] Along similar lines, the ECJ recognised that States should be offered a wide margin of discretion in the interpretation of this clause given the differences in their national cultures.[8]

The morality exception has been criticised by some for introducing extra-legal considerations into this field.[9] By contrast, others praised the role of patent law as a steering tool to guide technological development away from immoral fields.[10] The patent system has thus been seen as a "*way of knowing which inventions the state considers deserving of protection*".[11] Even in the case of the EPO, the granting of a patent can be seen as the sovereign act of an intergovernmental agency. Hence, the morality clause is used to avoid giving the impression that certain technologies are condoned or approved.[12] Yet, inventions can be developed and sold even in the

---

[5]Sommer (2013, p. 180). Outside of patent law, other organisations and normative instruments have affected this field. For example, the 1992 UN Convention on Biological Diversity and its Guidelines and Protocols consider the origin of genetic resources, informed consent and technology transfer (Sommer 2013, p. 182; *Convention on the unification of certain points of substantive law on patents for invention (ETS No. 047)*, 1963).

[6]Sommer (2013, p. 188).

[7]Stazi (2015, p. 181), Mills (2010, p. 26) and Armitage and Davis (1994).

[8]Ottolia (2011, pp. 324–325) and European Court of Justice (ECJ) (2001, § 35-49).

[9]Schneider (2014, p. 157). *Mills* was sceptical of whether a legislation designed to further economic interests could accommodate moral concerns (Mills 2010, p. 11).

[10]Liddell (2012, p. 159). On this, *Liddell* observed that: "*Since the purpose of the immorality exclusion is to ensure that patents incentivise socially beneficial inventions in a manner compatible with just and fair social organisation, it is obviously important to assess whether the protected technology and its likely uses are immoral. Failing to consider and exclude immoral technology would mean that the patent system was economically encouraging scientists to work on immoral forms of technology and to pressure governments to allow their use*" (Liddell 2012, p. 159).

[11]Derclaye (2009, p. 255).

[12]Schatz (1998, p. 12). Conversely, other authors argued that: "*The grant of a patent is by no means an authorization to implement the invention, let alone a stamp of moral approval. This is emphasized again by the wording of Recital 14 of the Directive*" (Torremans 2009a, p. 291). Other articles cited the statement of a British barrister, who held that: "*Granting a patent implies*

absence of patent protection, thus limiting the steering value of the morality clause. In fact, the issue of patentability is independent, albeit related, to the availability of a technology to the public.[13]

Identifying what constitutes an immoral invention is easier on paper than in reality. Therefore, doubts emerged over the ability of patent decision-makers to handle morality objections.[14] In particular, patent offices may not be equipped to examine those issues, as they lack both specific competences and mandate in this field.[15] Hence, it was argued that other organisations should be responsible for this.[16]

The possible unsuitability of the patent system to tackle morality questions was raised by the EPO and the ECJ.[17] In the *Netherlands* case, the ECJ explained that the Directive is solely concerned with the grant of patents and that it does not extend to other activities before and after the grant. Hence, the Directive is not meant to replace norms that guarantee compliance with ethical rules.[18] Equally, the EPO rejected the role of moral censor in a number of occasions.[19] This approach is founded on the idea that the patent system is not the ideal forum to discuss whether

---

*that the patent office, and hence the state, is 'putting its imprimatur' on the invention*" (Sterckx 2008, p. 478). In the *Michigan* case, the Technical Board of Appeal of the EPO held that: "*Article 53 (a) EPC is merely intended to prevent an invention, the publication or exploitation of which would infringe the fundamental principles of 'ordre public' or morality being given an appearance of approval through a patent issued by an international authority* (Technical Board of Appeal of the EPO (TBA) 2005, § 9.7). On this, see Sect. 5.1.3.6.

[13]As noted by *Moufang*: "*A patent is basically nothing more than a right to exclude others from doing certain acts. It does not give a positive right to use the invention without complying with all applicable rules of the legal system. Regardless of whether or not a patent is granted, the inventor has to respect all legal provisions that might possibly forbid the working of the invention*" (Moufang 1998, p. 70). For instance, a ban on the patentability of a technology could not be used to stop its production (Ottolia 2011, pp. 326–327).

[14]Moufang (1998, p. 72).

[15]Schneider (2014, p. 151).

[16]Moufang (1998, p. 72). *Sterckx* and *Cockbain* stated that: "*Patent law should not be seen as the area of law that bears the sole burden of avoiding or preventing abuses or risks associated with the exploitation of inventions*" (Sterckx and Cockbain 2012, p. 290). *Dworkin* adopted a similar line of thought when he affirmed that: "*Few would deny that there are major ethical issues relating to developments in biotechnology and genetic engineering; that there is a need to ensure that such ethical issues are properly addressed; that there should be adequate controls and monitoring of undesirable or questionable developments. The real question, though, is whether such control should be exercised in any significant way through the patent system. A rational answer must be 'no'*" (Curley and Sharples 2002, p. 570; Dworkin 1994).

[17]The dichotomy between law and morality before Courts was aptly summarised by the Law Officer in the case *In the matter for an application for a patent by A. and H.*, where he stated: "*I express no opinion as to whether the use of these articles is consistent with morality, because I am not aware that the law has laid down what the exact standards of morality are. I am a Court of law, and not a Court of morality*" (Mills 2010, p. 35; *In the matter for an application for a patent by A. and H.*, 1927).

[18]Bonadio (2012, p. 435) and European Court of Justice (ECJ) (2001, § 79-80).

[19]Bonadio (2012, p. 435), Opposition Division of the EPO (2001) and Technical Board of Appeal of the EPO (TBA) (1995).

a technology is intrinsically immoral.[20] Nevertheless, scholars have argued that the "*EPC requires the EPO to act as a moral censor*".[21] Since it would be impossible for the EPO to thoroughly analyse every patent application from a morality perspective, all that patent Examiners are required to do is to show a moral compass in line with society, rather than to demonstrate an expertise in this field.[22]

The morality assessment carried out by the EPO is integrated by the views expressed by third parties. The generous access granted by the EPC to its procedures has allowed NGOs and political groups to express their opinions on morality and thus to influence the rulings of the EPO. This has raised public awareness and sensitised authorities towards the ethical implications of patents.[23]

## *5.1.2 Normative Overview*

### 5.1.2.1 EPC

Article 53(a) EPC

Article 53 EPC outlines three categories of exceptions to patentability. The first type, contained in Article 53(a) EPC, is generally known as the morality clause.[24] It establishes that:

> European patents shall not be granted in respect of: (a) inventions the commercial exploitation of which would be contrary to 'ordre public' or morality; such exploitation shall not be deemed to be so contrary merely because it is prohibited by law or regulation in some or all of the Contracting States.

The purpose of this provision is to deny patent protection to inventions that would likely induce riots, public disorder or lead to criminal and offensive behaviour.[25] By

[20]Warren-Jones (2008a, p. 209).

[21]Sterckx and Cockbain (2012, p. 289).

[22]Sterckx and Cockbain (2012, p. 290) and Harmon (2006, p. 392).

[23]Bakardjieva Engelbrekt (2009, pp. 252–253).

[24]In the original version (EPC 1973), the clause read: "*Inventions the publication or exploitation of which would be contrary to 'ordre public' or morality, provided that the exploitation shall not be deemed to be so contrary merely because it is prohibited by law or regulation in some or all of the Contracting States*" (*Convention on the grant of European patents of 5 October* 1973, 1973). The norm was then amended in 2007. For an overview of the interpretation of the older version, Beyleveld and Brownsword (1993, pp. 48–52). Authors argued that the current version excludes both more and less in comparison to the older one. For an overview, Sterckx and Cockbain (2012, p. 294). This clause was based on Article 2 of the Strasburg Convention, although the latter provided optional exclusions instead of obligatory ones (*Convention on the unification of certain points of substantive law on patents for invention (ETS No. 047)*, 1963). The *travaux préparatoire* of the EPC show that little attention was paid to the morality clause during the drafting procedure (Stazi 2015, pp. 184–185; Mills 2010, p. 26).

[25]EPO (2018, § G.II.4.1).

the EPO's own admission, the norm will be invoked only in rare and extreme cases.[26] What constitutes such an extreme case has been the subject of much discussion. It has been argued that this proviso would apply only when the non-permissible use can be inferred from the nature of the invention itself, as in the typical letter bomb example.[27] The mere possibility that an invention might be abused does not constitute a sufficient ground for rejection under Article 53(a) EPC. Indeed, if the invention can be exploited in a way (even only one) that does not infringe the morality clause, patentability will not be denied.[28] Therefore, the exception would operate solely where an invention cannot be used in any socially fruitful manner.[29] This approach is particularly relevant in the field of dual-use technologies. Applications directly mentioning uses contrary to the morality clause would not be tolerated and those references would need to be deleted.[30]

The exception set in Article 53(a) EPC is aimed at hindering the patentability of inventions whose exploitation would be contrary to *ordre public* and morality. Commentators have dwelled on the possible meaning of the word exploitation and its significance in this context.[31] First, it could be read as encompassing objections to the technology and its use; in this case, exploitation would be equivalent to utilisation. A second reading would concern the consequences of patenting, thus encompassing the commercial use and financial benefits connected to the invention. A third option relates to the research needed to achieve such technology and could be equated to taking advantage of an immoral deed. Lastly, it could relate to protestations that such a technology is not worthy of patent protection. Authors noted that, thus far, the EPO has taken into consideration only the first type of objection. This shifts the focus to the morality of using and performing an invention rather than

---

[26]EPO (2018, § G.II.4.1). Classic examples are letter bombs or landmines (Schneider 2014, p. 150). In 1995, the Nuffield Council on Bioethics expressed a similar view, as it recommended that: "*The EPO maintain the so-called 'light-approach', namely that the immorality exclusion should only be considered by the EPO in extreme cases of obviously abhorrent inventions, leaving a more considered determination of the scope of Article 53(a) to national Courts*" (Nuffield Council on Bioethics 1995, p. 21; Bently and Sherman 1995, p. 286). Furthermore, the two scholars observed that: "*By recommending the so-called 'light approach' as an interim measure, the Report (like the EPO itself) implicitly approves the exclusion of ethical values from patent law*" (Bently and Sherman 1995, pp. 290–291).

[27]Schatz (2000, p. 218). Interestingly, it was noted that the most profound consequences of a technology will emerge only in the long run and are generally unforeseeable at the time of the patent assessment. This applies both to positive and negative repercussions of a new technology (Mills 2010, p. 188).

[28]EPO (2018, § G.II.4.1). The assessment can only be based on the claimed use and not on a merely possible one. For example, the cloning process mentioned in the application for "Dolly the Sheep" could not be refused merely because it could also be applied to humans ("The Life of Dolly | Dolly the Sheep," n.d.; Schatz 1998, p. 13). Patents were granted in Europe for the technology related to Dolly, covering both the cloning method and the cloned animal (Hally 2014; Campbell and Wilmut 2001; Campbell et al. 2000; Mayor 2000).

[29]Schneider (2014, p. 151).

[30]EPO (2018, § G.II.4.1.2).

[31]Liddell (2012, pp. 148–152).

to its socio-economic repercussions.[32] Equally, the language of the proviso suggests that it is only the exploitation that counts and not the patenting of the invention itself. Indeed, as highlighted by *Schatz* with reference to the *Relaxin* case, "*the morality of the act of patenting as such is not at issue in Art. 53(a) EPC*".[33] The EPO confirmed this approach by noting that it has not been vested with the task of considering the economic impact of granted patents in a specific technological sector.[34]

The second prong of Article 53(a) EPC establishes that the exploitation would not infringe the morality clause merely because it is prohibited in some or all contracting States.[35] Likewise, the fact that an invention can be exploited in one or all contracting States cannot *per se* be seen as proof that it does not infringe Article 53(a) EPC. Those aspects could only be part of the wider spectrum of evidence used in reaching a decision.[36]

---

[32] Liddell (2012, pp. 148–152).

[33] Schatz (2000, pp. 218–219). The "*no patents on life*" movement was founded in Munich in 1992 to facilitate patent oppositions and patent legislation (Curley and Sharples 2002, p. 565).

[34] EPO (2018, § G.II.4.1.3).

[35] Authors have examined what is meant by law and regulation in this context. It was argued that law comprises: "*Positive rules that are identified within the Contracting States as the laws of their legal systems. These rules may be of a constitutional or non-constitutional nature, rules belong to public or private divisions of the law... and so on. To qualify as 'law', such rules simply have to satisfy the source tests that de facto operate within the Contracting States*" (Beyleveld and Brownsword 1993, p. 76). While regulation could be seen as a synonym of law, it was argued that the best approach would be to give this term its own distinctive meaning. This would include a reference to rules (e.g. codes of practice, circulars) that cannot be considered as orthodox legal sources, but that are still fundamental for States to control behaviour (Beyleveld and Brownsword 1993, pp. 76–77).

[36] Warren-Jones (2007, p. 841). The proviso allows a product to be manufactured under a European patent and later exported in countries where its use is not prohibited (EPO 2018, § G.II.4.1.1). Specifically, it was held that laws and regulations that represent moral judgements in some States should be given evidential value, but cannot on their own prove such morality. Likewise, prohibitions hold a mere evidentiary value. Indeed, prohibitions found in national laws and regulations are neither a necessary nor a sufficient condition for a finding of immorality (Beyleveld and Brownsword 1993, pp. 79–80). Interestingly, it was observed that: "*The bearing of non-prohibition (or express permission) will vary from case to case. For example, where the conduct in question has been the subject of serious and prolonged debate within the history of the culture, and non-prohibition or permission is the explicit answer arrived at, this is surely relevant evidence. Similarly, if the conduct has gone unchallenged over a very long period, during which its existence has been common knowledge, there must be some sort of presumption that it is to be judged permissible. On the other hand, if the conduct in question is new, and not well publicised, or publicised in way that might distort its actual nature, then little or nothing may be read into its non-prohibition or express permission*" (Beyleveld and Brownsword 1993, p. 81).

### Rule 28 EPC

The EPC contains an additional norm on morality: Rule 28 of the Implementing Regulations.[37] The illustrative and non-exhaustive list contained in this proviso gives concrete form to the morality clause in the biotechnological field.[38] The norm states that:

> Under Article 53(a), European patents shall not be granted in respect of biotechnological inventions which, in particular, concern the following:
>
> (a) processes for cloning human beings;
> (b) processes for modifying the germ line genetic identity of human beings;
> (c) uses of human embryos for industrial or commercial purposes;
> (d) processes for modifying the genetic identity of animals which are likely to cause them suffering without any substantial medical benefit to man or animal and also animals resulting from such processes.

This proviso is amongst the Rules inserted by the Administrative Council of the EPO to ensure an approach to patentability of biotechnological inventions in line with the Biotechnology Directive. Specifically, the exceptions contained in Article 6 of the Directive were transported in this Rule.[39] The chosen categories represent technologies that were relevant at the time the Directive was passed and that were considered morally unacceptable from a patent perspective. New techniques that are not listed therein might be regarded as similarly unacceptable.[40]

The content and scope of those exceptions has been highly debated. The first exception is limited to the cloning of human beings. Processes for cloning non-humans would instead be acceptable, even if they claimed mammals and primates.[41] Those claims have often been connected to a disclaimer (e.g. "*provided that the process is not for cloning humans*") in order to overcome the Rule 28(a) EPC objection.[42] Under point (b) patents for transgenic humans and related processes would be denied. The exception in letter (c) relates to the commercial or industrial use of human embryos and has been particularly debated in the literature and in Courts.[43] Lastly, point (d) establishes that transgenic animals would be acceptable as long as the criteria set therein are met.[44]

---

[37] The list was previously found in Rule 23(d) and was later moved to Rule 28 (*Convention on the grant of European patents of 5 October 1973, as revised on 29 November* 2000 *(European Patent Convention - EPC)*, 1973).

[38] EPO (2018, § G.II.5.3).

[39] Torremans (2009b, p. 142).

[40] Intellectual Property Office (IPO) (2016, p. 39).

[41] Covone (2015, p. 22).

[42] Macchia (2012, p. 39).

[43] For an overview, Plomer (2009). The *WARF* and *Brüstle* cases famously refer to this exception. On this, see respectively Sects. 5.1.3.3 and 5.1.3.5.

[44] Covone (2015, pp. 23–24).

Following Article 164 EPC, Rules are an integral part of the EPC. However, in case of conflict, the articles of the Convention will prevail upon them. This means that a conflict between Article 53(a) EPC and Rule 28 EPC should be solved in favour of the former. Therefore, the Rule could not overcome conclusions reached on the basis of Article 53(a) EPC.[45] Equally, in comparing the scope of the morality clause set by Article 53(a) EPC and Rule 28 EPC, the Enlarged Board of Appeal remarked that "*it has not been argued that Rule 28... EPC took away the possibility to patent anything which had previously been regarded as patentable under Article 53(a) EPC, nor that the Directive did so*".[46]

By adopting this approach, the Board signalled that Rule 28 EPC is within the scope of Article 53(a) EPC and that this addition did not lead to legal innovation.[47]

### Relationship Between Article 52 and 53 EPC

The exclusions presented in Article 52(2) and 53 EPC express the intention of the legislator to remove certain items from the realm of patentability. Some noted that

---

[45]Torremans (2009b, p. 143). The connection between Rule 28 and Article 53(a) EPC has been challenged, as the two provisions could be seen in contrast with each other. *Sterckx* and *Cockbain* noted that: "*By incorporating provisions of the EBD* [European Biotechnology Directive] *into the EPC Rules – that is by seeking to force certain subject-matter to be found to be contrary to the morality provisions of Art. 53(a) EPC because it is EU law that such subject-matter shall be unpatentable – the Administrative Council of the EPO appears to have acted against this qualification of Art. 53(a) EPC. Thus the Administrative Council seems to have acted ultra vires*" (Sterckx and Cockbain 2012, p. 291). This argument was raised in the *Oncomouse* case (Technical Board of Appeal of the EPO (TBA) 2004). There, the Board dismissed such allegation by arguing that: "*An administrative action or rule of subsidiary legislation is ultra vires if it falls outside the scope of a law which precludes or limits the legal power of the person or body doing the act or making the rule which is consequently invalid – the term ultra vires denotes an 'excès de pouvoir'. That is quite clearly not the case here. The law in question, Article 53(a) EPC, contains nothing which precludes or limits its own subsequent interpretation whether by case-law... or by legislation (as in Rule 23d EPC)... Ultra vires requires an inconsistency but there is none - Article 53(a) EPC... remains unaffected by Rule 23d(d) EPC save that... the Rule deems four limited categories of inventions to fall within Article 53(a) EPC. That has been achieved in a perfectly valid – i.e. intra vires – manner*" (Technical Board of Appeal of the EPO (TBA) 2004, § 7.3). On this, see Sect. 5.1.3.1. Scholars criticised this conclusion by noting that, at the time the Rule was enacted, it would have excluded more from patentability than the Article, thus altering the scope of the exclusion. Hence, it was argued that the Rule should constitute a recommended rather than a binding interpretation of Article 53(a) EPC (Sterckx and Cockbain 2012, pp. 292–293).

[46]Enlarged Board of Appeal of the EPO (EBA) (2008, § 13). The Board referred to a 1984 article of *Dolder* to prove its point (Dolder 1984). Commentators argued that this statement might be incorrect. As some authors pointed out, WARF had been granted an equivalent patent in the US. Equally, a patent for a human embryo liver cell line had been granted in the UK. The patent that subsequently gave rise to the *Edinburgh* case was also initially granted by the EPO, despite raising issues very similar to those mentioned in *WARF* (Sterckx and Cockbain 2010, p. 100; Opposition Division of the EPO 2003a). On *WARF*, see Sect. 5.1.3.3.

[47]Treichel (2009, p. 454).

both exemptions serve the public interest and common good, but differ in their logic, function and regulatory goals. Article 52(2)(a) EPC is based on an "*access for all*" rationale, whereas Article 53(a) EPC adopts a "*no one should have access*" logic. The morality exception aims at suppressing certain technologies by exercising an *ex ante* control on their social desirability. This might, in turn, remove the legal incentive for the development of those inventions.[48] Others considered the difference between the two to lie in the fact that one covers items which "*might be inventions but should not have patents granted in respect to them*", whereas the other is directed at things that are not "*'inventions' in the sense that the word invention must be understood within patent law*".[49]

As for the relationship between the morality clause and Article 52(1) EPC in general, it has been suggested that the former constitutes a higher rank norm in comparison to Article 52(1) EPC due to its role and purpose. Hence, in case of conflict, the latter must yield to the former.[50]

#### 5.1.2.2 Biotech Directive

Norms

The Biotechnology Directive expressly addresses exceptions from patentability for *ordre public* and morality reasons both in its text and Recitals.[51] This structure constitutes a safety valve built in the patent system and reinforces Article 53

[48]Schneider (2014, p. 148). On this, *Stazi* maintained that Article 53(a) EPC: "*Can be held to be a veiled attempt to regulate innovation, including biotechnology, on the one hand by withdrawing the incentive to be dedicated to creating and distributing certain inventions by denying relative patents on moral grounds… on the other precluding the patentability of certain living organisms*" (Stazi 2015, p. 186).

[49]Sterckx and Cockbain (2012, p. 309). This would be the case even if they were to fulfil all other patentability criteria.

[50]Sterckx (2008, p. 488). The statement was made by the Rapporteur to the Enlarged Board of Appeal in the *WARF* case (Enlarged Board of Appeal of the EPO (EBA) 2008) On this, see Sect. 5.1.3.3.

[51]The implementation of the Directive was not identical across all member States (Schneider 2014, p. 158). Pursuant to Article 7 of the Directive, the ethical aspects of biotechnology will be assessed by the European Group on Ethics in Science and New Technology of the Commission (EGE). This advisory group issued reports in this sector, such as the ones on human stem cell patents or synthetic biology (European Group on Ethics in Science and New Technologies to the European Commission 2009; Crespi 2006, p. 571; European Group on Ethics in Science and New Technologies to the European Commission 2002). Significantly, the EPO has not always implemented the opinions of EGE, as for example in the *Edinburgh* case (Bakardjieva Engelbrekt 2009, p. 258; Opposition Division of the EPO 2003a). By analysing the different Opinions drafted by EGE, *Viens* compiled a list of the fundamental values and principles common in Europe. Those included the respect of human life and dignity, the relief of human suffering, justice and beneficence, freedom of research, individual autonomy and proportionality. Occasionally, references were made to the principle of non-commercialisation of the human body. Amongst the principles listed above, priority should be assigned to the respect of human life. No explanation was given on the origin and foundation of those principles (Viens 2009, pp. 90–91).

(a) EPC.[52] The Directive purposely employs vague and open concepts in order to be able to accommodate the diverse viewpoints of member States on morality.[53]

Article 6 establishes that:

> Inventions shall be considered unpatentable where their commercial exploitation would be contrary to ordre public or morality; however, exploitation shall not be deemed to be so contrary merely because it is prohibited by law or regulation.
>
> On the basis of paragraph 1, the following, in particular, shall be considered unpatentable:
>
> (a) processes for cloning human beings;
> (b) processes for modifying the germ line genetic identity of human beings;
> (c) uses of human embryos for industrial or commercial purposes;[54]
> (d) processes for modifying the genetic identity of animals which are likely to cause them suffering without any substantial medical benefit to man or animal, and also animals resulting from such processes.[55]

Article 6(1) of the Directive shares a number of similarities with Article 53 (a) EPC. As for Article 6(2), the examples show the clear priority given to inventions concerning the human body, advantaging humans or relating to human dignity.[56] The importance of human dignity is underlined by the references to this concept found in both the Directive and its Recitals.[57] In the context of the morality clause, human dignity has been understood as the intrinsic value of humans.[58]

---

[52] Torremans (2009a, pp. 289–291).

[53] Torremans (2009a, p. 291).

[54] *Plomer* observed how the use of human embryos is permitted in the EU, even if it is destructive. For example, the Directive on Human Tissues and Cells regulates some of those cases (Plomer 2009, p. 183). This Directive belongs to a group of three that is known as the European Union Tissue and Cells Directives (EUTCD). The main Directive provides the framework in this field (Directive 2004/23/EC). Two technical Directives complete the regulation in this sector (Directives 2006/17/EC and 2006/86/EC). In 2004, no European consensus could be found on whether the destructive use of embryos should be outlawed (Plomer 2009, p. 183). Due to this, *Plomer* stated that: "*If the EBA's construction of Article 6(2)(c) is correct, then there is a system conflict within EU law between legislation which permits the conduct of activities involving (destructive) uses of human embryos, including product development on an industrial and commercial basis, yet precludes the grant of property rights on the related 'inventions' as 'immoral'*" (Plomer 2009, p. 186).

[55] Article 6(2)(d) of the Directive was influenced by the test developed in the *Oncomouse* case (Porter 2009a, p. 15). On this, see Sect. 5.1.3.1. For an overview of the issues raised by this list, Lykkeskov and Jørgensen (2005).

[56] Sommer (2013, p. 256) and Macchia (2012, p. 39). Interestingly, Article 6(2) of the Directive does not explicitly mention plants. Similarly, environmental protection is not directly addressed (Stazi 2015, pp. 197–198). *Mills* raised a similar point and held that: "*There is no reference to plants in the provision. This suggests genetic modification of plants is acceptable and will, presumably, be determined on normal patentability criteria. The implication here is that there exists a higher standard of morality in relation to animal-related inventions than for plant-related inventions... on the other hand, EPO jurisprudence suggests that plant and animal varieties should be equated*" (Mills 2010, pp. 140–141).

[57] Sommer (2013, p. 185). In particular, reference is made to Recitals 20, 21 and 28 and to Article 5 and 6 of the Directive.

[58] Despite the difficulty in defining it, human dignity was interpreted in three ways, none of which was necessarily connected to a religious perspective. The first concerns the non-instrumentalisation

The list contained in Article 6(2) of the Directive is not exhaustive and does not limit the scope of Article 6(1).[59] Therefore, inventions could be held immoral even if they fall outside the realm of the four cases listed therein.[60] For the technologies listed in Article 6(2) of the Directive, there would be no need to assess their

---

or commodification of humans and entails a prohibition to use humans as means and objects. Commodification has been defined as the: "*Social process by which something comes to be apprehended as a commodity, as well as to the state of affairs once the process has taken place*" (Viens 2009, p. 99). This means that an instrumental value is associated to those items, which can be understood in commercial terms (Viens 2009, p. 100). As for objectification, it was stated that: "*To objectify something is to treat an entity with moral status as something to be used as an instrument to satisfy the interests or desired of oneself or others with impunity. Some people might think that objectification is the same as exploitation and/or commodification. Again, it is important to recognize that they are different notions. Objectification is similar to exploitation in that it is an instance of wrongful use and it is similar to commodification in that it is an instance of wrongful valuation, but objectifying something need not be the same kind of use or valuation, nor does exploiting or commodifying something seem to be an essential component to objectification*" (Viens 2009, p. 105). From a second perspective, human dignity can be intended as the non-commercialisation of the human body. This prohibits both the sale of human bodies and parts as well as using them as a form of financial gain. Lastly, the concept can be seen from a human rights perspective. In this case, human dignity is seen as the basis of human rights in the moral sense and independently of whether they have been codified (Sterckx and Cockbain 2012, pp. 306–307). A number of legislative instruments have codified the respect of human dignity and other fundamental rights. The Universal Declaration on Human Rights of 1948, the Oviedo Convention on Human Rights and Biomedicine, and the UNESCO Universal Declaration on the Human Genome and Human Rights might also be relevant (UN General Assembly 1948; *Convention for the protection of human rights and dignity of the human being with regard to the application of biology and medicine: Convention on human rights and biomedicine (ETS 164)*, 1997; UN Educational, Scientific and Cultural Organisation (UNESCO) 1997). For a complete list, Stazi (2015, pp. 211–2012) and European Group on Ethics in Science and New Technologies to the European Commission (2009, § 2.3, 3). The Charter of Fundamental Rights of the European Union, which assumed the same legal status as EU treaties pursuant to the Lisbon Treaty, is particularly important in this field, as it covers human dignity, the right to life and to individual integrity (*Charter of fundamental rights of the European Union (2007/C 303/1)*, 2000). In the biotechnological field, this resulted in the ban of eugenic practices, of the commodification of the human body and of human cloning (Stazi 2015, p. 203). Article 52(3) of the Charter of Fundamental Rights states that: "*In so far as this Charter contains rights which correspond to rights guaranteed by the Convention for the Protection of Human Rights and Fundamental Freedoms, the meaning and scope of those rights shall be the same as those laid down by the said Convention*". In light of this proviso, it has been argued that the ECHR has been incorporated into the Charter and thus placed at the head of the EU legal system (*European convention for the protection of human rights and fundamental freedoms - Council of Europe (ETS 5)* 1950). This way, they would play a role in the interpretation of European norms (Stazi 2015, p. 205). Nonetheless, it was noted that, despite the balance in the ECHR between public and private rights, this legal instrument might not be ideal in the field of patents. First, the application of those standards might affect the case law of the EPO and, consequently, legal certainty. Second, since the ECHR focuses on human rights, it might disregard other relevant aspects (i.e. social and economic impact) (Mills 2010, p. 140).

[59]Torremans (2009a, p. 290). Equally, it does not replace the morality clause of Article 53(a) EPC.

[60]Liddell (2012, p. 143).

patentability based on the balancing or rebuttable presumption tests, as those would apply only to the general morality clause.[61] Before the list of immoral inventions was added to Article 6(2) of the Directive, there was no specific guidance on which inventions would be considered to infringe Article 53(a) EPC.[62] Still, it is debatable whether the insertion of this list shed doubts on the application of the morality clause in this field or rather aggravated them.[63]

*Bostyn* argued that this proviso was added as the lack of a common notion of *ordre public* and morality would be problematic in the very sensitive field of patents for biotechnologies. Hence, the legislator provided a list of technologies that would be deemed contrary to the morality clause, thus setting a common *ordre public* and morality concept in this sector.[64] The list represents a "*static snapshot of the technologies and activities thought to be too immoral for patent protection at the time of the debates*".[65] Interestingly, some of the techniques listed therein had not been achieved at the time the Directive was passed.[66] These exclusions are not unified by a specific underlying rationale; hence, their guidance in assessing the morality of new controversial technologies is limited.[67]

In *WARF*, the Board of Appeal suggested that the list of exclusions should be construed pursuant to the Vienna Convention, which requires an interpretation based on the wording of the text and the intention of the legislator.[68] As for the scope of manoeuvre granted by it, in the *Netherlands* case the ECJ observed that:

> It is common ground that this provision allows the administrative authorities and Courts of the Member States a wide scope for manoeuvre in applying this exclusion... that scope for manoeuvre is necessary to take account of the particular difficulties to which the use of certain patents may give rise in the social and cultural context of each Member State, a context which the national legislative, administrative and Court authorities are better placed to understand than are the Community authorities.[69]

---

[61]Stazi (2015, pp. 199–200). On this, see Sect. 5.1.2.5.

[62]Schneider (2014, p. 152).

[63]Mills (2010, p. 136). The list of subject matter excluded from patentability on morality grounds presented in Article 6(2) of the Directive was supposed to clarify the application of the general morality exclusion of Article 6(1). However, this intent may have backfired, as the list proved particularly controversial (Porter 2009a, p. 5). Initially the list had a different content (Sommer 2013, pp. 185–186).

[64]Bostyn (2011, p. 236).

[65]Liddell (2012, p. 143). During the drafting of the Directive, some members of the European Parliament suggested to extend this list to, amongst others, patents on DNA sequences, animals and plants (Liddell 2012, p. 143).

[66]Schneider (2014, p. 158). For example, the cloning of human embryos and germline intervention in humans could not be achieved even more than a decade later.

[67]Stazi (2015, p. 200).

[68]Torremans (2009b, p. 145) and Technical Board of Appeal of the EPO (TBA) (2006, § 25 and 33).

[69]European Court of Justice (ECJ) (2001, § 37-38).

Numerous references to *ordre public* and morality are found in the Recitals of the Directive. Some of those explicitly address the morality clause, while others relate to principles that could have a bearing on it.

Recitals 36 to 45 expressly concern the morality clause. Recital 36 refers to TRIPS and employs the same language used in the morality clause of Article 27 (2) TRIPS. Recital 37 and 38 illustrate and follow the structure of Article 6 of the Directive. The first refers to the general principle, outlined in Article 6(1), that inventions contrary to *ordre public* and morality shall be denied patentability. The second addresses instead Article 6(2) of the Directive and its non-exhaustive list of technologies that infringe the morality clause. It is expressly mentioned that the list is supposed to be of guidance for national Courts and patent offices and that human dignity plays an important role in some of the exclusions. Therefore, it has been argued that Article 6 of the Directive should be read in light of Recital 38. Likewise, since the list in Article 6(2) of the Directive is not exhaustive, it has been argued that technologies that prejudice human dignity should be excluded from patentability. Those would include, for example, the production of human and animal hybrids as well as germinal and totipotent cells.[70] Equally, authors maintained that claims concerning chimeras mixing human and animal embryonic stem cells or germ cells could be refused by reading Article 6(1) of the Directive in light of Recital 38.[71]

The definition of *ordre public* and morality is briefly addressed in Recital 39. There, the legislator noted that those two concepts "*correspond... to ethical or moral principles recognized in a Member State*". Such *renvoi* to the ethical and moral principles recognised in *a* member State is particularly interesting, given that the formulation refers to *a* member State in singular form. This has been understood to mean that "*it is for each individual Member State to identify the dominant moral and ethical perception and take these into consideration*".[72]

Letter (a) and (b) of Article 6(2) of the Directive are addressed in Recital 40, where it is stated that exceptions for human cloning technologies and human germ line alterations derive from a consensus within the Community that such inventions violate the morality clause. An *a contrario* reading of Recital 40 hints to the fact that, on many points, there is no consensus on the content of the concepts of *ordre public* and morality.[73] Letter (a) is also relevant in the next Recital. At point 41, the European legislator explained the scope of processes for cloning human beings. Those include "*any process, including techniques of embryo splitting, designed to create a human being with the same nuclear genetic information as another living or deceased human being*". Recital 42 relates instead to Article 6(2) (c) of the Directive. It establishes an exclusion to what would otherwise be an absolute exception to patentability. Indeed, it holds that "*such exclusion does not*

[70] Stazi (2015, p. 205).

[71] Sommer (2013, p. 211).

[72] Bakardjieva Engelbrekt (2009, p. 246).

[73] Torremans (2009a, p. 290).

*affect inventions for therapeutic or diagnostic purposes which are applied to the human embryo and are useful to it*".

Significantly, the Recitals contain a number of references to other treaties and norms. Those include the TEU, the ECHR and the constitutional traditions common to member States.[74] References to such legal instruments are contained in Recital 43, alongside a reference to the fact that the EU needs to respect the fundamental rights contained therein. The role of EGE (European Group on Ethics in Science and New Technologies) is explained in Recital 44, whereas Recital 45 relates to the exception found at letter (d) of Article 6(2) of the Directive. While the formulation of the Recital is almost identical to that of the Article, it differs in one important aspect. The Recital describes what constitutes a substantial medical benefit. Specifically, it describes medical benefits in terms of research, prevention, diagnosis and therapy for men and animals.

Lastly, the Recitals refer to topics that have an indirect relevance in the field of patent morality. Those are Recitals 10, 11, 14 and 56. The first two concern the role of biotechnology in the environmental and health sectors. From a synthetic biology perspective, references to the utility of technologies providing less polluting and more efficient cultivation methods are significant. Likewise, the encouragement of the patent system to address health, hunger and epidemics would point to the social desirability of this technology. Recital 56 contains a reference to the Biodiversity Convention. Lastly, Recital 14 presents a cornerstone of patent law, which has often been cited by Courts and commentators alike in deciding upon the morality of an invention. It explains that a patent does:

> Not authorize the holder to implement that invention, but merely entitles him to prohibit third parties from exploiting it for industrial and commercial purposes... substantive patent law cannot serve to replace or render superfluous national, European or international law which may impose restrictions or prohibitions or which concerns the monitoring of research and of the use or commercialization of its results, notably from the point of view of the requirements of public health, safety, environmental protection, animal welfare, the preservation of genetic diversity and compliance with certain ethical standards.

### History

The focus on morality of the Directive is the result of an uneasy political compromise between EU institutions.[75] The complex legislative *iter* that led to the approval of the Directive shows the impact that morality concerns had on its development. Equally, the commitment of the European Parliament to insert morality provisions in the Directive is indicative of the uncomfortable feelings of MPs and the public towards the types of inventions covered by the Directive.

---

[74] Torremans (2009b, p. 147).

[75] Porter (2009a, p. 5) and Vandergheynst (1998, p. 92).

Issues emerged soon after the Commission presented its first draft Directive.[76] In 1989, the Economic and Social Committee objected to it by stating that it failed to draw ethical boundaries between what can be commodified and what not and that it did not explicitly state that humans are unpatentable.[77] The Parliament was also critical of the draft, since it believed that it did not take ethical and moral aspects enough into consideration. Because of this, the Parliament requested a number of amendments, which were partially implemented by the Commission.[78] Critiques were moved against this revised version as well, as the Parliament proposed further amendments.[79] In a Common Position from 1994, the Council accepted some of those alterations.[80] Despite this, the Parliament was not satisfied with the draft Directive due to ethical concerns; therefore, it suggested further changes, which were rejected by the Council. This led to a conciliation procedure between the Council and the Parliament that resulted in a new joint text. In spite of this, the Parliament rejected the draft in 1995 with a 240 to 188 vote.[81] This was the first time that a Directive was rejected by the European Parliament.[82] After this setback, the Commission presented a revised proposal focusing more on ethical concerns and morality.[83] After further proposals, amendments and Committees' work, the Directive was finally signed in July 1998 by the Parliament and the Council.

After the adoption of the Directive at European level, the norms needed to be transposed into national law.[84] During this process, significant asymmetries emerged. A high margin of discretion was conferred to national Courts in deciding

[76]The first draft of October 1988 did not directly address the notions of *ordre public* and morality (Vandergheynst 1998, p. 85). For an overview of the legislative history, Vandergheynst (1998, pp. 85–92).

[77]Porter (2009a, p. 10) and *Proposal for a Council Directive on the legal protection of biotechnological inventions (COM/88/496FINAL - SYN 159)* (1988).

[78]Porter (2009a, p. 10), *Amendments suggested by the European Parliament to the Draft Directive on the legal protection of biotechnological inventions ((*1992b*) OJ C 305/160)*, 1992 and *Amendments suggested by the European Parliament to the Draft Directive on the legal protection of biotechnological inventions ((*1992a*) OJ C 125/112)*, 1992.

[79]Porter (2009a, p. 13) and *Amendments suggested by the European Parliament to the Draft Directive on the legal protection of biotechnological inventions ((1993) OJ C 342/2)*, 1993.

[80]Porter (2009a, p. 13) and *Common position (EC) No 4/94 adopted by the Council with a view to adopting European Parliament and Council Directive 94/.../EC of... on the legal protection of biotechnological inventions (94/C 101/01)* (1994).

[81]Porter (2009a, p. 13) and *European Parliament decision on the joint text approved by the Conciliation Committee for a European Parliament and Council Directive on the legal protection of biotechnological inventions* (1995).

[82]Bently and Sherman (1995, p. 281).

[83]*Proposal for a European Parliament and Council Directive on the legal protection of biotechnological inventions* (1995). At this point, the highly controversial topic of the exclusion from patentability of human embryos had not yet entered the debate. The first explicit reference to this issue in relation to the Directive was raised in July 1996 by the Economic and Social Committee. For an overview of the history of this inclusion, Porter (2009a, pp. 18–21).

[84]The majority of States implemented Article 6 of the Directive in their national norms without any alterations. Nonetheless, there have been States that did not strictly follow the formulation of the

upon Article 6(1) of the Directive. Conversely, harmonisation was required for the exceptions in Article 6(2). Scholars interpreted this as a confirmation that the exclusions listed in Article 6(2) of the Directive should be interpreted according to their literal meaning, without resorting to extensive or restrictive readings.[85]

#### 5.1.2.3 Relationship Between Different Norms and Systems: The EPC and the Biotech Directive

The possibility of conflicts between the morality provisions in the EPC and the Directive was considered by scholars.[86] This analysis extended also to the impact that those norms can have on the balance and structure of the patent system in Europe.

Scholars argued that the insertion in the EPC of a Rule tailored upon Article 6 of the Directive does not guarantee a uniform practice amongst member States. Indeed, it is possible to interpret the notions of *ordre public* and morality in a different manner at both European and national level, thus giving rise to diverging approaches.[87] Moreover, conflicts can occur due to the structure of the patent system in Europe. The contracting States of the EPO do not coincide with those of the European Union. Therefore, differences in the case law of the EPO and of the ECJ might impact the national validity of patents granted at European level. Equally, problems could arise if the EPO adapts to the line traced by the ECJ on morality, since it is doubtful whether non-EU member States would accept ECJ decisions contradicting their national perceptions.[88]

---

Directive, especially with regards to Article 6(2)(c). Such differences concerned either the wording of the norm or the reference to medical legislation (Hellstadius 2009, pp. 121–122).

[85]Stazi (2015, p. 194). This different treatment was attributed to the different moral and cultural traditions of member States (Ottolia 2011, p. 325; European Court of Justice (ECJ) 2005, § 78; European Court of Justice (ECJ) 2001).

[86]The issue of the compliance of the EPC and the Directive with Article 27(2) TRIPS was raised in several occasions. As the Directive was still in a drafting stage, it was noted that there were two main differences between Article 27(2) TRIPS on the one side and Article 53(a) EPC and draft Article 9 of the Directive on the other. The first was the reference to the protection of human, animal and plant life or health, which was introduced upon suggestion of Japan and of a number of developing countries. The second dissimilarity concerned the reference to environmental issues (Ford 1997, p. 316). Furthermore, TRIPS leave States free to implement exceptions from patentability based on *ordre public* or morality grounds. The EPO and the EU have taken advantage of this option in the EPC and in the Directive respectively.

[87]Stazi (2015, p. 196). *Torremans* argued that: "*The exclusions are effectively non patentable subject matter. Morality is then a general safeguard clause that exists even for those inventions that cover patentable subject matter. If that turns out to be the case, then Rule 23 cannot be said to implement Article 53*" (Torremans 2009b, p. 142).

[88]Schuster (2012, pp. 636–637). Along these lines, *Sommer* noted that the Directive is addressed to EU member States and influences the relationship between the EPO and its own contracting States, at least those that are also part of the EU. Hence: "*Since the Directive complements Article 53 EPC, as implemented in Rule 28... as well as determining substantive EU patent law, conflicts can be*

A further layer of complexity is added by the fact that no single deciding body is responsible for the interpretation of the list of exclusions found in the Directive and in Rule 28 EPC. Indeed, parties challenging the interpretation given by national Courts could bring their critiques to the ECJ. Conversely, challenges to the interpretation given by the EPO could only culminate in a proceeding before the Enlarged Board of Appeal.[89] Such conflict could be aggravated by the fact that there is no formal hierarchical relationship between the EPO and the ECJ and by the absence of a system to solve conflicts between them.[90] Because of this, it was noted that:

> If this non-exclusive list expands in the future, because of new concerns of the Court of Justice or of the EPO Boards of Appeal, perhaps adopted in the Implementing Regulation (EPC) or in the Directive (EU), who will then be the final arbiter? The language of the Court of Justice in Opinion 1/09 on the EU patent, as well as in Case C-34/10... Brüstle . . . clearly supports the EU legal order at any price.[91]

#### 5.1.2.4 TRIPS

A proviso on the morality of inventions is contained Article 27(2) TRIPS. This norm, which so far has not been interpreted by the WTO, states that:

> Members may exclude from patentability inventions, the prevention within their territory of the commercial exploitation of which is necessary to protect ordre public or morality, including to protect human, animal or plant life or health or to avoid serious prejudice to the environment, provided that such exclusion is not made merely because the exploitation is prohibited by their law.

This article was the result of lobbying efforts during the Uruguay round.[92] In that occasion, Europe and a number of developing countries wished to exclude from patentability immoral technologies.[93]

Much like its European equivalents, Article 27(2) TRIPS does not define the notions of *ordre public* and morality. In interpreting those concepts, it is likely that the notions of public morals and public order found in Article XIV(a) GATS and in the *US gambling* case would be of relevance.[94] In the latter case, the WTO Panel addressed both concepts and observed that their content is not fixed, but can vary in time and space depending on a wide range of factors (e.g. social, religious and

---

*expected between the EU, the European Patent Organisation and the Member States/Contracting States subject to both regimes*" (Sommer 2013, p. 191).

[89] Liddell (2012, p. 143).

[90] Bakardjieva Engelbrekt (2009, p. 230).

[91] Sommer (2013, p. 221).

[92] Bonadio (2012, p. 441).

[93] Ford (1997, p. 315).

[94] Porter (2009b, p. 354), WTO Panel (2004) and *General Agreement on Trade in Services (GATS) (1869 U.N.T.S. 183)*, 1994.

cultural values).[95] At the same time, the Panel recognised a certain discretion of States in interpreting those issues based on their own values.[96] Lastly, the Panel provided its interpretation of both notions. In its opinion, public morals related to "*standards of right and wrong conduct maintained by or on behalf of a community or nation*", whereas public order encompassed "*the preservation of the fundamental interests of a society, as reflected in public policy and law. These fundamental interests can relate, inter alia, to standards of law, security and morality*".[97]

The interpretation and application of this Article have been considered unclear, especially in relation to Article 27(1) TRIPS.[98] *Porter* noted the existence of four different understanding of Article 27(2) TRIPS in the academic word. These four lines of thought can be described as follows:

> One group of commentators suggests that a WTO Member must first prohibit the commercial exploitation of an invention within its territory before it is entitled to exclude the invention from patentability on moral grounds. A second group takes the diametrically opposed view that the wording of Article 27.2 only requires the decision of a competent authority... that a ban is necessary to protect morality, but does not impose any legal requirement for the Member to also take positive steps to prohibit the actual commercial exploitation of the invention. According to the third view... the interpretation of Article 27.2 would be influenced by the 'morality jurisprudence' of the European Patent Office... that has been built up in relation to Article 53(a)... Finally, the fourth group suggests that when reviewing a WTO Member's decision to exclude inventions... a WTO panel or Appellate Body would apply the 'necessity test' used to scrutinize measures adopted under the 'exceptions clauses' found in other WTO Agreements; namely, Article XX of GATT and also Article XIV of the General Agreement on Trade in Services (GATS).[99]

Scholars argued that patentability can be refused on morality grounds only if the sale and distribution of the product are also prohibited;[100] that is, only if its commercial exploitation is excluded.[101] If this position were to prevail, the fact that in Europe there is no prohibition towards human treatments involving Human Embryonic Stem Cells (HESCs) would be problematic for the application of the morality clause to those inventions.[102] This position has been opposed by those who

---

[95]WTO Panel (2004, § 6.461).

[96]WTO Panel (2004, § 6.461). In particular, it was held that: "*Members should be given some scope to define and apply for themselves the concepts of 'public morals' and 'public order' in their respective territories, according to their own systems of scales and values*" (WTO Panel 2004, § 6.461).

[97]WTO Panel (2004, § 6.465, 6.467).

[98]Schneider (2014, p. 151).

[99]Porter (2009b, p. 345). For further information, Porter (2009b, pp. 346–352).

[100]Bonadio (2012, p. 441).

[101]de Carvalho (2018, p. 295), Ugurlu (2014, pp. 64–65) and Rothley (1997). Counterarguments based on Article 27(2) TRIPS were overcome by stating that: "*Only prohibitions by law having the purpose of protection of ordre public and morality should be determinative for the patent Examiner when deciding for the exclusion*" (Ugurlu 2014, p. 65).

[102]There are no prohibitions towards the commercial and industrial exploitation of products obtained from human embryos (Ugurlu 2014, p. 64).

believe that a ban is not necessary and that a State should simply prove that there is the necessity to prevent the commercial exploitation of that technology.[103]

The fact that Article 27(2) TRIPS must be applied in a non-discriminatory or unjustified manner may have an impact on the above scenario. *Porter* argued that it would be:

> Extremely difficult to see how a WTO Member could argue convincingly that the prevention of the commercial exploitation of an invention in its territory was 'necessary' if it in fact permitted the commercial exploitation of that invention.[104]

Additionally, scholars asserted that the principles of non-discrimination, justifiability and reasonableness would be infringed if Article 27(2) TRIPS were to be interpreted in a way that allows a mismatch between the morality standards expressed in the patent laws of a State and those expressed within its domestic regulations.[105] In light of this and given that Article 27(2) TRIPS must be read together with Article 27(1) and 30 TRIPS, it was suggested that the morality clause in TRIPS should be applicable only when a State has first prevented the commercial exploitation of that technology within its territory. Otherwise, a State could not justify deviating from its obligations under Article 27(1) TRIPS, as it could be argued that it is not providing patent protection without discrimination.[106]

Lastly, the reference to serious prejudice to the environment in Article 27 (2) TRIPS might suggest a broader spectrum of application of this norm in comparison to Article 53(a) EPC. Nonetheless, despite the different formulations, the fact that the EPO case law has taken into consideration environmental concerns in its decisions would undermine this conclusion.[107]

#### 5.1.2.5 *Ordre Public* and Morality

Despite the centrality of the notions of *ordre public* and morality, their definition remains vague, and so does their relationship to one another.[108] The complexity derives from a number of factors. First, the two concepts are interpreted differently in national and European patent law.[109] Second, the two notions have been employed in other ambits.[110] In particular, the concept of *ordre public* has a long tradition in

---

[103]Bonadio (2012, p. 441).

[104]Porter (2009b, p. 359).

[105]Porter (2009b, pp. 359–360).

[106]Porter (2009b, pp. 366–367).

[107]Sommer (2013, pp. 201–202).

[108]Beyleveld and Brownsword (1993, p. 53). For an assessment of whether the notion of *ordre public* is redundant given the reference to morality, Sterckx and Cockbain (2012, p. 298).

[109]Schatz (2000, p. 219).

[110]Over time, this included Article 36 TFEU, Article 56 CEE Treaty, Article F(d) TEU, as well as statements of the EGE and of the European Parliament (*Treaty on the Functioning of the European Union (TFEU) - 2008/C 115/01*, 2007; *Treaty on European Union (TEU) - 2008/C 115/01*, 2007);

the field of international law.[111] Lastly, the specific content of both notions could vary across different cultures and times.[112]

Over time, scholars, Courts and patent offices tried to delineate the content and boundaries of those two concepts. The Board of Appeal of the EPO defined both terms in the *Plant Genetic Systems* case.[113] *Ordre public* was read with reference to the protection of public security, of the environment and of the physical integrity of individuals.[114] Conversely, morality was interpreted as the totality of accepted

---

Vandergheynst 1998, pp. 82–85; *Treaty establishing the European Economic Community (EEC Treaty)* 1957). Similarly, it was cited in the *CIWF* case before the ECJ (Thomas and Richards 2004, p. 100; European Court of Justice (ECJ) 1998). *Ordre public* and morality have also been mentioned in other IP treaties (e.g. Community Design Directive) (Bakardjieva Engelbrekt 2009, pp. 242–243; *Directive 98/71/EC of the European Parliament and of the Council on the legal protection of designs* 1998).

[111]Torremans (2009a, pp. 291–293) and Moufang (1998, p. 71).

[112]Macchia (2012, pp. 38–39).

[113]Technical Board of Appeal of the EPO (TBA) (1995).

[114]The interaction between patent law and other legislative instruments was assessed by authors. On the one hand, the environmental provisions of the EC Treaty (ECT) demand patent law to take into account environmental concerns and global warming (*Treaty establishing the European Economic Community (EEC Treaty)* 1957). Indeed, Article 6 of the ECT explicitly states that: "*Environmental protection requirements must be integrated into the definition and implementation of the Community policies and activities referred to in Article 3, in particular with a view of promoting sustainable development*". Article 3 refers to policy goals that are particularly relevant for IP (i.e. approximations of laws to improve the functioning of the internal market, strengthening the competitiveness of the industry and promoting research and technological development). Yet, so far, patent law does not directly integrate environmental principles. However, it was argued that patent law might indirectly cater for the protection of the environment via Article 53(a) EPC (Derclaye 2009, pp. 271–272). In defining the concept of *ordre public*, the Board of Appeal of the EPO cited scholarly works. Following the position of *Singer, Stauder* and *Schatz*, it held that: "*'Ordre public' is formed by the ethically based constitutional or other rules, usually backed up by penal provisions, that reflect the basic values prevailing in society and trade. These protected values can in particular include public safety, the integrity of the individual and nowadays, certainly also the protection of the environment*" (Technical Board of Appeal of the EPO (TBA) 2005, § 6.9). Numerous scholars defined the two terms and delineated the differences between them. *Schneider* observed that *ordre public* relates to threats to the social structures that tie a society together. By contrast, morality represents the degree of conformity to moral principles and values prevailing in a society. Considering that those values change over time and across different cultures, the notions of *ordre public* and morality were seen as fluid (Schneider 2014, p. 151). Hence, from *Viens*' perspective, there is no single true morality notion. Morality represents cultural normative relativism and, as a result, there is no general framework to determine what is moral by using non-arbitrary or non-bias methods (Viens 2009, pp. 88–89). According to *Schatz*, the concept of *ordre public* encompasses laws that express the fundamental values of society. These include, *inter alia*, criminal laws, environmental criminal laws, constitutional protection of basic rights and the protection of animals and embryos. Conversely, morality was considered a question of: "*Actual, ethically substantiated codes of behaviour which through general acceptance are binding on society*" (Schatz 1998, p. 14). He took as an example the cloning of humans and noted that, even in the absence of laws against it, the idea is still clearly rejected by scientists (Schatz 1998, p. 14).

norms of right and wrong within a specific culture. For the purposes of the EPC, the relevant culture would be the European one.[115]

In her analysis, *Warren-Jones* focused on the treatment of the two terms and observed that the EPO views them conjunctively, so that one notion is a borderline clarification of the other.[116] For this reason, it was argued that a definition might be redundant, as the two concepts cover all aspects of public morality.[117] Similarly, *Liddell* believed that there was no need to draw a clear distinction between the two terms, since they both relate to whether a technology incentivises socially beneficial inventions in a fair manner.[118] This approach conflicts with the one adopted by other commentators, who argued that *ordre public* and morality constitute two distinct tests pursuant to the EPO case law.[119] This view was reinforced by the EPO, as it was held that they constitute two separate objections that can be raised either alone or together.[120]

#### Scope of Interpretation

Uncertainties exist over the scope of interpretation of the morality clause. In particular, it is debated whether the clause should be interpreted in a narrow way or in a broad one. The consequences of this choice are significant, as a broad reading would expand the scope of non-patentable subject matter. Such consequences would be particularly felt in the field of biotechnology, where oppositions on morality grounds have often been raised.

Traditionally, the morality clause has been applied restrictively. In the *Lubrizol, Plant Genetic Systems* and *Oncomouse* cases,[121] the EPO embraced a narrow reading

---

[115]Technical Board of Appeal of the EPO (TBA) (1995, § 6).

[116]Warren-Jones (2008b, p. 644), Warren-Jones (2008b, p. 642) and Technical Board of Appeal of the EPO (TBA) (1999). By contrast, in the *PGS* case, the Technical Board diverged from this approach (Warren-Jones 2008b, p. 644; Technical Board of Appeal of the EPO (TBA) 1995). On this, see Sect. 5.1.3.2. In another article, *Warren-Jones* observed that: "*There is a disparity between the way in which the EPO assesses the morality provision (arguably a conjunctive construction in which 'ordre public' represents a borderline clarification of the wider 'morality') and the way in which the Appeal Boards interpret the provision (arguably a disjunctive reading with 'ordre public' and 'morality' forming repetitious terms). Given that the Boards very often refer to the EPO guidelines in their decisions, this type of disparity is somewhat inconceivable*" (Warren-Jones 2008b, p. 646).

[117]Warren-Jones (2007, p. 834).

[118]Liddell (2012, p. 163).

[119]Torremans (2009b, p. 156). Other authors believed instead that the EPO treated the terms as synonyms (Warren-Jones 2008b, p. 642).

[120]Technical Board of Appeal of the EPO (TBA) (2004, § 10.2).

[121]Technical Board of Appeal of the EPO (TBA) (1988, 1990, 1995). On those, see respectively Sects. 5.1.3.6, 5.1.3.2 and 5.1.3.1.

and cited the *travaux préparatoire* in support of its approach.[122] A number of scholars shared this opinion. *Warren Jones* argued that, since the patent system is supposed to encourage innovation, exceptions should be construed narrowly.[123] Similarly, according to *Armitage* and *Davis*, the exceptions in Article 53 EPC were narrowly conceived.[124] Support for this approach derives also from *Moufang*, who spoke of this exception as a narrow gate, and from the Guidelines of the EPO, which state that Article 53(a) EPC is reserved for rare and extreme cases.[125] The patent community has also considered this clause a narrow gate that should be kept as closed as possible.[126] This narrow interpretation derives from the fact that it concerns fundamental legal and ethical principles.[127]

However, recent decisions have challenged this approach, albeit not openly. The Enlarged Board of Appeal rejected the existence of an a priori principle of narrow interpretation of patent exclusions in two cases and held instead that the EPC should be interpreted pursuant to the Vienna Convention.[128] The *Edinburgh* case followed a similar pattern and was criticised by scholars for its broad interpretation.[129] In the *WARF* case, although the Technical Board of Appeal explicitly discussed the scope of interpretation, the question was not addressed by the Enlarged Board. Nevertheless, its interpretation was broad.[130] Interestingly, in its submission to the Enlarged Board, the then President of the EPO, *Alain Pompidou*, stated that the EPC does not enshrine a principle of narrow interpretation of exclusions. In so doing, he suggested

[122]Technical Board of Appeal of the EPO (TBA) (1995, § 8, 18.5), Technical Board of Appeal of the EPO (TBA) (1990, § 4.5) and Technical Board of Appeal of the EPO (TBA) (1988).

[123]Warren-Jones (2007, p. 843).

[124]O'Sullivan (2012, p. 683) and Armitage and Davis (1994, pp. 16–26).

[125]EPO (2018, § G.II.4.1), O'Sullivan (2012, p. 684) and Moufang (1998, p. 69). Similarly, with regard to plant and animal varieties, the preparatory documents of the EPC were read as a hint that Article 53(a) EPC must be construed narrowly (Visser 2019, p. 69; Technical Board of Appeal of the EPO (TBA) 1995, § 6).

[126]Schneider (2014, p. 152).

[127]Moufang (1998, p. 70).

[128]O'Sullivan (2012, p. 681), Enlarged Board of Appeal of the EPO (EBA) (2010a) and Enlarged Board of Appeal of the EPO (EBA) (2010b). The ECJ adopted a broader interpretation in the *Brüstle* case. Article 31 and 32 of the Vienna Convention on the Law of Treaties provide guidance on the interpretation of treaties (*Vienna convention on the law of treaties (1155 UNTS 1155 331)*, 1969). In particular, Article 31 recommends interpreting norms in good faith and in accordance with their ordinary meaning in light of their context, object and purpose. The proviso in Article 31(3)(b) of the Vienna Convention was seen as a possible help in understanding the scope of the exception. Still, it was noted that national patent offices have often interpreted the Rule in a different manner (Torremans 2009b, p. 147). The reference was limited to the patentability of human embryos, but its rationale could be extended to other morality-based exclusions. Equally, resorting to the Vienna Convention as a means of interpretation was suggested by the Technical Board of Appeal in the *WARF* case (Torremans 2009b, p. 144; Technical Board of Appeal of the EPO (TBA) 2006, § 33-35). On this, see Sect. 5.1.3.3.

[129]O'Sullivan (2012, p. 686) and Opposition Division of the EPO (2003a).

[130]O'Sullivan (2012, p. 687) and Enlarged Board of Appeal of the EPO (EBA) (2008, § 16). On this, see Sect. 5.1.3.3.

an individual case-by-case interpretation as well as an evolutive and dynamic approach.[131] Support for this argument could also be found in the *BST* decision of the WTO.[132] There, it was decided that exceptions should not be narrowly interpreted without reason and that norms need to be read in light of the Vienna Convention.[133]

Based on the above, commentators argued that there is no longer an a priori principle of narrow interpretation of Article 53(a) EPC, but instead the proviso should be examined in light of the Vienna Convention.[134]

#### Common European Concept

Questions have arisen over whether a common European morality standard exists.

It has been argued that morality and *ordre public* should be read as common European concepts, since otherwise the unity of the EPO system would be put in jeopardy by diverging national approaches.[135] The European Convention on Human Rights was cited in support of this view, as it outlines a number of morality principles common in Europe.[136] References to a common approach can be also found in the case law. The EPO raised similar points in the *Oncomouse* and *Plant genetic* cases, where it referred to a conventionally accepted standard of conduct in Europe.[137] The Advocate General in the *Brüstle* case before the ECJ also considered the existence of a concept of *ordre public* shared by all member States.[138]

This approach is not unanimously accepted. For instance, at the first Patents Working Party meeting held in Brussels in April 1961, it was recognised that "*there was no European definition of morality*".[139] Additionally, this conclusion would be supported by the fact that morality and *ordre public* are open concepts that can be interpreted differently based on the local culture. Lastly, Recital 39 of the Directive

---

[131]O'Sullivan (2012, p. 689), Treichel (2009, p. 471) and Enlarged Board of Appeal of the EPO (EBA) (2008) (Comments by the President of the European Patent Office, § 36-38).

[132]WTO Appellate Body (1998).

[133]O'Sullivan (2012, p. 689), Thomas and Richards (2004, pp. 100–101) and WTO Appellate Body (1998, § 104).

[134]O'Sullivan (2012, p. 689).

[135]Bonadio (2012, p. 439) and Torremans (2009b, p. 159).

[136]The Convention is expressly cited in Recital 43 of the Directive.

[137]Technical Board of Appeal of the EPO (TBA) (1995, § 17) and Opposition Division of the EPO (1989, § 13.2). On those, see respectively Sects. 5.1.3.1 and 5.1.3.2.

[138]Advocate General Bot (2011, §114).

[139]Mills (2010, pp. 31–32) and EPC Working Party (Patent Working Party) (1961, § 5). Still, in the same context, it was: "*Unanimously agreed that interpretation of the concept of morality should be a matter for European institutions*". *Mills* reports that the parties believed that it was enough to mention the concept of morality and that no further explanations were needed in the norm. This was taken as a sign of the limited relevance of this issue in the drafting of the EPC (Mills 2010, p. 32).

would seem to confirm this, as it speaks of moral and ethical principles recognised in *a* member State.

Both Courts and the EPO have embraced this latter approach in multiple occasions. The ECJ reached similar conclusions in the *Netherlands* case and in other non-IP related ones.[140] Similarly, in *Leland Stanford*, the EPO held that there was no European consensus on that technology.[141] Equally, in *Edinburgh*, it noted that there were no uniform moral standards.[142] *Transgenic animals* was no different, as there the EPO observed that there was no single definition of morality based on economic or religious principles that represents an accepted standard within the European culture.[143]

Different approaches have been suggested to overcome the uncertainty connected to a morality assessment spanning across numerous countries. A maximalist test would hinder the grant of a patent if the application would be considered immoral in just one of the examined countries. On the other end of the spectrum is instead the minimal approach, which predicates that a patent should be refused only if all designated countries would consider it morally unacceptable.[144] Scholars have suggested a third option. If the EPO is aware of the immorality of the invention in one State, it should warn the applicant. This way, he could withdraw its application in that country. If he does not, the patent could still be granted, but it might be subject to revocation in a national proceeding pursuant to Article 138 EPC.[145] Since all of the approaches have pros and cons, it was held that "*there is no easy way out of the dilemma*".[146]

---

[140]European Court of Justice (ECJ) (2001, § 37-38), European Court of Justice (ECJ) (1979, § 15) and European Court of Justice (ECJ) (1974, § 18). Equally, the European Court of Human Rights maintained that there was no European-wide consensus on when life begins (European Court of Human Rights 2004, § 84).

[141]Opposition Division of the EPO (2001, § 51). On this, see Sect. 5.1.3.6.

[142]Opposition Division of the EPO (2003a, § 2).

[143]Technical Board of Appeal of the EPO (TBA) (2004, § 6.1). On this, see Sect. 5.1.3.1.

[144]Bonadio (2012, p. 440). It was held that: "*The view expressed by Straus in 1990 appears still compelling today. When an invention is morally acceptable in some, or even one, of the state parties to the EPC depriving an applicant of the advantage of the European patent system would seem ill founded. Indeed, as observed by Moufang, by opting for the standard in the states with the most restrictive views on morality the EPO would take the uneasy position of condemning the morality standards in states where the respective inventions are not considered immoral or against ordre public*" (Bakardjieva Engelbrekt 2009, p. 259). In commenting on the view of *Straus*, another scholar noted that: "*Straus has defended this position by arguing that it shows respect for sovereignty of the Contracting States to uphold their own morality concepts. In his view, the acceptability of an invention in even one contracting country would constitute evidence of absence of a European-wide morality and ordre public. On this ground he considers that such acceptability should eventually lead to at least dismissing objections on grounds of morality*" (Torremans 2009a, p. 298).

[145]Ugurlu (2014, p. 41), Torremans (2009b, p. 159) and Bakardjieva Engelbrekt (2009, pp. 260–261).

[146]Bakardjieva Engelbrekt (2009, p. 259). For a brief overview of the pros and cons, Bakardjieva Engelbrekt (2009, pp. 257–261). *Torremans* suggests adopting the minimalistic approach or, even

### Standards of Morality

Over time, two different standards have been devised to assess whether an invention fails the morality clause: the balancing exercise and the rebuttable presumption test.

The balancing exercise requires moral and immoral aspects to be balanced against each other. Based on which one weights more heavily, the decision will be taken. By contrast, the rebuttable presumption approach establishes that the presence of moral aspects raises a favourable presumption that can be rebutted only in case the immoral aspects are such that a positive decision would be untenable.[147] As noted by *Warren-Jones*, the rebuttable presumption test is aimed at detecting stronger immoralities, since in every other case there will be a favourable decision. Conversely, the balancing test might lead to a negative decision even when both aspects are closely tied. Therefore, while the rebuttable presumption approach tackles cases where the decision would be irrefutable, the balancing one might be more contentious.[148]

The balancing test is inextricably linked to the so-called unacceptability standard, which entails a conscious decision-making process to assess whether a conduct is acceptable or not.[149] By contrast, the rebuttable presumption approach is connected to the abhorrence standard, which is judged more intuitively.[150] The latter refers to

better, the distributive one (Torremans 2009a, pp. 300–301). Following this latter approach, the patent would be granted in those member States where the patent is considered moral. This would require the EPO to compile an overview of morality standards across its contracting States (Torremans 2011, pp. 301–302).

[147]Warren-Jones (2007, p. 834).

[148]Warren-Jones (2007, pp. 834–835).

[149]The *Oncomouse* case represents an example of a balancing standard of unacceptability (Technical Board of Appeal of the EPO (TBA) 1990). On this, see Sect. 5.1.3.1. To judge whether a patent should be rejected, attention was paid to the careful weighing of the suffering of animals and risks to the environment against the usefulness of the invention to mankind (Technical Board of Appeal of the EPO (TBA) 1990, p. 90, § 5). Soon after, this test was used in the *Upjohn* case relating again to patents for mice (Min 2012, p. 262; McNab et al. n.d.). On this, see Sect. 5.1.3.6. The Board considered also the balancing exercise in the *Relaxin* case (Warren-Jones 2007, p. 837; Opposition Division of the EPO 1995b, § 18.8). On this, see Sect. 5.1.3.6.

[150]By contrast, other decisions used the abhorrence test, which requires patents to be denied only in extreme cases. For example, this would apply when an invention would universally be regarded as outrageous (Technical Board of Appeal of the EPO (TBA) 1995, § III., X.). In *WARF*, the President of the EPO advocated the use of a rebuttable presumption approach in lieu of the balancing one, at least when human life is involved (Enlarged Board of Appeal of the EPO (EBA) 2008; Warren-Jones 2008a, p. 205) (Comments of the President, § 53-55). Similarly, authors argued that the abhorrence standard is likely to be applied to humans, whereas the unacceptability approach would be used for animals and plants (Min 2012, p. 264). Equally, other scholars positively considered the application of the abhorrence approach to human biotechnologies, since it would permit a greater amount of innovation (Warren-Jones 2007, p. 839). However, this division might be impracticable, since the norms do not distinguish between humans and animals (Warren-Jones 2007, p. 839). Moreover, this conclusion is contradicted by the fact that the EPO adopted the abhorrence standard in a plant case (*Plant Genetic Systems*) and the unacceptability one by a balancing exercise in a human case (*WARF*). On this, see respectively Sects. 5.1.3.2 and 5.1.3.3. Equally, there seems to be no valid reasons for using two different tests (Min 2012, p. 264).

cases considered immoral beyond reasonable doubt. Because of this, the unacceptability test is seen as more stringent, since it can lead to a finding of unpatentability even if the invention is "merely" unacceptable instead of clearly abhorrent.[151]

The EPO and the Boards have adopted different approaches on this matter. The first has embraced the abhorrence rebuttable presumption test, whereas the Boards have swung between the two.[152] A clear sign of the approach of the EPO is contained in its Guidelines for Examination, according to which morality would be infringed if "*it is probable that the public in general would regard the invention as so abhorrent that the grant of patent rights would be inconceivable*".[153]

Proof

Following the *Edinburgh* and the *Plant Genetic Systems* rulings, it was noted that even if all the contracting States of the EPC were to judge a situation uniformly, this would still not be automatically relevant under Article 53(a) EPC.[154] This left commentators wondering which other types of evidence should be taken into consideration by the EPO.[155]

Currently, surveys, legal standards, religious teachings, human rights and opinions of ethical committees could all be examined in determining whether an

[151]Warren-Jones (2007, pp. 836–837).

[152]Warren-Jones (2007, pp. 843–844).

[153]EPO (2018, § G.II.4.1). From a scholarly perspective, it has been argued that the EPO adopted such a high standard of consensus due to its reluctance to decide on morally grey areas (Ottolia 2011, p. 316). In light of this, authors argued that the only way in which the interpretation given by the EPO Guidelines could be reconciled with that of the Boards of Appeal would be to consider the two notions as: "*Truly independent options: the moral assessment requires the decision-maker to select either 'ordre public' or 'morality' as the applicable standard. This means that one term must fit the other to represent a narrower meaning*" (Warren-Jones 2008b, p. 646). From this perspective, the EPO appears to have opted for the morality standard (Warren-Jones 2008b, p. 651). Scholars sided with the abhorrence approach, not only because numerous decisions embraced it, but also because it would fit a narrow construction of exclusions from patentability (Warren-Jones 2007, pp. 843–844). Additionally, if inventions were to be assessed by examining their possible benefits and risks, many technologies in the chemical, pharmaceutical and military field would not receive patent protection (Ugurlu 2014, p. 44). Therefore, it was argued that a finding of non-patentability pursuant to the morality clause should be made only when States unanimously abhor that technology; otherwise, patents should be issued. In light of this, exceptions from patentability should be taken by relying upon incontrovertible evidence, so that development is not hindered without a serious reason. A wide scope of evidence could be employed to reach this conclusion (e.g. laws, regulations, reports, practice and empirical evidence) (Warren-Jones 2008b, p. 660).

[154]Opposition Division of the EPO (2003b) and Technical Board of Appeal of the EPO (TBA) (1995). On this, see Sect. 5.1.3.2.

[155]Torremans (2009b, p. 158).

invention infringes the morality clause. Those indicia could operate at the same time and none of them would be decisive, since they all have shortcomings. Yet, if examined together, they may be able to provide an overview of the conventionally accepted standards of conduct.[156]

History of the Invention

The relevance of the history of an invention is a contentious matter. Scholars have criticised approaches where patent protection is withheld if the immoral steps occurred during the initial creation of the invention.[157] Indeed, it was observed that the EPO never considered the origin of a material in its decisions.[158] Similarly, in the *Netherlands* case, the ECJ argued that the scope of the Directive did not extend to "*activities before and after the grant, whether they involve research or the use of the patented products*".[159]

Nevertheless, following the *Brüstle* case before the ECJ, it would seem that the full history of the invention is relevant for the morality assessment.[160] However, it is still not clear how far would patent offices and judges need to investigate in order to establish the morality of an invention.[161]

### 5.1.3 European Case Law Overview

The EPO as well as the ECJ have been confronted with the application of the morality clause pursuant to the EPC and the Directive in a number of cases. An

[156]Liddell (2012, p. 162).

[157]Warren-Jones (2007, p. 833). Others considered this task as beyond the realm of the EPO (Ugurlu 2014, p. 53).

[158]Torremans (2009b, p. 166).

[159]European Court of Justice (ECJ) (2001, § 79).

[160]European Court of Justice (ECJ) (2011). On this, see Sect. 5.1.3.5.

[161]Bonadio (2012, p. 442) and Sterckx and Cockbain (2012, p. 304). This past-looking approach was compared to the one adopted in TRIPS and in the Directive towards the origin of genetic resources. However, the two systems were considered to be different, given the different consequences connected to them (mere administrative sanction vs. invalidity of the patent) (Schuster 2012, p. 637). The analysis of the full history of the invention is connected to the complicity of the patent applicant in the prior morally problematic step. This issue emerged in the field of human embryos, where the relationship between the patent applicant and the person who carried out the destruction of the human embryo was assessed (Ugurlu 2014, pp. 61–62; Sterckx and Cockbain 2010, pp. 101–104). To show the rationale behind this position, examples were made referring to the possibility that an invention was developed by using slaves as guinea pigs. Permitting a patent in such cases would effectively allow the manufacturer to profit from his immoral behaviour (Beyleveld and Brownsword 1993, p. 51).

overview of the most relevant ones from a synthetic biology perspective is presented below.

#### 5.1.3.1 Oncomouse

In 1985, the University of Harvard presented a patent application before the EPO for, *inter alia*, a rodent that had been genetically modified to carry a gene that increased its susceptibility to cancer. The application concerned both method and product claims and would be used in cancer research.

First, the application was rejected by the Examining Division in 1989 on the basis of Article 53(b) and 83 EPC. In that occasion, the relevance of Article 53(a) was discussed, as it was held that ethical questions needed to be considered.[162] Issues were raised because the animals were objectified and because of the possible destructive consequences of having transgenic animals escaping into the environment. Concerns over the interference with natural evolutionary patterns were also raised. On this, the Examining Division held that patent law was not the right tool to address those issues.[163]

On appeal, the case was decided by the Technical Board.[164] In assessing the application of Article 53(a), the Board noted that:

> There is also a danger that genetically manipulated animals, if released into the environment, might entail unforeseeable and irreversible adverse effects... Considerations of precisely this kind have also led a number of Contracting States to impose legislative control on genetic engineering. The decision as to whether or not Article 53(a) EPC is a bar to patenting the present invention would seem to depend mainly on a careful weighing up of the suffering of animals and possible risks to the environment on the one hand, and the invention's usefulness to mankind on the other.[165]

The Board remitted the case to the Examining Division for further prosecution. The Division then granted the patent, which was subsequently opposed by seventeen parties on the basis, amongst others, of Article 53(a) EPC. In reaching its decision, it

---

[162] Sterckx and Cockbain (2012, p. 244), Technical Board of Appeal of the EPO (TBA) (1990, § II) and Opposition Division of the EPO (1989).

[163] Technical Board of Appeal of the EPO (TBA) (1990, § II). In the view of the Examining Division, technologies could not be excluded from patentability merely because they could be dangerous, since the regulation of hazardous materials is assigned to specialised authorities rather than the EPO (Mills 2010, p. 59).

[164] According to *Warren-Jones*, the Board did not separate the issues of *ordre public* and morality (Warren-Jones 2008b, p. 641).

[165] Technical Board of Appeal of the EPO (TBA) (1990, § 5). It was noted that, at least on paper, the Board advocated for a substantive rather than procedural approach to evaluate morality (Sterckx and Cockbain 2012, p. 296). On the release of GMOs into the environment, Mills (2010, pp. 148–152) and *Council Directive 90/220/EEC on the deliberate release into the environment of genetically modified organisms* (1990).

held that arguments holding that patents on living matter or animals are not allowed *per se* must fail as a matter of principle.[166]

In its judgement, the Opposition Division confirmed the patent, albeit in an amended form. Dissatisfied with this conclusion, six opponents appealed this conclusion and the case landed once again before the Technical Board of Appeal in the *Transgenic animals/Harvard* case. To decide on the patentability of this technology, the Board assessed the relationship between Rule 23(d) EPC and Article 53 (a) EPC.[167] This could be seen as a "two-stage test"; inventions passing the hurdle set by the Rule could still be caught by the Article 53(a) EPC exception.[168] As for the test set by Rule 23(d)(d)(now Rule 28(d)), the Board considered it a "balancing test", aimed at weighing the suffering of animals against the resulting medical benefits to humans and animals.[169]

Under Rule 28(d) EPC, it was necessary to assess whether benefits could be obtained for all the animals that were likely to suffer. Since the benefit claimed in the patent could only be obtained for mice and not for rodents in general, as instead alleged by the applicant, such claims would fail the balancing test set by Rule 28(d). The same conclusion would be reached in case Article 53(a) EPC were to apply.[170] Instead, for claims referring exclusively to mice, the Board found that they would pass both the Rule and the Article test, since the animals that would suffer and those yielding medical benefits would overlap.[171]

Lastly, concerning the environmental risks connected to genetically modified animals, the Board dismissed them as having, at the most, a neutral effect on the case. In particular, it maintained that:

> While a risk of release or escape exists, just as there is such a risk with zoo or circus animals, the risk can only be regarded as minimally more than hypothetical when one considers the secure conditions under which laboratory mice are kept and the level of regulation of the use and keeping of animals for experimental purposes... Further, in the event of release or escape, it must be questionable whether oncomice would cause any damage, let alone any lasting damage, to the environment. The only perceivable threat is that, by mating with mice already in the wild, the oncogene would be spread. Against that, there must be the possibility

[166]Sterckx and Cockbain (2012, p. 245).

[167]Technical Board of Appeal of the EPO (TBA) (2004, § 1.3).

[168]Technical Board of Appeal of the EPO (TBA) (2004, § 6.3). The possibility of a triple test has also been suggested (Visser 2019, p. 70; Ottolia 2011, pp. 318–319; Technical Board of Appeal of the EPO (TBA) 2004, § 13).

[169]Technical Board of Appeal of the EPO (TBA) (2004, § 6.2). The presence of harm was considered apparent from the patent application. On the other hand, proof of substantial benefit required additional information to show that, at the time the patent was filed, there was a substantial medical gain. Such evidence could be submitted until the time the decision was made (Warren-Jones 2008a, p. 203; Technical Board of Appeal of the EPO (TBA) 2004, §9).

[170]Technical Board of Appeal of the EPO (TBA) (2004, § 12.2).

[171]As far as evidence was concerned, the Board considered opinion pools of very limited relevance (Technical Board of Appeal of the EPO (TBA) 2004, § 10.4).

that, because of their manipulated state, oncomice would not survive as long in the wild as non-manipulated mice.[172]

In discussing the repercussions of this ruling on environmental protection, scholars noted that those types of concerns are generally directed to the consequences and impact on nature of technologies or concern the dominion exercised by mankind on it. According to *Mills*, the EPO viewed those issues as against morality and not against *ordre public*. He also considered it unlikely that a patent would be rejected on morality grounds on the basis of environmental issues alone.[173] This conclusion seems to be confirmed by EPO precedents, as lower Boards have uniformly rejected environmental concerns, often by noting that the patent system was not the appropriate forum for this type of enquiry.[174]

In conclusion, the Board considered the claims relating to mice as patentable and remitted the case to the Opposition Division. However, the patent was subsequently revoked, as the applicant failed to take the steps necessary to maintain it.[175]

Scholars examined the possible impact of public opinion on the patentability of this technology. Since at the time of the patent application the European public was barely informed about the oncomouse, it had no opportunity or reason to form an opinion about its morality. Because of this, the morality clause could not be overcome by arguing that there was no public concern at the time the application was filed.[176]

#### 5.1.3.2 Plant Genetic Systems

In 1990, a patent was granted to Plant Genetic Systems (PGS) for the genetic engineering of herbicide-resistant plants. The patent was opposed by Greenpeace on a number of grounds, including Article 53(a) EPC. The environmental organisation opposed the idea that life forms could be seen as monopolisable items and was worried about the impact of this technology on farmers and genetic diversity.

[172]Technical Board of Appeal of the EPO (TBA) (2004, § 13.2.9). On environment issues, the Board noted that: "*The same factors are considered except that the agreed suffering to animals is replaced by possible threats to the environment if oncomice were to escape (or to be released, deliberately or accidentally) into the wild. As is only to be expected of a danger yet to materialise, there was no evidence to support such environmental arguments which played very little part in the appeal proceedings. The environmental arguments are thus if anything weaker in this case than in T 356/93*" (Technical Board of Appeal of the EPO (TBA) 2004, § 13.2.7, 13.2.8; Technical Board of Appeal of the EPO (TBA) 1995).

[173]Mills (2010, p. 60).

[174]Warren-Jones (2008b, p. 645). Other authors commented on the centrality of environmental concerns in the morality clause. Those issues are an incontrovertible part of the moral assessment. Hence, the Board required the Examining Division to consider environmental issues when reassessing the case (Warren-Jones 2008b, p. 646; Technical Board of Appeal of the EPO (TBA) 1990, § 5).

[175]Sterckx and Cockbain (2012, p. 252).

[176]Thomas and Richards (2004, p. 101).

Equally, risks connected to the development of new pests and weeds were mentioned. Greenpeace sought to revoke the patent, as it argued that its granting "*facilitates and, in effect, acts as enabling legislation for the release of genetically engineered organisms*" in the environment.[177] By contrast, PGS held that patents do not constitute a positive right to exploit the invention and that inventions are assessed via regulatory and controlling systems at national and EU level. Lastly, the company alleged that its technology could increase genetic diversity as well as help developing nations.[178]

In its decision, the Opposition Division maintained the patent. It objected to the issues raised by Greenpeace by noting that the alleged risks were unproven and that they could change during the lifetime of the patent. Significantly, it maintained that "*the possibility of risks traditionally has no bearing on whether a patent is granted or not*".[179] Furthermore, it held that the Opposition Division was not the right forum to discuss those matters.[180]

Greenpeace appealed this ruling and the case was brought before the Technical Board of Appeal. Despite recognising that there was no European definition of *ordre public* and morality, the Board outlined its understanding of both concepts. *Ordre public* covered the protection of public security as well as the physical integrity of individuals as members of society. The protection of the environment was also included in this concept.[181] Morality referred instead to the idea that "*some behaviour is right and acceptable whereas other behaviour is wrong, this belief being founded on the totality of the accepted norms which are deeply rooted in a particular culture*".[182]

On environmental threats, the Board stated that the claims did not concern the misuse or destructive use of the technology, given that they concerned activities and products (i.e. protection of plants from weeds as well as plants and seeds) that do not violate the conventionally accepted standards of conduct existing in Europe. The argument of Greenpeace that those techniques represent a dominion of mankind over the natural world was also rejected. This technology was seen as the application of

[177]Sterckx and Cockbain (2012, pp. 255–256).

[178]Sterckx and Cockbain (2012, p. 256).

[179]Opposition Division (Technical Board of Appeal of the EPO (TBA) 1995, § 31).

[180]Sterckx and Cockbain (2012, p. 257).

[181]Technical Board of Appeal of the EPO (TBA) (1995, § 4-5). The classification of environmental threats as either belonging to *ordre public* or morality concerns was addressed in this case and a theoretical line was traced between the two scenarios. On the one hand, the appellant argued that morality would be called into question if the technology enabled mankind to control the natural world. On the other hand, the applicant believed that *ordre public* would come into play when the exploitation of the technology would seriously prejudice the environment (Mills 2010, pp. 66–67). *Warren-Jones* maintained that: "*On the rationale of the Technical Board of Appeal in the PGS case, dumping radioactive waste would be an issue of 'ordre public', but 'litter-bugging' a question of 'morality'. It is difficult to argue that this is an easy dividing line to predict or to justify, given that it centres upon the degree to which a single moral tenet is breached, rather than upon which moral tenet*" (Warren-Jones 2008b, p. 645).

[182]Technical Board of Appeal of the EPO (TBA) (1995, § 6).

modern scientific knowledge to advance the control and understanding of phenomena relating to plants. In the opinion of the Board, this was not enough to consider those activities as intrinsically wrong. The Board reaffirmed its point by noting that:

> Plant biotechnology per se cannot be regarded as being more contrary to morality than traditional selective breeding because both traditional breeders and molecular biologists are guided by the same motivation, namely to change the property of a plant by introducing novel genetic material into it in order to obtain a new and, possibly, improved plant. However, compared with traditional breeding techniques, genetic engineering techniques applied to plants allow a more powerful and accurate control of genetic modifications... The impressive potential of these techniques is at the origin of the concerns and apprehensions expressed by public opinion and generates considerable disagreement and controversy... Like any other tool, plant genetic engineering techniques can be used for constructive or destructive purposes. It would undoubtedly be against 'ordre public' or morality to propose a misuse or a destructive use of these techniques.[183]

As for the environmental hazards alleged by Greenpeace, in the view of the Board, a patent could be revoked under Article 53(a) EPC only if the risks to the environment have been sufficiently substantiated by the time the EPO reaches a decision. Here, this had not been the case, since Greenpeace could not document those hazards conclusively.[184]

Afterwards, the Board maintained that Article 53(a) EPC is not violated merely because a patent seeks to cover living matter.[185] Lastly, the Board addressed its role and that of the patent system in granting rights on these inventions. It pointed out that the function of the EPO is to issue patents. A number of regulatory authorities and bodies operate alongside the Office to control and supervise their exploitation. Those authorities are involved independently of whether the invention is patent protected or not and their intervention is needed, since the risks connected to exploitation are

[183]Technical Board of Appeal of the EPO (TBA) (1995, § 17). This approach was contested by noting that the fact that an activity has been performed in the past does not mean that it will be considered moral in the future. Specifically, in analysing the comparison drawn by the Board, *Warren-Jones* observed that traditional breeding techniques could be seen as acceptable domination, since they employ naturally occurring processes. On the other hand, rDNA techniques represent a stronger form of domination, given that they facilitate combinations that do not occur in nature (Warren-Jones 2008a, pp. 200–201).

[184]Technical Board of Appeal of the EPO (TBA) (1995, § 18). Judging from this, it would seem that the EPO does not adopt a precautionary principle (Sommer 2013, p. 202). On this, see Sect. 2.4.1.2. The precautionary principle and the risks connected to genetically modified animals were discussed with regards to a patent for a genetically modified salmon. As reported by *Sommer*, the application concerned an all-fish chimeric growth hormone gene construct. The Norwegian Advisory Board on Ethical Aspects of Patenting noted that this growth hormone presented problematic side-effects, especially from a behavioural, physiological and genetic perspective. This could have had an impact not only on transgenic fishes, but also on wild ones. Notwithstanding this, the patent was granted (Sommer 2013, p. 204; Fletcher and Hew 2001).

[185]Technical Board of Appeal of the EPO (TBA) (1995, § 10-11).

generally impossible to foresee on the mere basis of what is disclosed in the patent application.[186]

In light of the above, the patent was found in compliance with Article 53(a) EPC and the case was thus remitted to the Opposition Division, which upheld the patent, albeit in an amended form.[187]

#### 5.1.3.3 WARF

In 1996, the Wisconsin Alumni Research Foundation (*WARF*) filed a patent application that would become a *cause célèbre* in the application of the morality clause before the EPO. The invention concerned primate embryonic stem cells, including human ones, as well as processes for obtaining them.

The Examining Division was first confronted with this application and rejected it on the basis of Article 53(a) EPC and Rule 28(c) EPC. In breaking with prior EPO practice, the Division adopted a broad reading of the exceptions, despite holding that it was not doing so. Specifically, it maintained that the use of embryos as raw materials for creating items for industrial application equalled an industrial use of that embryo. This inclusion was possible, since the Examining Division noted that the morality clause referred to inventions. This could be understood to cover all aspects of the invention that were made available to the public, rather than limiting it only to the patent claims.[188]

The decision was appealed by WARF.[189] In its ruling, the Technical Board of Appeal observed that all the techniques that were mentioned in the application implied the destruction of embryos.[190] The matter was then referred to the Enlarged

[186]The position of the Board raised questions on the role of the EPO towards technological hazards. *Schatz* noted that the position of the EPO is unclear. Considering that threats to the environment will be assessed only when they are sufficiently substantiated at the time the decision is made, he concluded that the: "*Examining Divisions of the EPO do not have any particular duty of investigation and that they are allowed to base their assessment of technological risks on what is disclosed in the patent application as filed*… *if the assumption is correct*… *then it is very likely that Opposition Divisions and Boards of Appeal will arrive at an assessment of risks which is different from that of the Examining Divisions, since, as a rule, relevant arguments and evidence will be presented by the opponent, especially in cases in which the opponent is a non-industrial entity*" (Schatz 2000, p. 221).

[187]Sterckx and Cockbain (2012, p. 261).

[188]Sterckx and Cockbain (2012, pp. 280–281).

[189]Sterckx and Cockbain (2012, pp. 282–284).

[190]Technical Board of Appeal of the EPO (TBA) (2006). On the interpretation of the Rule, the Board noted that there was only one decision of the Enlarged Board of Appeal on the matter (*Diagnostic methods* case) (Enlarged Board of Appeal of the EPO (EBA) 2005). It then referred to the statement of the Enlarged Board of Appeal that: "*The frequently cited principle according to which exclusion clauses from patentability laid down in the EPC were to be construed in a restrictive manner, did not apply without exception*" (Technical Board of Appeal of the EPO (TBA) 2006).

Board of Appeal for the clarification of a number of open points.[191] In that circumstance, the Board embraced the broader interpretation already adopted by the Technical Board and considered the morality clause to refer to the invention as a whole rather than to the claimed subject matter.[192] Since the application involved the destruction of human embryos and considering that this was an essential aspect of the industrial and commercial exploitation of the invention, it violated the Rule.[193]

The approach adopted by the Board constituted a radical change compared to the line of decisions previously set by the EPO for the morality clause. Until then, the EPO followed the so-called balancing approach, which revolved around whether the benefits deriving from the invention outweighed its disadvantages. However, the Board noted that this approach may not be advisable when human dignity is involved.[194]

---

[191]The Board was asked to shed light on the following questions: *"(1) Does Rule 23d(c) EPC apply to an application filed before the entry into force of the rule? (2) If the answer to question 1 is yes, does Rule 23d(c) EPC forbid the patenting of claims directed to products (here: human embryonic stem cell cultures) which - as described in the application - at the filing date could be prepared exclusively by a method which necessarily involved the destruction of the human embryos from which the said products are derived, if the said method is not part of the claims? (3) If the answer to question 1 or 2 is no, does Article 53(a) EPC forbid patenting such claims? (4) In the context of questions 2 and 3, is it of relevance that after the filing date the same products could be obtained without having to recur to a method necessarily involving the destruction of human embryos (here: eg derivation from available human embryonic cell lines)?"* (Technical Board of Appeal of the EPO (TBA) 2006, § VII). In addressing the first point, the Enlarged Board of Appeal noted that Rule 28 (c) EPC applied to all pending applications, including those filed before its entry into force. The second question was pivotal in the analysis of the Board, as it clarified that: "*Rule 28(c) EPC… forbids the patenting of claims directed to products which - as described in the application — at the filing date could be prepared exclusively by a method which necessarily involved the destruction of the human embryos from which the said products are derived, even if the said method is not part of the claims*" (Enlarged Board of Appeal of the EPO (EBA) 2008, § Headnote). For an overview of the case, Rowlandson (2010).

[192]Specifically, it stated that: "*This Rule (as well as the corresponding provision of the Directive) does not mention claims, but refers to 'invention' in the context of its exploitation. What needs to be looked at is not just the explicit wording of the claims but the technical teaching of the application as a whole as to how the invention is to be performed. Before human embryonic stem cell cultures can be used they have to be made. Since in the case referred to the Enlarged Board the only teaching of how to perform the invention to make human embryonic stem cell cultures is the use (involving their destruction) of human embryos, this invention falls under the prohibition of Rule 28 (c)… To restrict the application of Rule 28(c)… EPC to what an applicant chooses explicitly to put in his claim would have the undesirable consequence of making avoidance of the patenting prohibition merely a matter of clever and skilful drafting of such claim*" (Enlarged Board of Appeal of the EPO (EBA) 2008, § 22).

[193]Enlarged Board of Appeal of the EPO (EBA) (2008, § 25). Accordingly, the Board stated that: "*The argument raised by the Appellant, namely that the exclusion from patentability would go much too far if one would consider all the steps preceding an invention for the purposes of Rule 28(c)… EPC, is not relevant*" (Enlarged Board of Appeal of the EPO (EBA) 2008, p. 06, § 23). Scholars criticised such reading (Torremans 2009b, p. 164).

[194]Sterckx and Cockbain (2012, pp. 285–286). The issue of research funding was considered both by the applicant and the Board. Specifically, it was noted that: "*The very fact that the Community*

In conclusion, the decision of the Enlarged Board was seen as a fundamental step towards the "constitutionalisation" of European patent law, given that it identified the high-ranking norms behind Rule 28(c) EPC. These include the respect of human dignity and the prohibition of the commodification of human embryos.[195] This decision was seen as a confirmation that the purpose of Article 53(a) EPC is to incorporate fundamental ethical and legal principles into European patent law.[196]

#### 5.1.3.4 Myriad

Article 53(a) EPC concerns were raised for BRCA-related patents before the EPO.

In *Breast and ovarian cancer/University of Utah*, the Board decided on a patent related to the Myriad saga.[197] Initially, before the Examining Division, the application had been objected mostly on novelty grounds. Morality concerns emerged after the patent was granted in 2001, as eight parties opposed it, *inter alia*, based on Article 53(a) EPC. The concerns related to the lack of valid consent from the donors and the adverse effects that the patent could have on research and healthcare. Those arguments were dismissed by the Opposition Division. After the patent was maintained in amended form, the patentee and an opponent appealed this decision. In 2007, the Board dismissed the appeal. It noted that Article 53(a) EPC relates to inventions whose exploitation would be contrary to *ordre public* and morality. Instead, in the case at hand, the Board considered the objection of Greenpeace to relate to the exploitation of the patent rather than to the exploitation of the invention. Because of this, the objection was outside of the realm of Article 53(a) EPC.[198]

One year later, the Board of Appeal was confronted again with two Myriad-related cases. In *Mutation/University of* Utah, it dealt with a patent granted in 2001 and later opposed, *inter alia*, on Article 53(a) EPC grounds.[199] It was argued that

---

*funds such research shows that the legislator did not want to exclude activities such as those underlying the present invention and which include the use (and destruction) of human embryos. However, Council press release 11554/06 (Presse 215) of 24 July 2006, states on page 7 that as regards Community Research '... the Commission confirmed that it will continue the current practice and will not submit to the Regulatory Committee proposals for projects which include research activities which destroy human embryos, including for the procurement of stem cells. The exclusion of funding for this step of research will not prevent the Community funding of subsequent steps involving human embryonic stem cells'*" (Enlarged Board of Appeal of the EPO (EBA) 2008). For an overview of other contexts that "use" human embryos as well as of conflicting notions of their status in the regulatory framework, Odell-West (2020).

[195]Treichel (2009, p. 451). The importance of human dignity was mentioned by Advocate General *Jacobs* in the *Netherlands* case (Advocate General Jacobs 2001, § 6, 70, 76).

[196]Treichel (2009, p. 455).

[197]Sterckx and Cockbain (2012, pp. 275–277) and Technical Board of Appeal of the EPO (TBA) (2007).

[198]Technical Board of Appeal of the EPO (TBA) (2007, § 53).

[199]Sterckx and Cockbain (2012, pp. 277–279) and Technical Board of Appeal of the EPO (TBA) (2008a).

human genes belonged to the common heritage of mankind and were therefore not patentable. Similarly, worries about the inhibitive effects of the patent on research and the diagnostic choices of patients were expressed. Those arguments were rejected by the Opposition Division, which maintained the patent in an amended form. Before the Board, all objections pertaining to Article 53(a) EPC were rejected once again.

A third Myriad patent was decided in 2008 in *Method of diagnosis/University of Utah.*[200] The granted patent was opposed, amongst others, on Article 53(a) EPC grounds. In its decision to revoke the patent, the Opposition Division did not mention the morality clause. When faced with this argument, the Board of Appeal rejected it on the basis of its prior findings in the *Breast and ovarian cancer* case, just as it had done in the *Mutation* one.

#### 5.1.3.5 Brüstle

*Brüstle* represents a landmark case for the interpretation of the morality clause in Europe. The case was brought to the attention of the ECJ upon referral of the German Supreme Court. The case originated from a German patent granted to Dr. Brüstle of the University of Bonn for isolated and purified neutral precursor cells obtained from HESCs (Human Embryonic Stem Cells) that could be used in the treatment of patients afflicted by dementia and Parkinson's disease.[201] The patent was initially challenged by Greenpeace before the German Federal Patent Court on morality grounds.[202] In particular, it was argued that it would be unethical to grant a patent for an invention that employed a human embryo that was later destroyed.[203] As the Court ruled in favour of Greenpeace, Dr. Brüstle appealed to the German Supreme Court, which in turn referred the matter to the ECJ.[204]

---

[200] Sterckx and Cockbain (2012, pp. 279–280) and Technical Board of Appeal of the EPO (TBA) (2008b).

[201] A patent having the same claims had been granted by the EPO even before the Board of Appeal reached its decision in the *WARF* case (Ugurlu 2014, pp. 70–71). On this, see Sect. 5.1.3.3. Similarly, it was noted that the German Federal Patent Court invalidated the patent pursuant to the morality clause, despite the grant of the EPO patent only 6 months before (Plomer 2009, p. 194).

[202] For a brief overview, Treichel (2009, p. 462) and Hellstadius (2009, pp. 127–128).

[203] Questions regarding the interpretation to be given to the concept of human embryo were raised also in another case before the ECJ (European Court of Justice (ECJ) 2014). There, the Judges had to decide whether an unfertilised human ovum whose division and further development have been stimulated by parthenogenesis and is incapable of developing into a human being should be included within this definition. The Court replied in the negative and held that such unfertilised human ovum would not fit within this definition if, in light of the current scientific knowledge, it does not in itself have the inherent capacity of developing into a human being.

[204] German Federal Patent Court (BPatG) (2006). The referral concerned the following three questions: "*1. What is meant by the term 'human embryos' in Article 6(2)(c) of [the Directive]? 2. What is meant by the expression 'uses of human embryos for industrial or commercial purposes'? Does it include any commercial exploitation within the meaning of Article 6(1) of [the*

In considering whether "*the exclusion from patentability concerning the use of human embryos for industrial or commercial purposes also covers the use of human embryos for purposes of scientific research*", the Court noted that "*the grant of a patent implies, in principle, its industrial or commercial application*".[205] Therefore, the ECJ held that Article 6(2)(c) of the Directive covered also uses for scientific research if they constitute the object of a patent application.[206]

Afterwards, the Court considered whether the destruction of embryos would hinder the patentability of the invention even if such destruction was not mentioned in the specifications of the patent. In the opinion of the ECJ, this circumstance would not affect the unpatentability of the invention. The ECJ noted that, otherwise, this proviso would be redundant, given that applicants could circumvent it by skilfully drafting the claims. In reaching this conclusion, the Judges referred to the *WARF* ruling of the Enlarged Board of Appeal of the EPO.[207] Despite this reference to the EPO, the two systems remain separate. Therefore, although this ECJ judgement can be seen as persuasive from an EPO perspective, it is not binding on the latter.[208]

---

*Directive], especially use for the purposes of scientific research? 3. Is technical teaching to be considered unpatentable pursuant to Article 6(2)(c) of the Directive even if the use of human embryos does not form part of the technical teaching claimed with the patent, but is a necessary precondition for the application of that teaching: because the patent concerns a product whose production necessitates the prior destruction of human embryos, or because the patent concerns a process for which such a product is needed as base material?*" (European Court of Justice (ECJ) 2011, § 23). In replying to the first query, the Court adopted a wide interpretation of the concept of embryo, which encompassed any ovum, as soon as fertilised (European Court of Justice (ECJ) 2011, § 38). Significantly, the Judges stated that the Directive does not define the concept of human embryo or refer to national norms on this. Hence, for the purposes of the Directive, this notion shall be read: "*As designating an autonomous concept of European Union law which must be interpreted in a uniform manner throughout the territory of the Union*" (European Court of Justice (ECJ) 2011, § 26). This argument had been supported by Advocate General *Bot*, who held that concepts found in Article 6(2) of the Directive shall be subject to a common understanding within all EU States. Conversely, States argued that they should retain their discretion in interpreting this concept (Sommer 2013, p. 193). In particular, the Advocate General stated that: "*If it were left to the Member States to define the concept of a human embryo, in view of the differences which exist in this regard, this would mean, for example, that an invention like that of Mr Brüstle could be granted a patent in some Member States, while the patentability of such an invention would be excluded in others. This would run counter to the main objective of the directive, which is to establish effective and harmonised legal protection of biotechnological inventions*... *As regards the scope accorded to the Member States by Article 6(2) of the directive, the Court ruled that that provision allows the Member States no discretion with regard to the unpatentability of the processes and uses which it sets out. This binding aspect of one of the key provisions of the directive would also seem to call for a uniform interpretation of the concept of a human embryo within the Union. I cannot see how such a categorical prohibition, applying to all the Member States, could exist on the basis of concepts which were not common*" (Advocate General Bot 2011, § 56, 60).

[205] European Court of Justice (ECJ) (2011, § 41-44).

[206] Bonadio (2012, p. 434) and European Court of Justice (ECJ) (2011, § 46).

[207] European Court of Justice (ECJ) (2011, § 50-51). On this, see Sect. 5.1.3.3.

[208] EPO (2018, § G.II.5.3).

Human dignity was central in this decision. Both the Court and the Advocate General referred to this notion, although they held that their conclusions were not dictated by moral beliefs.[209] In particular, the Advocate General *Bot* explained that:

> The Union is not only a market to be regulated, but also has values to be expressed. Before it was even enshrined as a fundamental value in Article 2 of the EU Treaty, the principle of human dignity had been recognised by the Court as a general legal principle.[210]

Similarly, the European Charter of Fundamental Rights adopted with the Lisbon Treaty and the Oviedo Declaration of the Council of Europe were seen as a codification of the norms and values enshrined in European culture.[211]

The decision caused a stir and was criticised for negatively affecting the biotech industry in Europe.[212] Nevertheless, other commentators believed that the scope of the decision might be more limited than it appeared at first glance, given the existence of new techniques that no longer require the destruction of embryos.[213] However, not everyone was critical of this decision. For example, the European Parliament welcomed this ruling and the *WARF* decision.[214] The EPO adopted a middle ground position on the matter.[215]

---

[209]Bonadio (2012, p. 434), European Court of Justice (ECJ) (2011, § 30) and Advocate General Bot (2011, § 39-40, 46, 76, 96). Commentators have criticised this approach of the Court, by noting that: "*The ECJ's denial in Brüstle that it was dealing with moral issues, when clearly... it was, was disingenuous and disappointing*" (Harmon et al. 2013, p. 100).

[210]Advocate General Bot (2011, § 46).

[211]Schneider (2014, p. 161), (*Charter of fundamental rights of the European Union (2007/C 303/1)*, 2000) and (*Convention for the protection of human rights and dignity of the human being with regard to the application of biology and medicine: Convention on human rights and biomedicine (ETS 164)*, 1997). The Advocate General traced the importance of human dignity to the Charter of Fundamental Rights of the European Union (Article 1 and 3) (Advocate General Bot 2011). Other authors cited the Preamble of the 1997 Oviedo Convention as encoding the protection of human dignity in the European legal system (Zullo 2012, p. 90) (*Convention for the protection of human rights and dignity of the human being with regard to the application of biology and medicine: Convention on human rights and biomedicine (ETS 164)*, 1997).

[212]Harmon et al. (2013) and Bonadio (2012, p. 434).

[213]Harmon et al. (2013, p. 106). Likewise, it has been argued that the impact of the decision is relativised by narrow national and EPO practices (Schneider 2014, p. 161). For a narrow reading of Article 6(2)(c) of the Directive by national patent offices, Ottolia (2011, p. 322).

[214]European Parliament (2012, § 6). On this, see Sect. 5.1.3.3.

[215]In a blog post on patents and biotechnology, the then President of the EPO *Benôit Battistelli* stated that the Office is: "*Well aware of the sensitivity of the issues involved, and although biotech accounted for under 5% of all European patent applications received in 2010 has set up a dedicated specialist taskforce and applies the rules very strictly (the grant rate in biotech is 28%, compared with 42% overall). Even so, every so often one of these cases becomes a cause célèbre... If the judges rule in favour of a restrictive interpretation of biotech patentability provisions, the EPO will immediately implement it. The European legislator may also decide that further changes to the law are needed. Whatever is decided, I welcome evidence-based debate and invite all interested parties, especially associations active on these issues, to discuss them with us on the basis of objective data. It should also be kept in mind that changes to our legal framework are likely*

#### 5.1.3.6 Other Cases[216]

Lubrizol

In 1992, the Opposition Division was asked to decide on the morality of a patent for a genetically modified plant. In this controversy, the common heritage objection was raised. The Opposition Division however dismissed it, by noting that the notion of heritage could only apply to something that already existed, which was not the case in this situation. In commenting on this, authors noted that this objection can be classified as "*raising general developmental issues of the technology, rather than of the specific invention, because what is being objected to is the existence of the technology itself*'.[217]

That same year, in another case regarding again a Lubrizol patent controversy, the Opposition Division explained that the invention could be used to generate new plants with an increased nutritive value that could alleviate global food shortage. In light of this, the Division argued that the invention would not violate the morality clause.[218]

Relaxin

The patent was initially granted without having encountered any Article 53(a) EPC objections. However, soon after, an opposition was filed. The morality of the invention was challenged by the Green Party, which considered genes as common human heritage. In its ruling, the Opposition Division discussed the standard needed under Article 53(a) EPC and held that:

> Only in those very limited cases in which there appears to be an overwhelming consensus that the exploitation or publication of an invention would be immoral may an invention be excluded from patentability under Article 53(a).[219]

The Greens argued that the significance of DNA lies in its informational content and that the use of the genetic information of a person constitutes an exploitation of the human body, which would infringe the morality clause. To this, the EPO Opposition Division replied that the patenting of single human genes is not tantamount to patenting human life. DNA is not life, but rather a chemical substance that carries genetic information. Indeed, even if every gene were cloned, it would still be impossible to reconstruct a human being on that basis alone.[220]

---

*to have economic consequences, including possible deterrent effects for the siting of research centres in Europe*" (Battistelli 2011).

[216]For an overview of other relevant cases, Sterckx and Cockbain (2012, pp. 252–264).

[217]Warren-Jones (2008a, p. 195) and Opposition Division of the EPO (1992a).

[218]Grosheide (2010, p. 250) and Opposition Division of the EPO (1992b).

[219]Opposition Division of the EPO (1995b, § 6.5).

[220]Milius and Townend (2008, p. 6) and Opposition Division of the EPO (1995a, § 6).

After the Opposition Division rejected the opposition and maintained the patent, the opponent appealed. Also this time, the move was not successful and the appeal was dismissed.[221]

### Hairless Mouse

The company Upjohn applied for a patent for a hairless mouse, which could be used in finding a treatment for hair loss. The patent was refused by the EPO and later withdrawn. Amongst the critiques raised against it, Article 53(a) EPC was cited. It was considered that the benefit that could be derived from the animal (i.e. cure for hair loss) did not justify its suffering.[222]

Equally, commentators noted that the term "substantial" should be understood to mean that it alleviates or prevents a serious condition. In light of this, a cure for boldness, while desirable, would not suffice.[223]

### Leland Stanford

The case was brought before the Opposition Division and concerned an animal-human chimera. The invention provided significant medical benefit, since the chimeric animal could be used as a model for HIV infection. In its decision, the Division held that the patenting of animals cannot be considered *per se* as unethical within European society. In particular, while chimeras could be instinctively seen as distasteful or even immoral at first glance, there was no consensus under Article 53(a) EPC or Rule 28 EPC that would render them unpatentable. In reaching this conclusion, the Division seems not to have taken into consideration Recital 38 of the Directive. As for the level of proof required, it was noted that threats and hazards that are not conclusively documented are not relevant under Article 53(a) EPC. Moreover, the Division observed that, as long as the invention has a legitimate use, the EPO cannot act as a moral censor and hinder patentability on the basis of Article 53 (a) EPC.[224]

The Opposition Division decided on whether the patent would negatively impact future research or access to the technology. In dismissing those issues, it claimed that "*the EPO had not been vested with the task of taking into account the economic effects of the grant of patents in specific areas and of restricting the field of*

[221]The Division did not set any moral divide between patenting genes and other human substances (e.g. adrenalin) (Sterckx and Cockbain 2012, pp. 271–274).

[222]Thomas and Richards (2004, p. 98).

[223]Thomas and Richards (2004, p. 103).

[224]O'Sullivan (2012, p. 685), Rothschild and Newman (2002, p. 77) and Opposition Division of the EPO (2001, § 8).

*patentable subject-matter*".[225] The risks connected to xenotransplantation were also considered. The Opposition Division viewed them as unsubstantiated and therefore not enough to deny patentability. In reaching this conclusion, the Division noted that other regulatory bodies might be better equipped to assess those risks.[226]

Michigan State

In 1991, the Michigan State University presented a patent application for a euthanasia method in mammals.[227] The patent was granted by the EPO and soon after opposed, amongst others, on morality grounds. The reason was that the composition was aimed at killing living beings, including humans.

In its decision, the Opposition Division held that euthanasia of animals could not be seen as infringing *ordre public* and morality. Hence, once the patent was amended to exclude humans, it was considered in compliance with Article 53(a) EPC. The case was then brought before the Technical Board of Appeal. The Board interpreted Article 53(a) EPC restrictively.[228] On the basis of this, it then interpreted the concept of exploitation by noting that:

> If the intended exploitation of the invention does not infringe the principles of 'ordre public' and morality it is not sufficient that one out of several or more or even all other conceivable exploitations or uses of the invention's teaching would be or could be regarded as breach of the principles of 'ordre public' or morality, even if that exploitation constituted a serious breach of the principles of 'ordre public', for example a criminal offence. The mere possibility of abuse of the invention is not sufficient to deny patent protection pursuant to Article 53(a) EPC, if the invention can also be exploited in a way which does not and would not infringe 'ordre public' and morality.[229]

## 5.2 The Morality Clause and Synthetic Biology Inventions

### *5.2.1 Introduction*

Morality issues have been central in the debate on the patentability of biotechnological inventions. It is therefore unsurprising that they could have a pivotal role also

[225]Warren-Jones (2008a, pp. 205–206) and Opposition Division of the EPO (2001, § 49).

[226]Warren-Jones (2007, p. 838) and Opposition Division of the EPO (2001, § 45-47). *Mills* noted that the: "*Assessment of the hazards stemming from exploitation of a given technology was one of the more important functions of regulatory authorities, which are in a position to carry out a realist evaluation of risks on the basis of regulations in force, objective criteria and scientifically valid parameters*" (Mills 2010, p. 67).

[227]Sterckx and Cockbain (2012, pp. 264–271) and Technical Board of Appeal of the EPO (TBA) (2005).

[228]Technical Board of Appeal of the EPO (TBA) (2005, § 5).

[229]Technical Board of Appeal of the EPO (TBA) (2005, § 5.8).

in the development of synthetic biology.[230] Those issues span from the ethical, safety and regulatory ones examined in Chap. 2 to purely legal ones. In commenting on them, authors argued that "*SynBio inventions may be subject to ethical or moral exclusions*"[231] and stated that:

> There is no doubt that, at least theoretically, these general exclusion clauses could serve to objection to some inventions derived from synthetic biology through their being especially alarming in the broadest ethical-social terms.[232]

This section will examine the impact of the morality clause on the patentability of SynBio inventions and will examine whether specific inventions or fields of research will be excluded from patent protection because of it. The analysis will begin with a general assessment and will move on to individual research areas. These findings will be integrated with the assessment of a number of selected SynBio patent applications presented before the EPO.[233]

## 5.2.2 *General Assessment*

### 5.2.2.1 Relevant Moral Concerns

Article 53(a) EPC limits the operation of the morality clause to inventions contrary to *ordre public* and morality. This means that ethical objections that do not fall within those two definitions will not be relevant from a legal morality perspective.[234] For this reason, before proceeding to the juridical evaluation, it is necessary to establish which issues raised by synthetic biology might be within the scope of *ordre public* and morality and would thus be relevant in a legal morality assessment.[235]

---

[230]Commentators noted that the impact of the morality clause in synthetic biology will vary on the basis of the forum in which such issues are discussed. Specifically, if synthetic biology were discussed only in the administrative arena, it would probably be handled just like any other biotechnology. This means that the morality clause will be interpreted restrictively. On the other hand, if this discipline were negotiated in the legislative arena, exemptions would probably be interpreted broadly (Schneider 2014, pp. 162–165). For an overview of some SynBio ethical issues, Deutsche Akademie der Technikwissenschaften et al. (2009, § 4.4).

[231]Agovic (2014, p. 108).

[232]Casabona (2014, p. 178) and Spanish Bioethics Committee and Portuguese National Ethics Council for the Life Sciences (2011, p. 23). Authors noted that synthetic biology might: "*Represent the last chance for recovering the relationship between the biotech industry and patent morality*" (Agovic 2014, p. 115). In particular, the approach adopted by patent offices might shape the way and the conditions in which SynBio research is performed in Europe (Agovic 2014, p. 115).

[233]The patent applications selected for this analysis are the same ones that were used in Chap. 4.

[234]This does not mean that those wider ethical concerns and objections are irrelevant or unjustified; rather, that the legislator has not considered them important from a legal morality perspective.

[235]Many of those concerns have already emerged in other biotechnological fields. This is unsurprising, as the debate on whether SynBio represents a new ground-breaking technology or a mere

Relevant concerns include environmental issues, the impact on biodiversity, the misappropriation of genetic resources, biosecurity and biosafety threats as well as health issues caused by SynBio inventions.[236] Lastly, critiques concerning human dignity, the commodification of life as well as the idea that genes belong to the common heritage of mankind are also relevant.[237]

By contrast, a number of objections could not be taken into consideration when establishing the morality of SynBio inventions. Those include labelling and regulatory issues as well as arguments against the general use of patents in this field.[238] The reason for this is that those concerns are far removed from the legal interpretation of the concepts of *ordre public* and morality, even though they are often presented under the guise of ethical and moral principles.

Amongst those legally irrelevant objections, the critique to the use of patents in this sector acquired a prominent position and started a complex policy debate. Despite the high level of interest attracted by this debate, this dispute relates to policy matters rather legal morality ones. This conclusion is confirmed by commentators, who held that the act of patenting itself is not relevant under the morality clause.[239]

improvement of already existing techniques is still ongoing. In the course of the SYBHEL project, doubts were expressed on whether synthetic biology will raise new or different ethical questions compared to other biotechnologies. Yet, even identical concerns could be raised more strongly in this field, given the greater distance between SynBio and nature. For example, *Rutz* stated that: "*All the known objections against genetic engineering and GMOs are thus also raised against synthetic biology and might even be articulated more strongly in view of its greater distance to 'nature'*" (Rutz 2007, p. 1073). Equally, authors noted that synthetic biology is likely to revive questions on the role played by ethics in patent procedures (Agovic 2014, p. 102).

[236]This non-exhaustive overview presents the issues that could be raised against synthetic biology. It includes arguments already raised in the past before the EPO and Courts as well as issues that might emerge for the first time in synthetic biology. It does not imply that any of those arguments will be successful.

[237]Schneider (2014, p. 168). In its opinion on synthetic biology, EGE stated that: "*As far as the patenting and common heritage issue is concerned, the Group acknowledges the complexity of the topic... the Group stresses that general ethical issues involved in patent applications have to be addressed properly in the patent allocation system*" (European Group on Ethics in Science and New Technologies to the European Commission 2009, p. 54). Furthermore, synthetic biology has been accused of impacting the relationship between humans and the natural world in a morally undesirable way more than any other biotechnology (Kaebnick 2009, p. 1106).

[238]*Beriain* noted that: "*As it is commonly known, some people oppose to the use of the patent system in the case of synthetic biology as they consider it as contrary to morality or public order*" (de Miguel Beriain 2010). This hints to concerns over the use of patents *per se* in this field rather than the issuing of a patent for a specific invention.

[239]Schatz (2000, pp. 218–219).

#### 5.2.2.2 Applicable Norms

After having established which concerns could be relevant for a legal morality assessment, it is important to determine which norms would be applicable.

As discussed in Chap. 4, the Directive might not be applicable to all SynBio research areas. In particular, xenobiology and protocells might not fit the definition of biological material set by the Directive and the Implementing Regulations of the EPC. If this were the case, morality would be assessed only pursuant to Article 53 (a) EPC, given that Article 6 of the Directive and Rule 28 EPC would not apply. The repercussions of such inapplicability described in Chap. 4 would extend to the moral enquiry as well. However, the inapplicability of these norms may not impact the results of the legal assessment. Protocells and xenobiology inventions do not fall within the scope of Article 6(2) of the Directive.[240] Indeed, since those SynBio research areas do not relate to either humans or animals, this proviso is not directly applicable to them. The same conclusion would be reached by examining the rationale of this norm, as those two research areas do not put at risk either human dignity or the wellbeing of animals. Hence, the inapplicability of this Article is unlikely to alter the legal assessment, at least as long as those techniques are confined to lower life forms.

Equally, current xenobiology and protocell research raises only limited morality and *ordre public* concerns under Article 53(a) EPC. In the absence of controversies, the impact of the lack of ECJ jurisdiction on them might be less perceived. However, it cannot be excluded that the impossibility to trigger the referral mechanism to the ECJ might become problematic in the future and that contrasting interpretations might emerge at national level.[241]

EPC

Article 53(a) EPC constitutes the pivotal norm against which the morality of SynBio inventions will be judged.

Given its general formulation, the norm adapts well to future developments and therefore also to disciplines like synthetic biology. Scholars confirmed this view by holding that Article 53(a) EPC is a "*provision that is future-proof in the sense of being technology neutral*".[242] Similar statements were made by the Advocate General in the *Netherlands* case and by *Alain Pompidou*, former President of the EPO.[243]

---

[240]References to Article 6(2) of the Directive should be read to extend also to the list found in Rule 28 EPC.

[241]On this, see Sect. 4.2.2.1.

[242]Sommer (2013, p. 198).

[243]The Advocate General held that: "*The Directive thus provides an essential safeguard against the issue of such a patent. That safeguard is moreover so framed as to accommodate future developments: the generality of the standard ensures that it can be applied to inventions in this fast evolving field the detail of which cannot at present be foreseen. It is no doubt for that reason also that the*

A number of objections moved against synthetic biology would fit the concepts of *ordre public* and morality, especially if those were to be interpreted extensively. From a practical perspective, the framing of an issue under the morality or *ordre public* heading should not affect the result of the legal assessment. For instance, the classification of an issue as either contrary to *ordre public* or to morality would not impact the finding of unpatentability and its consequences.

Directive

The Directive will be applicable to most SynBio inventions. Nevertheless, even in those cases, synthetic biology inventions are likely to fall outside the list set in Article 6(2) of the Directive, at least for now. Possible exceptions concerning synthetic life and de-extinction will be analysed in the specific sections dedicated to them.[244]

The limited relevance of Article 6(2) of the Directive is connected to the type of inventions currently developed by synthetic biology. For the moment, SynBio inventions relate to microorganisms. Therefore, provisions relating to humans and animals do not affect them. This conclusion would be confirmed by an analysis of the rationale of this non-exhaustive list. Indeed, if one were to consider the protection of human dignity and animal welfare as the basis for this norm, SynBio inventions would generally fall outside of its scope, given their focus on microorganisms. In light of this, it is unlikely that the list found in Article 6(2) of the Directive will play a role in the morality assessment of most SynBio inventions, at least in the short term.[245]

#### 5.2.2.3 Balancing Exercise and Rebuttable Presumption Approaches

Two trends have emerged in the application of Article 53(a) EPC, namely the balancing exercise and the rebuttable presumption approach.

---

*legislature chose not to lay down in Article 6(2) an exhaustive list of examples of inventions which are to be considered unpatentable by virtue of Article 6(1). A case-by-case evaluation of patent applications in the light of moral consensus is the surest guarantee that the right to human dignity will be respected, and that is the framework established by the Directive*" (Advocate General Jacobs 2001, § 201). *Alain Pompidour* stated that: "*The generality of Article 53(a) EPC ensures that it can be applied to inventions in the fast evolving field of biotechnology that cannot at present be foreseen. It also ensures that possible changes in fundamental legal and ethical principles can be taken into consideration and immediately incorporated into patent law*" (Sommer 2013, p. 198; Pompidou 2006).

[244]On this, see respectively Sects. 5.2.3.2 and 5.2.3.6.

[245]Given the two-phase structure of the morality test, even inventions that passed this first threshold could still be found unpatentable under Article 53(a) EPC, as stated in *Oncomouse* and *WARF*. On this, see respectively Sects. 5.1.3.1 and 5.1.3.3.

Following the latter, the EPO established in its Guidelines that Article 53(a) EPC could be successfully invoked only in rare and extreme cases. This high threshold would generally be hard to reach in the field of synthetic biology. Indeed, SynBio inventions are not generally and univocally seen as abhorrent, even by those who consider this technology undesirable.

Similar conclusions have been supported by scholars. If Article 53(a) EPC applies to rare and clearly abhorrent cases, it would only prevent the patentability of the SynBio equivalent of letter bombs and landmines. This would include viruses and bio-weapons threatening health, safety and security.[246] However, criminals and terrorists are very unlikely to file for patent applications. From this perspective, problems are more likely to be raised by dual-use research.[247]

The situation would be more complex under the balancing test. If Boards were to adopt this approach, any immoral feature of an invention would play a role in the patent morality assessment. This would include characteristics that do not reach the level of abhorrence required for the rebuttable presumption test. Because of this, even less marked immoral aspects could tilt the scale towards the rejection of patent protection. This might have repercussions on SynBio patents and lead to an increased number of immorality findings. Such number might even be higher, if one considers that the balancing approach has often been advocated for inventions not concerning human life.[248]

#### 5.2.2.4 Environment

Environmental risks have played an important role in the application of Article 53 (a) EPC. This is likely to be the case also with synthetic biology, as this new discipline presents a number of environmental hazards, which could be even more significant than those raised by prior biotechnologies.[249]

For example, in the *Oncomouse* case, the EPO dealt with questions regarding the destructive consequences of the escape of genetically modified animals and their interference with evolutionary patterns. In its decision, it dismissed environmental

[246] Schneider (2014, p. 163). *Rutz* argued that SynBio biological weapons would squarely fall within the Article 53(a) EPC exception (Rutz 2009, p. 16).

[247] On this, see Sect. 2.4.1.2.

[248] The conclusions reached on the basis of the above could vary between countries, as the notions of *ordre public* and morality could be interpreted differently at European and national level. Equally, the use of a maximalist, minimalist or distributive approach would affect the patentability of SynBio inventions at EPO level.

[249] The patent system would not be the right forum to raise questions on the approach to environmental risks (e.g. whether a precautionary approach should be adopted) and on the release into the environment of SynBio products. These issues are within the purview of regulatory bodies. On this, *Rutz* noted that: "*Patents are not responsible for everything and they also cannot solve everything. This applies to ethical and social, as well as, economic, questions. One should not forget that there is 'life after grant', and that other regulatory means are available*" (Rutz 2009, p. 17).

concerns as having a neutral impact on the case. Specifically, it held that the risk of release was hypothetical (much like in a zoo) and that, even in the event of escape, the modified animals would not survive in the wild given their predisposition to diseases.

Those conclusions would not hold up in the case of synthetic biology. First, many SynBio inventions will be released deliberately into the environment. Hence, their risk of release is not hypothetical. Second, the survival rate of those organisms in the wild may be higher than that of an oncomouse. Indeed, a mouse with a predisposition for tumours is obviously at disadvantage in the wild, as far as its survival chances are concerned. By contrast, many SynBio inventions are not unfit for survival. Creations requiring a deliberate release are generally capable of surviving in nature, thus increasing their possible effects on the environment. Lastly, microorganisms that are currently not robust enough to survive in nature (e.g. minimal genomes) might acquire increased resistance, as those techniques evolve over time and via evolutionary pressure.

In light of this, environmental risks could gain increased significance in this sector, especially after the decision of the Board of Appeal in *Plant Genetic Systems*, where the Examining Division was urged to consider environmental issues as an incontrovertible part of the moral assessment. Nevertheless, it is important to remember that those risks must necessarily be defined and proven, as mere allegations would not suffice. In particular, the alleged hazards would need to be sufficiently substantiated and materialised by the time the decision is made. This requirement may strongly impact the outcome of the assessment. Indeed, risks are generally impossible to foresee on the basis of the patent application alone; yet, the Examining Division is required to make a decision on patentability. Its impossibility to make a thorough morality evaluation at the granting stage would most likely result in the patentability of SynBio inventions, except in clearly abhorrent cases. By contrast, in proceedings before the Opposition Division and the Board, opponents would be able to present more arguments and evidence to substantiate their claims, thus increasing their chances of success.[250] So far, no conclusive proof of environmental risks has been presented for SynBio. However, this situation may change rapidly, as this discipline develops further and additional studies are undertaken.

#### 5.2.2.5 Safety, Dual-Use and Risks

Synthetic biology has sparked numerous safety questions. Nonetheless, their impact on patent morality is not univocal.

SynBio bacteria exclusively engineered to wipe out the world's population would obviously not be patentable on morality grounds. This conclusion would equally

[250]This conclusion could be extended to proceedings before Courts. Also, just as for other biotechnologies, it is to be expected that political, environmental and NGO groups will launch the majority of those oppositions.

apply to bioweapons. Those inventions represent the SynBio equivalent of a letter bomb and would therefore be excluded from patentability. Given their abhorrence, it is unlikely that the EPO would grant a patent on them pursuant to its Guidelines. Nevertheless, this situation might be relevant only in theory, as it is highly unlikely that the developers of such weapons would apply for patents.

In less extreme cases, when a moral use for the invention is shown in the claims, the patent application will not be rejected on morality grounds. However, the immoral claims will need to be amended or eliminated in order to comply with *ordre public* and morality. This was confirmed by the EPO, which observed that, as long as an invention has a legitimate use, the Office could not act as a moral censor and hinder patentability on the basis of Article 53(a) EPC.[251]

The possibility to envision moral uses alongside immoral ones often emerges in synthetic biology, given the high number of dual-use research in this field.[252] In those situations, the following parameters would apply. First, immoral patent claims would need to be expunged from the patent application. By contrast, merely theoretical immoral uses that are not claimed in the patent would not have a bearing. In fact, most applications in the chemical and pharmaceutical field could theoretically be misused; yet, this does not prevent their patentability on the mere chance of an immoral use not claimed in the patent. Even in situations where acceptable uses coexist alongside problematic ones (e.g. SynBio microorganisms for bio-remediation that could also be used for bio-terrorism purposes), "*the refusal of an application under Art. 53(a) on the basis of this possible offensive use would be unjustified, as long as there are acceptable purposes for which the invention can be used*".[253] Therefore, risks of misuse are not *per se* sufficient to establish the contrariety of a SynBio invention to *ordre public* and morality.

This is confirmed by the *Oncomouse* decision, where it was held that technologies could not be excluded from patentability on morality grounds because they could be dangerous.[254] In light of this, SynBio inventions could not be dismissed *per se* on morality grounds merely because they might create hazards.

The type of applicant applying for the patent will play no role in the morality assessment. Patent morality will be examined in the same way, independently of whether SynBio inventions are presented by companies or by biohackers.[255] The circumstance that biohackers might have developed problematic products is not a concern of the patent system, but should rather be examined by the regulatory one. In

[251] O'Sullivan (2012, p. 685), Rothschild and Newman (2002, p. 77) and Opposition Division of the EPO (2001, § 8).

[252] Schneider (2014, pp. 163–164). On this, see Sect. 2.4.1.2. *Schneider* noted that: "*It will be difficult to discern 'good use' from 'abuse', as research on harmful viruses, biochemical and other agents must not be intrinsically malicious. Therefore, it is doubtful whether and to what extent barring such inventions from patentability would actually have an impact on such research and investment*" (Schneider 2014, p. 169).

[253] Fernandez y Branas (2014, pp. 193–194).

[254] On this, see Sect. 5.1.3.1.

[255] On this, see Sect. 2.3.2.5.

fact, the regulatory framework is likely to play a pivotal role in questions of safety, security and release of SynBio inventions, as those issues generally fall within the purview of regulatory bodies and governments.

#### 5.2.2.6 Commercialisation

So far, synthetic biology inventions have not been banned as such from entering the market.

Yet, even if this were the case, decisions excluding the commercialisation of SynBio inventions in Europe would not *per se* impact the patent morality assessment. At the most, they could be taken into consideration as evidentiary value.[256] Equally, the possibility to exploit SynBio inventions is not *per se* a sign that they do not infringe Article 53(a) EPC. This is especially the case, since the possibility to exploit SynBio inventions on the market is not the result of prolonged debates and considerations on a topic of common knowledge, but it is rather based on regulations set for other biotechnologies.[257] Indeed, public debate and awareness of synthetic biology have been quite limited so far. Hence, no specific relevance should be attached to the possibility to exploit SynBio inventions in Europe. However, this might become relevant, in case the idea were to prevail that non-patentability must be preceded by the prevention of the commercial exploitation of an invention.

On the one hand, it is always important to remember that the granting of a patent for an invention does not mean that a product can be rendered available to the public.[258] Therefore, in the case of synthetic biology, the granting of a patent does not overcome the regulatory hurdles needed to bring inventions to the market. This includes, *inter alia*, provisions on their deliberate release in the environment, their labelling as well as safety. On the other hand, the possibility to put products on the market without having obtained patent protection first seems unlikely in this field,

[256]As noted by *Beyleveld* and *Brownsword*, the probatory value of a prohibition or permission would vary if the conduct had been subjected to a serious and prolonged debate and the decision on commercialisation was the result of it. This does not seem to be the case for SynBio inventions, as their commercialisation was not at the centre of a wide debate (Beyleveld and Brownsword 1993, p. 81).

[257]On this, see Sect. 2.4.2.

[258]The second period of Article 53(a) EPC confirms that exploitation shall not be deemed contrary to the morality clause merely because laws or regulations prohibit it. In the *Netherlands* case, Advocate General *Jacobs* offered a practical example concerning GMOs. He noted that: "*There is a general moratorium on the use of these in the European Union at the moment, but it will not necessarily be indefinite. Similarly at national level an inventor may anticipate a change of government. Alternatively, an inventor may wish to manufacture an invention in a Member State where the exploitation (but not the manufacture) of the invention is prohibited, with a view to exporting it to States in which its exploitation is not prohibited*" (Advocate General Jacobs 2001, § 106). This reasoning could be extended to SynBio inventions and is an example of the disassociation between the norms on patentability and those on commercial exploitation.

given the high costs connected to research and product development as well as the pattern usually followed in other biotechnologies.

Lastly, it is important to note that the support provided by the EU and governments towards synthetic biology research is not *per se* a sign of the morality of SynBio inventions, especially in light of the *WARF* case.[259]

#### 5.2.2.7 Impact of the Benefits Promised by Synthetic Biology

Synthetic biology has been connected to numerous advantages across the medical, food, agricultural, energy and environmental sectors. If a balancing test were adopted, the moral aspects and benefits provided by an invention would need to be taken into consideration. Given the advantages connected to synthetic biology, proving that these inventions have moral features should not be difficult. The fact that some of those benefits have already been mentioned in the Directive is also a plus. These include biotechnologies for the environment (e.g. cultivations that are less polluting and more economical in their use of the ground) as well as techniques for the improvement of life in developing countries in the fields of health, epidemics and hunger and the conservation and sustainable use of biological resources and diversity. Interestingly, the statements of the Opposition Division in the *Lubrizol* case could be helpful here.[260] In that occasion, the increased nutritive value of the invention and its ability to alleviate global food shortage were seen as a sign that the invention did not violate the morality clause. Since SynBio inventions often promise to bring similar advantages, this could grant more support to the opinion that they do not infringe the morality clause set in Article 53(a) EPC. Equally, following *Plant Genetic Systems*, SynBio inventions would be judged as moral, if they do not concern a misuse or destructive use of a technology.[261] Given that SynBio inventions are generally targeted to moral goals, they would normally fall outside of this category.

#### 5.2.2.8 Public Perception

Opinion polls and public perception cannot be taken as the sole indicator of the abhorrence or acceptance of an invention from a morality perspective. Although several decisions have dismissed their relevance, scholars maintained that:

> It is only where there are no means of assessing an intuitive understanding of society's moral sensibilities towards a particular technology, such as where the general technology has not been considered by the public and there are no close comparators, with similar groupings of technology, that the patent office or Courts may need recourse to more direct evidence in order to inform their intuitions. Opinion polls which identify 'public abhorrence' to at least

[259]On this, see Sect. 5.1.3.3.

[260]On this, see Sect. 5.1.3.6.

[261]On this, see Sect. 5.1.3.2.

general categories of technology could become applicable in forming part of this more direct evidence.[262]

Scholars argued that if those surveys were to show an extreme public reaction against such technologies, the patent should be denied. In all other cases, the patent should be granted and further analysis should be left to the relevant regulatory bodies.[263]

Whether synthetic biology could fall within the above criteria is debatable. On the one hand, this technology has not gathered much public attention, especially in comparison to other biotechnologies.[264] Indeed, surveys showed that the public was generally unaware of this new technology. For instance, in 2010, only 17% of the people involved in a survey organised by the EU Commission knew of this discipline. On the other, the perception of synthetic biology as having close comparators depends on personal views. SynBio could be seen as either an extension of earlier biotechnologies or as a distinct technique due to its increased level of modification. From this last perspective, SynBio could pose new challenges that could not simply be solved by equating it with earlier technologies.

So far, synthetic biology encountered limited obstruction, with the exception of public backlash over the use of SynBio substitutes for palm oil in detergents.[265] Nevertheless, the absence of widespread opposition in the public should not be interpreted as a sign of its acceptance. During the *Oncomouse* saga, authors argued that the lack of concern in the public could not be used as an argument in favour of patentability, since the technology was not well-known at the time the patent application was presented.[266] The low level of awareness of synthetic biology in Europe could sustain such an argument. Indeed, it is currently hard to imagine a significant public backlash against the patentability of those inventions on moral grounds. However, the situation could evolve quickly. If this were the case, public perception might assume a more prominent role. Still, public opinion alone is not decisive in the morality assessment. As the Board noted in *Plant Genetic Systems*, polls showing opposition against a patent are not sufficient to establish its contrariety to morality and *ordre public*.[267] This includes cases in which the majority of the population of all contracting States would be against it.

In addition to public awareness of SynBio, surveys examined also the acceptance of this technology.[268] EU-wide surveys showed that 17% of the interviewees did not approve of this technology under any conditions and that 21% did so only under

[262]Warren-Jones (2007, pp. 844–845).

[263]Warren-Jones (2007, p. 845).

[264]On this, see Sect. 2.3.2.4.

[265]On this, see Sect. 2.2.4.6. Nevertheless, there are organisations that have expressed critiques against SynBio patent application from the very beginning. This is for example the case with the ETC Group. On this, see Sect. 2.2.2.1.

[266]Thomas and Richards (2004, p. 101). On this, see Sect. 5.1.3.1.

[267]On this, see Sect. 5.1.3.2.

[268]On this, see Sect. 2.3.2.4.

special circumstances. The biggest group (36%) approved only if strict norms were applicable and 23% did not reply, as they did not have a clear opinion on the matter. These results hint to a negative approach towards synthetic biology. National studies paint a partially different picture.[269] Despite variations across sectors, UK studies showed that citizens considered this field as acceptable with opinions ranging from about 55% to 80% (medical applications were the most accepted, while food/crop ones the least). The same study showed that interviewees saw synthetic biology as generally useful for society (64–79% across all applications). However, risks for the environment were considered problematic by the majority of citizens in the food and bioremediation sectors (56%).[270] Despite contrasting numbers, both surveys show that public objection is not strong enough to reach the level of abhorrence and that this discipline is likely to be considered morally acceptable in Europe.

Lastly, authors sustained the tolerability of synthetic biology for two additional reasons. First, it was noted that even those who object to SynBio inventions might often be willing to weigh them against possible benefits. Second, while SynBio could be perceived as an objectionable technology, it might still not be seen as completely undesirable.[271]

#### 5.2.2.9 Perception of Equivalence

Boards have considered an invention unproblematic from a morality perspective if a comparable technique had also been previously considered morally acceptable. This approach, known as the "equivalent approach", has been used in the *Relaxin* and *Leland Stanford* cases to reject moral objections moved against new technologies.[272] For example, in *Relaxin*, it was argued that patenting life was intrinsically immoral. This critique was overcome by comparing the invention with other patented products of human origin (e.g. proteins). Since the products covered by the Relaxin patent were conceptually no different from already patented ones, this moral objection was rejected. Equally, in the *Leland Stanford* case, moral objections to chimeras were overruled by drawing parallels with the accepted practice of medical donations.[273]

Parallels between technologies were drawn by the Board in the *Plant Genetic Systems* case.[274] The similar motivations that guided plant biotechnology and traditional selective breeding techniques meant that the former could not be considered more contrary to morality than the latter, since the difference between the two amounted to a mere variation in the level of control exercised over genetic

---

[269]The EU-wide survey showed differences in the levels of acceptance between countries. On this, see Sect. 2.3.2.4.

[270]Bhattachary et al. (2010, pp. 53–64).

[271]Kaebnick (2009, p. 1107).

[272]On this, see Sect. 5.1.3.6.

[273]Warren-Jones (2008a, p. 199).

[274]On this, see Sect. 5.1.3.2.

modifications. If this reasoning were applied to synthetic biology, it would be unlikely for those inventions to be rejected on morality grounds. Synthetic biology could be seen to share the same motivations of earlier biotechnologies and to represent an improvement in the ability to control and modify biological systems. Furthermore, the drawing of parallels between earlier biotechnologies and synthetic biology would not be difficult, considering the ongoing debate over whether SynBio is a mere advancement over prior technologies or instead a new paradigm shifting one. Indeed, if SynBio is seen as a continuation of gene transfer techniques, the conclusions reached for the latter could be applied to the former. Since there is no consensus as to whether GMOs should be banned, a restrictive approach to synthetic biology could not be implied from this.[275]

Nevertheless, this parallel may not be fully appropriate in this case. First, the control exercised by synthetic biology over biological systems is not comparable with selective breeding techniques and it is a stark improvement over the decades-old techniques mentioned in *Plant Genetic Systems*. Secondly, the fact that activities have been performed in the past does not mean that they will be considered moral in the future and vice versa; techniques evolve rapidly and so do sensibilities over what is abhorrent or acceptable from a morality perspective.

#### 5.2.2.10 Perspectives and Conclusions

The morality of SynBio inventions from a patent perspective has seldom been analysed. The few scholars that addressed this issue agreed on the lack of imminent legal moral concerns in this field. What they disagreed on is whether those issues are likely to arise in the future.

On the one hand, *Rutz* maintained that it is unlikely that synthetic biology inventions will be considered contrary to morality and *ordre public* pursuant to the EPC.[276] The OECD agreed with this position and noted that:

> While the ethical justification of DNA synthesis may be debatable, synthetic biology does not fall under the scope of general exclusion from patentability, such as 'inventions contrary to public order or morality'.[277]

On the other hand, scholars argued that SynBio inventions could become as disputed as HESCs (Human Embryonic Stem Cells).[278] However, for the moment, those fears seem exaggerated. Limits to the patentability of HESCs can be found in

---

[275]Kaebnick (2009, p. 1108).

[276]Rutz (2009, p. 16).

[277]OECD (2014, p. 96).

[278]Agovic (2014, p. 108). The same author argued that: "*If SynBio is to avoid the faith of HESCs patents in Europe… there needs to be a pro-active discussion between public, legislature and scientists on the issue of general ethical concerns raised by patentability of SynBio inventions and a strong will to have them addressed properly and in time, in the patent allocation system*" (Agovic 2014, p. 118).

the legislation and in the case law. By contrast, current SynBio inventions do not operate in areas specifically excluded by the legislator. Furthermore, the same author noted that many morality concerns in this discipline are still far-fetched. This is because synthetic biology is still in its infancy. However, as this technology progresses, the situation might change quickly.[279] Equally, other authors noted that, for the moment, synthetic biology might not be particularly problematic from a morality perspective, since it focuses on unicellular organisms. However, should this discipline transition to more complex organisms, concerns might arise.[280]

An accurate description of the moral issues raised by synthetic biology patents was offered by *Torremans*. In his opinion:

> One could be inclined to say that these morality issues do no arise in relation to synthetic biology inventions. This is clearly the result of the fact that at present no human genetic material and no embryonic material is involved in the early development phases of the science of synthetic biology. At present these early development phases are confined to rather primitive bacterial forms of life and the relevant genomes. The exclusion of human material makes the matter far less controversial, but it is no doubt also true that society has not yet woken up to this new form of biology that sets out to create artificial forms of life. Because if one puts it this way, morality arguments do not seem to be that far away. At present thought, the matter is more one of biosafety. We do not want these artificial forms of bacterial life to get out in nature and reproduce themselves in an uncontrolled way... these are in a way ethical concerns, but no doubt ethics and morality will become more important the closer this new discipline comes to higher forms of life.[281]

### 5.2.3 *Specific Assessment*

Having examined the general issues relating to the morality of synthetic biology inventions, it is now time to consider the morality of the main SynBio research fields, both current and futuristic.

In the analysis pursuant to Article 53(a) EPC, inventions will be assessed under both the rebuttable presumption and the balancing approach. While the former has been consistently employed by the EPO in the examination phase, Boards have oscillated between the two. Therefore, a double analysis is necessary, since it is unclear which approach will be adopted.

#### 5.2.3.1 Minimal Genome

In Europe, the morality of minimal genomes will be assessed under both the Directive and the EPC.

---

[279] Agovic (2014, p. 115).

[280] Kaebnick (2009, p. 1107).

[281] Torremans (2011, p. 307).

The four categories set by Article 6(2) of the Directive would not apply to this line of research. Indeed, both the examples presented therein and the rationale of the norm relate to human dignity and the suffering of animals. Since minimal genome creations are limited to microorganisms, they would fall outside this list and the rationale of the norm.

Having passed the Article 6(2) hurdle, those inventions will be considered pursuant to Article 53(a) EPC.

A first critique that could be moved against minimal genomes relates to the simplification and commodification of life.[282] This type of objection has usually been raised for human life and higher life forms. In the field of microorganisms, such concerns seem particularly weak and would find little support in the case law and in the public. In fact, while microorganisms are undoubtedly life forms, they do not raise comparable morality questions due to their perceived lower moral status. Furthermore, those critiques embody personal moral concerns, which would not be univocally perceived as problematic.

Likewise, minimal genomes are unlikely to raise environmental issues. Under current conditions, minimal genomes are not robust enough to survive in the wild and therefore their use is restricted to the laboratory. According to prior EPO case law, this would reduce the impact of possible environmental concerns. Moreover, the case law of the EPO shows that environmental issues have had *per se* very limited impact on the morality assessment. In light of this, it is likely that no morality issues will be raised against minimal genomes on environmental grounds. This situation may of course change once minimal genomes become strong enough to survive and reproduce in the wild.

Morality objections against minimal genomes would have few chances of success under both Article 53(a) EPC tests. Indeed, if the rebuttable presumption test were to apply, those inventions could not be considered immoral, since they do not express the level of abhorrence necessary to rebut such favourable presumption. Similar conclusions would be reached if a balancing exercise were used, since the immoral features would most likely be balanced out by the moral ones. This conclusion is confirmed by an analysis of the patent applications presented before the EPO, where no morality issues were raised.[283]

In light of the above, and as long as the situation doesn't drastically change (e.g. transition to higher life forms), minimal genome inventions are highly unlikely to be denied patentability on the basis of the morality clause.

---

[282] In its 25th opinion on the ethics of synthetic biology, EGE shortly addressed the question of morality. It observed that the idea of Venter to achieve a minimal genome was taken into consideration by ethicists in 1999. In their opinion: "*The prospect of constructing minimal and new genomes did not violate fundamental moral precepts or boundaries, but did raise questions about the possible consequences of synthetizing new free-living organisms in relation to the concept of life and our relation to it*" (European Group on Ethics in Science and New Technologies to the European Commission 2009, p. 40).

[283] Blattner et al. (2010), Glass et al. (2008b) and Blattner et al. (2003).

#### 5.2.3.2 Synthetic Life

Synthetic life is one of the SynBio fields that poses the most questions from a legal morality perspective. The accusations moved against these inventions span from "playing God" arguments to the objectification of life and its wrongful patenting. Such critiques are bound to become stronger and more widespread, as synthetic biology moves into the creation of eukaryotes.[284] This would ideally culminate in a synthetic human genome, which is currently under discussion.

Some of the above critiques have already been addressed in the case law. For instance, the objection against patenting life is not new and has already been dismissed by the EPO in the *Plant Genetic Systems* case.[285] There, the Board rejected the idea that Article 53(a) EPC would be violated solely because a patent covered living matter. Commentators took a similar stance, as it was argued that "*the claim 'no patents on life' – to the extent that it implies that patenting living matter is immoral per se – cannot possibly be satisfied under Art. 53(a) EPC*".[286]

The "playing God" metaphor has been most often associated with the work of Craig Venter in the field of synthetic life.[287] However, also this objection is unlikely to lead to a finding of unpatentability on morality grounds. This has several reasons. First, the "playing God" argument was never successfully presented before the EPO. Second, this objection has been challenged by commentators and by religious groups for being inappropriate. For instance, the Commission of the Bishops' Conferences of the European Community (COMECE) rejected it and so did scholars, who viewed the work of Venter as a tour de force rather than an attempt to "play God". Lastly, this critique could be seen as a publicity tag rather than a legal argument. Indeed, this approach has mostly been adopted in the press to discuss the implications of synthetic biology. Hence, while this characterisation might impact the views of the general public, this objection still displays a personal ethics character rather than a legal one.

Furthermore, all the critiques presented above share a common weakness. Currently, synthetic life inventions are restricted to microorganisms and therefore they raise limited morality concerns, given that the morality of patenting microorganisms *per se* has been accepted for decades. By contrast, should synthetic biology move away from microorganisms and venture into human and higher life forms, numerous morality questions would arise.

An examination of the application of Article 6(2) of the Directive confirms this conclusion. So far, Article 6(2) of the Directive does not cover synthetic life inventions. Current work is focused on microorganisms (e.g. bacteria, yeast) and is

---

[284] This started with the project to synthesise the genome of yeast. On this, see Sect. 2.2.2.2.

[285] On this, see Sect. 5.1.3.2.

[286] Schatz (2000, pp. 218–219).

[287] Authors noted that this metaphor has been: "*Most often associated with overzealous Craig Venter but less so with the synthetic biology project in general*" (Schneider 2014, p. 167). For an overview and critique of the "playing God" and reductionist arguments, Douglas et al. (2013).

thus outside the scope of the norm, which covers instead human dignity and animal suffering. Hence, Article 6(2) of the Directive does not impact the moral assessment of synthetic life inventions concerning microorganisms.

Equally, limited morality questions would be raised under Article 53(a) EPC. Current SynBio inventions do not reach the level of abhorrence necessary to overcome the rebuttable presumption of morality. The same is true for the balancing approach, where the possible immoral aspects of synthetic life microorganisms would not be enough to tilt the balance towards their unpatentability.

Nevertheless, this approach leaves open one question. The end product of synthetic life inventions (e.g. yeast) might be similar if not identical to something found in nature. In this case, the difference between the two would be found in their production method: one via nature and one via synthesis. Recent decisions of the Board of the EPO and the ECJ have highlighted the importance of the origin and history of an invention. If this reasoning were to apply here, would it impact the morality assessment? Essentially, the question revolves around the possibility that those inventions could be considered against *ordre public* and morality because they were obtained artificially. So far, no clear stance has been taken by the EPO on this point. *Rutz* noted that:

> Whether the creation of 'artificial life' in itself, or its patenting, could be considered immoral has, to my knowledge, not been explicitly addressed by any legal authority to date. However, case law under the EPC has confirmed the patentability of microorganisms and higher life forms – that is, plants and animals – including genetically modified forms, which could also be considered 'artificial'.[288]

If this were the case, the artificial origin of those microorganisms should not play a role from a legal morality perspective.

In light of the above, it is unlikely that current synthetic life inventions would be found unpatentable on morality grounds. This conclusion is confirmed by the absence of morality critiques in the patent applications presented before the EPO in the field of synthetic life.[289] Nonetheless, the analysis presented above might drastically change in the future, given that the application of synthetic life techniques to humans and other higher life forms could be problematic under Article 53(a) EPC and Article 6(2) of the Directive.

From this perspective, the realisation of the HGP-write project is particularly challenging for a number of reasons. For starters, a line needs to be drawn between the synthesis of the human genome and the construction of a whole human organism.[290] The promoters of the HGP-write project held that their work would be limited to creating cells (including human ones) and not people.[291] The latter type

---

[288]Rutz (2009, p. 16). Similar conclusions have been reached by *Schneider*, who observed that morality issues connected to synthetic life have not been addressed so far (Schneider 2014, p. 155).

[289]Glass et al. (2008a, 2010, 2014), Venter et al. (2013), Benders et al. (2012) and Venter and Smith (2008).

[290]Bailey (2016).

[291]Pollack (2016).

of creations, which are still at the sci-fi stage at the moment, would surely raise serious morality concerns from both a philosophical and legal point of view.[292] In fact, they would raise *ordre public* and morality issues under Article 53(a) EPC as well as Article 6(2) of the Directive.[293]

The synthesis of the human genome in a cell is instead a grey-area. Human cells have been patented before and have been considered by the EPO as belonging to the definition of microorganism. Their inclusion within this notion provides an advantage for their patentability, as microorganisms have generally raised limited patent morality concerns. However, the synthesis of the human genome, albeit in cell form, might raise questions.

First, would this violate human dignity? For instance, would an exact copy of the genome of a human being be considered an affront to his dignity? This issue has already been raised, as scholars questioned whether: "*would it be O.K., for example, to sequence and then synthetize Einstein's genome?*".[294] This question has no easy answer. On the one hand, it could be argued that this affects the dignity of a person as well as his human rights. Genomes are specific to each person and their reproduction would be a violation of this uniqueness and be seen as intrinsically wrong pursuant to European culture. On the other hand, it has been argued that:

> Genomes are not some kind of special 'sacred' entities. If it is morally all right to synthetize the genome of a bacterium or even of a mammoth, then it is OK to do the same with the human genome. If it is morally OK to know what each and every base pair of a human genome is, there is no ethical reason not to reverse engineer it. It's the same entity.[295]

This argument might be objected to by noting that the moral value attached to the genome of a microorganism is not comparable to that of the human genome. Sensibilities on this point are different and so are norms. For example, Article 6 (2) of the Directive covers human dignity and does not consider bacteria. Therefore, both legal and value judgements seem to point to a rather incomplete equivalence between human and non-human genomes.[296]

Further questions are raised by the possibility to alter the synthetic human genome. If in the field of microorganisms this could be seen as an acceptable praxis

---

[292]In discussing enablement issues for a xenobiology patent application, an EPO Examiner noted that: "*There is no evidence that synthetic human beings or other higher organisms can be created by design in the way applicant understands the terminology. Further, there are moral issues insofar as the claims extend to this matter*" (Krishnakumar et al. 2010, Examination Document of 5 February 2013, p. 3). On this, see Sect. 5.2.3.3. This shows that the EPO is well aware that synthetic human beings could be problematic from a morality perspective.

[293]Those inventions would infringe Article 6(2) of the Directive, since they would affect human dignity and animal well-being.

[294]Pollack (2016).

[295]Bailey (2016).

[296]On a related note, it was also asked: "*Where does the ethical boundary lie? Is it the synthesis of 10 genes, 1,000 genes, or a whole genome? And why does it change at that boundary?*" (Boeke et al. 2016, p. 10).

(e.g. the Yeast 2.0 project), matters might be more complex for humans.[297] DNA synthesis is an error-prone technique and changes to the genome could cause unpredictable consequences. Also, a genome that has been changed could be accused of immoral eugenics attempts or of causing suffering. Furthermore, the inclusion of artificial chromosomes into human cells could also be denied patentability pursuant to Article 53(a) EPC.[298]

A report compiled by the promoters of the HGP-write project highlighted further contentious points. The first concerned whether there is a moral difference between natural and synthetic genomes.[299] This issue hints at the morality of artificial life and at the relevance of the history of an invention. As mentioned by *Rutz*, no cases have explicitly discussed the morality of artificial life and its patentability, but patents for synthetic life inventions involving bacteria have nonetheless been granted.[300] Recent trends have shown the importance of the history of the invention (e.g. *WARF* and *Brüstle*).[301] Hence, it cannot be excluded that this would be extended to cover the synthetic or natural origin of an invention. Furthermore, authors noted that *Brüstle* would most likely have an impact on this field, especially if scientists were to use SynBio tools to create pseudo-embryonic human life forms.[302] Nonetheless, for the moment, the absence of specific provisions and cases to the contrary would hint to the morality of artificial life inventions. This conclusion would be confirmed by the absence of morality critiques in the patent proceeding for JCVI syn1.0 before the EPO.[303]

In light of the above, it is unclear whether the synthesis of the human genome in cells would be considered contrary to *ordre public* and morality and thus be rejected pursuant to Article 53(a) EPC. Similarly, it is uncertain if Article 6(2) of the Directive would have any impact on this matter, since the list aims to protect human dignity, but makes no reference to human cells and their morality.

Given the complexity of the topic as well as its sensibility, it seems advisable for patent offices, Courts and scholars to start considering the possible moral implications of the patentability of such SynBio inventions. If this field continues to evolve

---

[297]On this, see Sect. 2.2.2.2.

[298]European Group on Ethics in Science and New Technologies to the European Commission (2009, p. 46).

[299]Boeke et al. (2016, p. 10).

[300]Rutz (2009, p. 16).

[301]On this, see respectively Sects. 5.1.3.3 and 5.1.3.5.

[302]Casabona (2014, p. 181).

[303]The reaction to a human genome synthesised via a computer should also be considered. At the time of the creation of JCVI syn1.0, *Venter* announced that this was the: "*First... species... whose parent is a computer*" (Wade 2010). If the same were true for a human genome, what would be the legal morality impact of such a situation? How would society and the legal world react to the use of a synthetic genome to create humans without biological parents (Pollack 2016)? In the case of JCVI syn1.0, this issue did not raise legal critiques. Yet, the situation may be diametrically different if a human genome is involved. On this, see Sect. 2.2.2.2.

at a fast rate and if synthetic human genomes are achieved in the near future, the patent system would need to start preparing for the challenges ahead.

#### 5.2.3.3 Xenobiology

The application of the Directive in the field of xenobiology is doubtful. Yet, even if the provisions of Article 6(2) were to apply, this would not lead to the exclusion of xenobiology inventions on morality grounds. The examples set in the norm relate to human and animal inventions, whereas xenobiology is currently limited to microorganisms. This would also exclude an extension of the rationale of the norm to xenobiology inventions.

Article 53(a) EPC would instead surely apply to those inventions. From this perspective, a number of issues could be raised. First, it could be argued that xenobiology could be damaging to the environment. The characteristics of those inventions are such that their impact on nature and biodiversity is hardly predictable and, exactly because of this, their effects could be destructive. The situation is rendered more problematic by the fact that those inventions might be deliberately released into the environment. This means that problematic interactions with nature are more likely to occur. Lastly, the high level of manipulation required by those techniques raises "playing God" concerns.

In spite of this, an analysis of the morality of those inventions shows that those arguments are unlikely to succeed under the current legal and scientific conditions. First, the alleged environmental risks would need to be proven by the time the decision on the patent is reached. If serious environmental problems could be demonstrated by then, xenobiology inventions might be considered problematic from a morality perspective. However, for now, such risks are unsubstantiated. Hence, they would not be significant in an Article 53(a) EPC assessment. This applies also to health hazards.

Furthermore, concerns over the destruction of biodiversity could be countered by arguing that xenobiology provides new diversity, albeit artificially. Lastly, "playing God" arguments are likely to be rejected, given their general weakness and the fact that they have never been accepted as such by the EPO.

The morality of xenobiology inventions could be argued by comparing this technology to prior moral ones. The EPO often drew parallels between current and prior technologies to prove the morality of the former. The idea is that, if one is an ideal continuation of the other, one cannot be more contrary to morality than the other. While this position could be criticised, it is obvious that if the EPO were to adopt it, it would lead to a finding of morality. However, this approach might not adapt well to xenobiology. Despite sharing goals with prior biotechnologies, scientists involved in xenobiology try to change biological paradigms rather than gain more control over them. From this perspective, an extension of the morality of earlier biotechnologies might be erroneous.

In light of the above, it can be argued that current xenobiology inventions would not be rejected under the rebuttable presumption approach, since they would fall short of the abhorrence standard. Similar conclusions could be reached if the balancing approach were to apply. In the absence of proof of serious threats, the advantages provided by xenobiology inventions would tilt the scale towards their morality.

This conclusion seems to be confirmed by the examined xenobiology patent applications presented before the EPO.[304] Even in the one case where morality issues were raised, no specific objections against this discipline could be detected. Indeed, in a document from February 2013, an EPO Examiner considered enablement issues under Article 83 EPC and stated that the:

> Applicant is claiming any organism comprising a synthetic genome. This covers eukaryotes including even human beings. There is no evidence that synthetic human beings or other higher organisms can be created by design in the way applicant understands the terminology. Further, there are moral issues insofar as the claims extend to this matter. Hence, the claims should be restricted to microorganisms to which the technology can reasonably be applied.[305]

Thus formulated, this objection seems to concern a matter of enablement and correspondence between the claims and the actual invention. Furthermore, the statement refers to the morality of the synthesis of human beings and higher organisms rather than to the use of xenobiology techniques. Hence, even this patent application and its examination seem to confirm the conclusions presented above.

#### 5.2.3.4 Protocells

The morality of protocells will most likely be assessed on the basis of Article 53 (a) EPC. As mentioned in the previous chapter, protocells do not currently fall within the definition of biological material. Hence, they will be regulated pursuant to the EPC alone.[306]

Under Article 53(a) EPC, three types of concerns might be raised for those inventions. On the one hand, it could be argued that protocells blur the definition of life. On the other, protocells could be accused of simplifying life. Lastly, the "playing God" argument might be put forward. This is because protocells conjure up the idea of creation, as one of the goals of this line of research is to identify how cells were first originated. The first two arguments do not carry a legal value, but rather a philosophical one. Hence, their relevance for a legal morality analysis is limited. More issues could instead be raised by the "playing God" critique. However, also

[304] Cho et al. (2007, 2012).

[305] Krishnakumar et al. (2010, Examination Document of 5 February 2013, p. 3).

[306] If the Directive were applied to them, those inventions would need to be assessed from a morality perspective under Article 6 of the Directive. Protocells would generally fall outside the list presented in Article 6(2). This line of research is so far restricted to cells and no work has been carried out in relation to humans and animals. Should this situation change in the future, for example with work related to embryonic cells, this assessment would need to be revised. Nevertheless, these considerations are confined to a distant future, since this field is still in its infancy.

these are likely to fail. This argument is generally not a strong one and has not been accepted *per se* by the EPO. Furthermore, it could be objected that researchers are not "playing God", but merely reconstructing cells using a different approach, namely the bottom-up one. Therefore, although this may be seen critically from a philosophical and personal morality perspective, it does not constitute a legally valuable argument.

As a result, protocells are likely to pass the morality test under both the rebuttable presumption and the balancing test. In the first case, protocells raise no issues that would lead to a rebuttal of the presumption and to a finding of unpatentability. Likewise, under the second test, the balance would be in favour of the morality of the invention, since no issues have been found that could be used to argue otherwise.

In conclusion, protocells are unlikely to raise morality issues on the basis of the current legal and scientific panorama.[307]

#### 5.2.3.5 Microorganisms

Microorganisms are not mentioned in the list set by Article 6(2) of the Directive. As noted before, this list is targeted to humans and animals and does not include references to lower life forms, such as microorganisms, which would therefore fall outside of its scope.

Equally, the morality clause contained in Article 53(a) EPC has never been applied to deny the patentability of microorganisms *per se* on morality grounds. This derives from two factors. First, microorganisms have a long tradition of patentability. Second, there is a different level of sensibility for morality issues for microorganisms compared to higher life forms.

In light of this, it can be concluded that there are no objections which, *per se*, would exclude SynBio microorganisms from patentability on morality grounds. The examined patent application presented before the EPO confirms this conclusion, as it did not raise any morality concerns.[308]

#### 5.2.3.6 De-extinction

De-extinction is problematic from a morality perspective. The possibility to bring back to life extinct creatures subverts evolution and raises questions on the purposes of this line of work. An assessment of its legal morality should be approached from two perspectives: the de-extinction of animals and that of human or human-like creatures. This separation is critical, as the second type of de-extinction (i.e. human or human-like creatures) raises different and more problematic morality questions.

---

[307]Given the absence of patents in this field, this conclusion cannot be confirmed by an analysis of patents and patent applications.

[308]Hansen (2013).

In the past years, experiments have successfully brought back to life extinct creatures.[309] Synthetic biology plans to expand this possibility and recreate a wider range of creatures, including the wholly mammoth. The recreation of the mammoth would involve its DNA and an elephant egg and birth mother. As the DNA is added to the egg, the latter would be developed into an embryo. The resulting organism would have the nuclear genotype of the DNA donor and the mitochondrial genome of the egg donor. This approach, which so far has not been implemented in practice, would lead to the creation of an animal/animal chimera. Indeed, despite the use of the mammoth DNA, the mitochondrial genome of the egg would still influence the development of the embryo.

This invention could then be caught by Article 6(2)(d) of the Directive and be rejected on immorality grounds.[310] These techniques would lead to an alteration of the genetic identity of the mammoth, thus fulfilling the first criterion set for the application of this provision. The second and the third requirements (i.e. the possibility of suffering and the lack of substantial medical benefits) would need instead to be assessed on a case-by-case basis. Nevertheless, it is foreseeable that they would both be fulfilled.

The introduction of mammoth DNA into a different species could lead to interactions and alterations that could cause physical pain to the animal. Also, the risk of abnormal births and genetic abnormalities should not be underestimated. If physical pain were present, this requirement would surely be met. There is however a second layer of complexity to be taken into consideration. So far, the focus has been on physical pain; yet, moral suffering cannot be disregarded. If this type of non-physical pain were to be taken into account, it could change the morality assessment. For example, recreated animals would come to life in an environment that is different from the one they originally inhabited. This transformation would be particularly felt for animals that have been extinct a long time (e.g. mammoths). Equally, the limited number of animals of that species available in the world would force them to a life of isolation. For instance, if a wholly mammoth were recreated, this animal would constitute an *unicum* and would not have any contacts with other mammoths. This might cause distress and inflict pain to the animal, albeit a non-physical one.[311]

Similarly, the requirement of substantial medical benefit is problematic. First, one could ask which kind of benefit could be derived from animals that have been extinct for a long time. The applicant would need to prove that a substantial medical benefit exists and must show a correspondence between the animal species that is suffering and the one offering the benefit. This requirement would thus hinder the patentability

[309]On this, see Sect. 2.2.3.1.

[310]Transgenic animals have been protected in the EU due to their usefulness. Still, a line was drawn between what is acceptable and what not. This distinction takes into consideration animal suffering and was necessary to avoid gratuitous experiments having non-proportional purposes (Sommer 2008, p. 157).

[311]Savulescu and Powell (2013).

of animals that have been recreated for non-medical purposes, thus including general scientific curiosity and recreational purposes (e.g. mammoths for zoos).[312]

In light of this, recreated animals could be considered immoral pursuant to Article 6(2)(d) of the Directive.

Creations that passed this threshold might nonetheless be caught by Article 53 (a) EPC, both under the balancing and the rebuttable presumption approach. Pursuant to the first, the applicant would need to show the benefits of its invention and prove that they would overcome its immoral features. Such elements would likely include the commodification of animals, which were recreated to fulfil human purposes and with disregard to their well-being.[313] The subversion of evolutionary patterns would also constitute a strong argument and so would their possible impact on the environment. Whether those critiques would be sufficient to tilt the scale towards unpatentability in the case of the balancing approach or overcome the rebuttable morality presumption will depend of the specifics of the patent. Nevertheless, a case could be made for their contrariety to *ordre public* and morality under the EPC.

The patent morality assessment of human and quasi-human creations would be more problematic. For example, the recreation of the Neanderthal poses difficult questions concerning human and human-like chimeras. To recreate the Neanderthal, scientists would need to start by synthesising its genome. They would then insert it into human stem cells, to be birthed by a human surrogate mother. While this prospect is still sci-fi, the moral implications of this line of work are evident.

On the one hand, this technique would employ human biological material, in the form of stem cells and a human birth mother. Conversely, the DNA would be derived from a human ancestor, belonging to the same genus (i.e. *homo*), but a different species (i.e. *homo neanderthalensis* instead of *homo sapiens*). Because of this, it is difficult to determine whether such creation would constitute a human/human chimera or instead a human/non-human one.[314]

---

[312]Lastly, even if those inventions were to pass the morality clause, they could still fail the industrial application requirement set by Article 57 EPC. For example, cloning dinosaurs to gain information on their evolutionary disparity from modern animals was seen as a scientific advance rather than an industrially applicable invention (Warren-Jones 1998, p. 449).

[313]Nonetheless, the well-being of animals has been of limited relevance at times. For instance, in a case concerning a genetically modified salmon, the issues caused by a chimeric growth hormone were disregarded in deciding on the patent before the Norwegian patent office. In that case, however, the animals had not been recreated, but rather genetically engineered. On this, see Sect. 5.1.3.2.

[314]Over time, philosophers and legal scholars have examined different concepts and criteria to define what is human. While an extensive analysis is impossible in this context, a few general indications can be presented. *Kant* defined humans in terms of their ability to set ends for themselves and to achieve them in a practical sphere. They were also seen to possess a distinctive dignity that could not be assigned a market price (Burke 2012, p. 241). On the other hand, *von Zeiller* argued that persons are those beings: "*Empowered with senses and reason, which in and of itself must be viewed as the subject of rights and obligations*" (Bernat 2008, p. 2). Others noted instead the need to resort to both quantitative and qualitative criteria to determine whether someone

Deciding whether a being can be classified as human is fundamental from a legal perspective. As noted by *Bernat*, "*every human being is a person, and being a person is a necessary and sufficient condition for having legal rights and obligations*".[315] He then considered what it is generally meant by "human". This term would seem to comprise members of the human species, *Homo sapiens*.[316] If this interpretation were to prevail, Neanderthals would not be considered humans.

---

can be considered a human being. From a qualitative perspective, humans can be distinguished by their higher faculties and the self-awareness of their own status. By focusing on the essence of humans, this approach raises a number of religious and philosophical questions. Matters are not easier from a quantitative perspective. Humans have a 95% genetic homology with chimpanzees, but the latter have never been considered human. On the other hand, human recipients of animal organ transplants are still considered fully human (Burke 2012, pp. 241–242). Scholars noted that this dichotomy could be solved by observing that: "*Chimeras contain a given percentage of cells that contain only human genes with the rest of the cells containing entirely nonhuman genetic material. Thus chimeras contain portions that are fully human while transgenic animals do not. Therefore, they might be considered more human than a transgenic animal with the identical percentage of human genetic material*" (Burke 2012, p. 241). Still, problems could remain, for example in the case of creatures having a human central nervous system, but who look like animals. In this case, there would be a clash between the qualitative and quantitative thresholds (Burke 2012, p. 242). Other scholars listed amongst the characteristics of humans the capacity to reason, to act for normative reasons and in autonomy, to engage in complex social relationships, to display empathy and sympathy and to have faith (i.e. believe in a God) (Bernat 2008, p. 8). Nevertheless, the author noted that there are times when members of the species *Homo sapiens* would not display those characteristics (i.e. as a foetus, a small child) (Bernat 2008, p. 9). Equally, non-human animals would acquire the status of moral agent (i.e. a person in the legal sense) when they display some of the meaningful characteristics listed above. In particular, the deciding factor would be the ability to act for normative reasons, including moral ones (Bernat 2008, p. 9). A middle-way approach was envisioned for creatures having human cells transplanted into a non-human animal. Those creatures might not be univocally considered non-human, since they might have a moral status that, albeit not equivalent to that of humans, might still be greater than that of laboratory animals (Baylis and Robert 2007, p. 44). In the United States, US Congressman *Rose* raised a similar issue and maintained that: "*Congress must have a better definition of how much genetic material constituted the legal definition of what constitutes a human being. Should an animal that contains one-half of a human code be considered human? How about one quarter human genetic material? Should genetically altered foetuses be considered patentable subject matter under current patent law?*" (Bently and Sherman 1995, p. 277). As previously seen, a patent for a human-animal chimera was presented to the USPTO. On this, see Sect. 4.2.3.6. The applicants were worried that the: "*Artificial creation and propagation of cloned, chimeric, and transgenic animals could mean the end of the wild and the substitution of a bio-industrial world*" (Burke 2012, p. 237). In its reply, the USPTO, while noting that it will apply patent law without discriminating between different technological fields, maintained that the utility requirement must be interpreted to exclude from patentability inventions that would be: "*Injurious to the well-being, good policy, or good morals of society*" (Burke 2012, p. 238). It also added that: "*Congress did not intend 35 U.S.C. 101 to include the patenting of human beings*" (Burke 2012, p. 238). In the US, there were many reactions towards patents on human chimeras. The USPTO Commissioner stated that: "*There will be no patents on monsters*", while President *Bush* asked Congress to prohibit the: "*Creation of animal-human hybrids… human life is a gift from our Creator, and that gift should never be discarded, devalued or put up for sale*" (Burk 2013, p. 237).

[315]Bernat (2008, p. 1).

[316]Bernat (2008, p. 5).

The recreation of the Neanderthal would likely fall within the scope of Article 6 (2) of the Directive. The first three items mentioned in that list share the same rationale, namely the protection of human dignity. Hence, the perceived humanity of those chimeras could be relevant from this perspective. While a specific analysis would need to be carried out ad hoc for each application, it is possible that Article 6 (2)(a)(b)(c) might be infringed by the recreation of the Neanderthal. For one, those techniques might employ human embryos for commercial and industrial purposes. Following the *Brüstle* and *WARF* cases, this might lead to a rejection of the patent application on morality grounds.[317] On this issue, authors maintained that there is: "*no prima facie good reason to give less protection to a non-human embryo that has the potential to develop into a human-like being, than to a human embryo that has the potential of becoming a born person*".[318] Equally, chimeras that require the destruction of human embryos for their production would be contrary to morality.[319]

Furthermore, some of those techniques may fall within the definition of human clones, especially if interpreted broadly. Following Recital 41 of the Directive and the indications of the EPO, this includes "*any process, including techniques of embryo splitting, designed to create a human being with the same nuclear genetic information as another living or deceased human being*".[320] In this case, the qualification of the Neanderthal as a human may be pivotal and so would the interpretation of the concept of "*same nuclear genetic information*".

Lastly, in case the nuclear genetic information of the Neanderthal is influenced by the human mitochondrial genome, this could lead to alterations of the germ line genetic identity. Hence, this might be caught by Article 6(2)(b) of the Directive. On the other hand, the fact that human cells have been used in an animal might not be contrary to morality.[321] Indeed, in the *Leland Stanford* case, the Opposition Division of the EPO considered the morality of a human/animal chimera, specifically an immunocompromised mouse implanted with human cells. In its Article 53(a) EPC assessment, the EPO observed that:

> It is undeniable that the production of chimeric animals containing human organs grown from human cells isolated from aborted foetuses or deceased persons, whether children or adults, instinctively appears distasteful, if not immoral, to many people at first glance. On the other hand, the medical benefits conferred by the invention are not in dispute among the parties... and the use of donated human material for research is widely accepted provided consent was given, which there is no reason to doubt in the present case.[322]

---

[317]On this, see respectively Sects. 5.1.3.5 and 5.1.3.3.

[318]Bernat (2008, p. 10).

[319]Bernat (2008, p. 10).

[320]EPO (2018, § G.II.5.3).

[321]This does not extend to cases when human embryos are used for industrial or commercial purposes.

[322]Opposition Division of the EPO (2001, p. 23).

Other commentators argued against the morality of human/animal chimeras. For instance, it was stated that claims concerning chimeras mixing human and animal embryonic stem cells or germ cells could be refused by reading Article 6(1) of the Directive in light of Recital 38.[323] Similarly, since the list in Article 6(2) of the Directive is not exhaustive, other technologies that prejudice human dignity could be excluded from patentability. Those might include the production of human and animal hybrids.[324] EGE reached a similar conclusion in its Opinion on synthetic biology, where it noted that the production of chimeras from germ cells should not be rewarded with patents, as it offends human dignity.[325]

---

[323]Sommer (2013, p. 211). *Warren-Jones* had a different take on the matter. She considered the response that would occur in Europe, if a patent for a human-animal chimera similar to that presented in the bogus application of Rifkin and Newman in the US were submitted. On this, see Sect. 4.2.3.6. In her view: "*Hypothetically, were such an application to be presented to individual Member States, determining patentability under Art. 6 (even within the intent outlined in Recital 38) would not automatically lead all Member States to reject such an application. Where the application is directed to the product and not the process; where it defines introduction of genomic, but not whole genome, human DNA this would take it outside of the strict interpretation of a chimera, but could factually amount to a genetic alteration akin to such a hybrid; where the human cellular contribution is minimal, or where it provides introduction of material during the process of differentiation of either the human or the animal, it could arguably fall outside of the prohibition in Recital 38. In addition, there is no clear divide between the morality of what is so obviously precluded and the morality of what slips through the definition. Even without the need for 'bogus' applications, technological development has the unfortunate habit of constantly testing the boundaries of permissibility and this is a fundamental reason why biotechnology is so contentious. This argues that there is no such thing as an exception that is so obviously immoral that 'it goes without saying', and this deficit permits different standards of regulation*" (Warren-Jones 2008b, pp. 649–650).

[324]Stazi (2015, p. 205).

[325]Burke (2012, p. 243) and European Group on Ethics in Science and New Technologies to the European Commission (2009, p. 46). *Burke* argued that the patent system does not have the right mechanisms and language to be able to deal with human/animal chimeras. Human/animal chimeras could be useful in research on human diseases or embryonic development. They could also be used for organ transplants (Burke 2012, p. 237). The notion of chimera encompasses: "*A mixture of cells from two or more genetically distinct organisms of the same or different species. They are mosaics at the cellular level; individual cells are derived from either the host or the donor but not both*" (Baylis and Robert 2007, p. 41). Chimeras need to be distinguished from hybrids, which are: "*Created by breeding across species. Hybrids are generally the result of combining an egg from one species with sperm from another to form a single embryo. Hybrids contain recombined genetic material throughout their genome and throughout all the tissues in their body*" (Baylis and Robert 2007, p. 41). In *Brüstle*, the Court stated that: "*Whereas processes, the use of which offend against human dignity, such as processes to produce chimeras from germ cells or totipotent cells of humans and animals, are obviously also excluded from patentability*" (European Court of Justice (ECJ) 2011, § 38). On this, see Sect. 5.1.3.5. *Warren-Jones* addressed the role of chimeras and hybrids in the Directive (in particular Recital 38) and held that: "*A 'chimera' is a term which means the product of cellular mixing between species... but it does not dictate the ratio of cells or determine which species comprise the mix... Clearly, what is intended to be precluded by this provision is much more narrow and can only accord with the main text where it is read as meaning that it would be immoral to patent a process to produce an animal-human hybrid. This then is stated to be achieved by combining either germ cells (sperm and ova) or totipotent cells... since this preclusion*

As for animal/animal chimeras, the recreation of the Neanderthal could be challenged under Article 53(a) EPC. In addition to the arguments presented for animal chimeras, human ones add a further layer of complexity. Those chimeras could be accused of exploiting humans and infringing their human rights. Those critiques would apply both if Neanderthals were seen as human/non-human chimeras or as purely human ones; in the latter case, those objections would obviously be stronger. In commenting on those issues, *Casabona* observed that:

> The hypothetical future creation of human beings would probably come under the limiting clause on morality and public ordre, particularly during the initial embryonic phase.[326]

#### 5.2.3.7 Data Processing and Encoding

Inventions concerning data processing and encoding are unlikely to be contrary to *ordre public* and morality pursuant to the EPC and the Directive.

To begin, those inventions do not fall within the four cases listed in Article 6(2) of the Directive and would not fit within the rationale of the norm. Consequently, they would pass this morality hurdle.

The situation is more complex under Article 53(a) EPC. One could argue that these techniques misuse and exploit genetic material for human purposes, hence immorally objectifying and commodifying it. This position is however untenable. The notion of commodification of life has usually been associated with human life and other higher life forms. Employing it for strands of DNA encoding a message and stored in non-human vectors seems far-fetched. Furthermore, commodification has been interpreted as the attribution of an instrumental value understood in commercial terms to items that should not be seen from that perspective. It traces back to an understanding of the human body and biological systems as means and objects. Clearly, this is not the case for data processing and encoding. This conclusion is confirmed by the interpretation of DNA given by the EPO in the *Relaxin* case, where it was argued that DNA is not *per se* life, but merely a chemical substance carrying genetic information.[327]

In light of the above, those inventions are unlikely to be considered immoral under either the balancing test or the rebuttable presumption one. As for the latter, no abhorrent issues could be identified; hence, the morality presumption could not be rebutted. Equally, a balance between moral and immoral features of those inventions would tilt the scale towards their morality, as moral aspects seem to be predominant.

---

*does not extend to the simple combination of smaller sections of DNA than a whole genome with its attendant cell, transgenic animals with human DNA fall outside of the prohibition, as do animal clones. What is, however, more worrying is that this prohibition does not extend to the product animal-human hybrid itself*' (Warren-Jones 2001, p. 181).

[326] Casabona (2014, p. 184) and Spanish Bioethics Committee and Portuguese National Ethics Council for the Life Sciences (2011, pp. 24–25).

[327] On this, see Sect. 5.1.3.6.

The conclusions presented above are confirmed by the patent applications presented before the EPO. No morality issues were raised in these circumstances, apart from one exception.[328] In examining an application for "*encoding text into nucleic acid*", the EPO noted that some of the claims referred to a synthetic organism. Because of this, the Examiner asked the applicant to amend the claims to insert the words "*non-human*". This was requested also for claims for animals, which should be transformed into "*non-human animal*", and for cells. For the latter, the suggested language referred to "*cell or non-human synthetic organism*".[329] The applicant included those changes, so that the patent covered "*a method of generating a non-human recombinant or synthetic organism or cell*" and "*non-human recombinant or synthetic organism or cell comprising the synthetic nucleic acid sequence*".[330]

The objection of the EPO referred to the use of human synthetic biological materials, rather than to the morality of using biological systems for encoding purposes. Therefore, it does not seem to contradict the conclusions reached above.

#### 5.2.3.8 Algorithms and Simulation Tools

The morality of algorithms and simulation tools will be assessed exclusively under Article 53(a) EPC, as the Directive will not be applicable to this kind of inventions. Yet, even in this case, it is improbable that this examination will lead to a finding of unpatentability on morality grounds, as those inventions do not present immorality features reaching the level of abhorrence or unacceptability required by the EPO.[331] This conclusion is supported by the analysis of the patent applications presented before the EPO, where no references to morality concerns were found.[332]

#### 5.2.3.9 Completely *ex Novo*

So far, no completely *ex novo* synthetic biology inventions have been devised. In spite of this, the following points could be made.

If the Directive were to apply, its provisions would not have an impact on the morality of SynBio *ex novo* inventions. Just as for other SynBio inventions, the complexity of altering and creating biological system implies that at the beginning

[328]Lu and Farzadfard (2016), Venter et al. (2015) and Lu and Siuti (2015).

[329]Hutchison et al. (2012 Email from the EPO to the applicant of 18 June 2015).

[330]Hutchison et al. (2016).

[331]Obviously, this does not take into consideration proverbially problematic cases, such as simulation tools determining the best way to spread a life-threatening epidemic or how to design highly contagious deadly viruses via synthetic biology. This would represent the SynBio equivalent of a letter bomb and would fall within the scope of Article 53(a) EPC.

[332]Sarmiento et al. (2015).

research will focus on simpler life forms (i.e. microorganisms). However, microorganisms are excluded from the scope of Article 6(2) of the Directive and therefore would not lead to a finding of unpatentability under this norm.

Similarly, Article 53(a) EPC is unlikely to lead to such conclusion under both the rebuttable presumption and the balancing approach. This is due to a number of factors. First, at the beginning those inventions would concern microorganisms, which tend to raise fewer morality concerns. Second, some of the arguments that could be made in this field have generally been unsuccessful before the EPO. For example, "playing God" critiques have usually been dismissed and so have objections to patents on life.

Environmental complaints have also enjoyed limited success so far. For now, it is too early to tell if and what impact those inventions would have on nature, especially since no inventions of this kind have been devised. However, just as in the field of xenobiology, serious and proven environmental risks could play a role in the legal morality assessment of those inventions.

In its 2009 opinion on synthetic biology, EGE reached a similar conclusion. It dismissed the idea that new genomes would be contrary to *ordre public* and morality and considered instead their possible relevance from a philosophical viewpoint. Specifically, it held that:

> The prospect of constructing minimal and new genomes did not violate fundamental moral precepts or boundaries, but did raise questions about the possible consequences of synthetizing new free-living organisms in relation to the concept of life and our relation to it.[333]

## 5.2.4 Conclusions

As it is to be expected when dealing with biotechnologies, morality issues are often raised against synthetic biology inventions. Papers and studies often refer to the ethical challenges posed by this discipline and thus indirectly support the idea that the morality clause will be pivotal in the assessment of the patentability of synthetic biology inventions. However, the examination presented above shows that this is not necessarily the case.

First, it became clear that not all concerns are legally relevant. Many of those critiques are connected to personal rather than legal morality, and are therefore irrelevant for a patent assessment. This is for example the case with critiques moved against the use of patents in this sector; opposing patents for SynBio inventions *per se* is a matter of general policy that is to be dealt with by the legislator and is not within the framework of the morality clause, as decided upon by Courts and patent offices.

[333] European Group on Ethics in Science and New Technologies to the European Commission (2009, p. 40).

At the same time, it is important not to confuse patent concerns with regulatory ones. As correctly noted by *Rutz*, patents are not responsible for everything and they also cannot solve everything.[334] This means that the patent system should not be called upon to deal with health, safety, and environmental issues of a general nature. Patents do not authorise the implementation or the sale of an invention, but simply regulate the right to exclude others from exploiting it. For this reason, a patent assessment before patent offices and Courts is not the right forum to address general regulatory issues and it is therefore unsurprising that such general topics do not play a role in the decision to grant patents for SynBio inventions. In fact, the patent system is tasked with assessing specific patent applications and not to deal with general regulatory or legislative issues. The situation is obviously different for concerns relating to a specific patent and not to a technology in general; for instance, detailed and proven environmental concerns raised against a patent application would be taken into consideration in deciding on its morality.

The review of the SynBio patent applications presented before the EPO shows however that such specific and legally relevant morality concerns against SynBio inventions are yet to be successfully raised. On this, one cannot but agree with the statement of *Torremans* describing how morality issues are unlikely to be raised at the moment given that synthetic biology is currently confined to rather primitive bacterial forms of life.[335] Nevertheless, he noted, it is undeniable that society has not yet woken up to this technology that sets out to create artificial life forms. Because, if one shapes the morality question from this perspective, critiques do not seem to be that far away.

Indeed, synthetic biology could become a victim of its own success. If this discipline were to achieve its goals pertaining to higher life forms, legal morality claims can be envisioned. In particular, morality critiques could become the most concrete way to exclude a range of inventions from patentability, as the exclusions set in Article 52 EPC will eventually become less prominent due to the technical progress of this discipline and the increasing gap between nature and SynBio inventions.

This view is reinforced by the fact that the morality clause has often been invoked by civil societies, who tend to be very active in the opposition of patents for biotechnologies before the EPO and Courts. Conceptually, this clause appears to have a specific affinity with the goals of NGOs and non-profit organisations. Therefore, it is to be expected that those civil society organisations will continue to be the ones raising the majority of morality questions against SynBio inventions. Whether those claims will be successful will depend on the technical evolution of this discipline and whether it would be able to transition to higher life forms.

Morality issues surrounding SynBio patents are more likely to emerge at the opposition, appeal and judiciary stage. The Examining Divisions of the EPO have adopted a rebuttable presumption approach with an abhorrence standard. Hence,

---

[334]On this, see Sect. 5.2.2.4.

[335]On this, see Sect. 5.2.2.10.

only evidently problematic issues would lead to the rejection of a patent application on morality grounds. Because of this, it is unlikely that SynBio patent applications will be rejected at the examination stage, at least as long as SynBio inventions do not concern humans and higher life forms. By contrast, morality arguments might have an increased chance of success if formulated at a later stage. This has three reasons. First, Boards have swung between the rebuttable presumption approach and the balancing one. If the latter were applied, the chances of rejection on morality grounds would be higher. Second, third parties opposing the patent gain a more prominent role at those stages, given the generous access granted by the EPC to its procedures. Third, objections presented at a later stage can rely on additional evidence. In order to successfully object to the patentability of an invention, the opponent would need to demonstrate the risks connected to it, as mere allegations would not suffice. Given the expanded timeframe connected to later examination stages, the opponent would have more time to gather the necessary information and proof.

In any event, for now, morality issues strong enough to be able to deny on their own the patentability of synthetic biology inventions are still years away. This is confirmed by the patent analysis conducted in the prior sections, where it was shown that no patents were rejected on morality grounds. At the same time, those patent samples showed the direction that SynBio might take in the next years. Claims relating to humans were clearly flagged by the EPO, which considered them problematic. Since synthetic biology is moving in that direction, it can only be expected that these issues might become more prominent and problematic in the future.

If until then no theoretical approach would have been developed on the morality of SynBio inventions, there is the risk that the first wave of applications concerning humans and higher life forms would represent a shock for the patent system and lead to morality conflicts, just like in the case of earlier biotechnologies. A way to mitigate this would be to begin considering those issues at a theoretical level and establish if and how Article 6(2) of the Directive and Article 53(a) EPC could apply to those cases.

## References

Advocate General Bot (2011) Opinion of Advocate General in Oliver Brüstle v Greenpeace eV (C 34/12)

Advocate General Jacobs (2001) Opinion of Advocate General in Kingdom of the Netherlands v European Parliament and Council of the European Union (C 377/98)

Agovic A (2014) Patenting SynBio in Anglo-America and Europe: chaos or opportunity. In: de Miguel Beriain I, Casabona CMR (eds) Synbio and human health: a challenge to the current IP framework? Springer, Dordrecht, pp 101–121

Amendments suggested by the European Parliament to the Draft Directive on the legal protection of biotechnological inventions ((1992a) OJ C 125/112), 1992

Amendments suggested by the European Parliament to the Draft Directive on the legal protection of biotechnological inventions ((1992b) OJ C 305/160), 1992

Amendments suggested by the European Parliament to the Draft Directive on the legal protection of biotechnological inventions ((1993) OJ C 342/2), 1993

Armitage E, Davis I (1994) Patents and morality in perspective. Common Law Institute of Intellectual Property, London

Bagley MA (2003) Patent first, ask questions later: morality and biotechnology in patent law. William Mary Law Rev 45:469–547

Bailey R (2016) Is it ethical to synthesize an entire human genome? Yes! [WWW Document]. http://reason.com. URL https://reason.com/2016/05/13/should-we-synthesize-a-human-genome/. Accessed 29 Feb 2020

Bakardjieva Engelbrekt A (2009) Institutional and jurisdictional aspects of stem cell patenting in Europe (EC and EPO): tensions and prospects. In: Plomer A, Torremans P (eds) Embryonic stem cell patents: European law and ethics. Oxford University Press, Oxford, pp 228–269

Battistelli B (2011) Patents and biotechnology – latest developments [WWW Document]. http://blog.epo.org. URL https://blog.epo.org/patents/patents-and-biotechnology---latest-developments/. Accessed 29 Feb 2020

Baylis F, Robert JS (2007) Part-human chimeras: worrying the facts, probing the ethics. Am J Bioeth 7:41–45

Benders GA, Glass JI, Hutchison CA, Lartigue C, Vashee S, Algire MA, Smith HO, Merryman CE, Noskov VN, Chuang RY, Gibson DG, Venter JC (2012) Methods for cloning and manipulating genomes. EP2403944 A2

Bently L, Sherman B (1995) The ethics of patenting: towards a transgenic patent system. Med Law Rev 3:275–291

Bently L, Sherman B, Borges Barbosa D, Basheer S, Visser C, Gold R (2010) Exclusions from patentability and exceptions and limitations to patentees' rights (No. SCP/15/3-Annex 1). WIPO Standing Committee on the Law of Patents

Bernat E (2008) Which beings should be entitled to human rights? Med Law Int 9:1–12

Beyleveld D, Brownsword R (1993) Mice, morality and patents: the onco-mouse application and Article 53(a) of the European Patent Convention. Common Law Institute of Intellectual Property, London

Bhattachary D, Pascall Calitz J, Hunter A (2010) Synthetic biology dialogue. Biotechnology and Biological Sciences Research Council (BBSRC), Swindon

Blattner FR, Posfai G, Herring C, Plunkett G, Glasner J (2003) Bacteria with reduced genome. US20030138937 A1

Blattner FR, Glasner JD, Herring CD, Plunkett G, Posfai G, Twose TM (2010) Bacteria with reduced genome. EP1483367 B1

Boeke JD, Church G, Hessel A, Kelley NJ (2016) HGP-write: testing large genomes in cells - meeting summary. Presented at the HGP-write: testing large genomes in cells workshop, Boston, pp 1–13

Bonadio E (2012) Biotech patents and morality after Brüstle. Eur Intellect Prop Rev 34:433–443

Bostyn SJR (2011) A decade after the birth of the biotech directive: was it worth the trouble? In: Arezzo E, Ghidini G (eds) Biotechnology and software patent law: a comparative review of new developments, new directions in patent law. Edward Elgar, Cheltenham, pp 221–259

Burk DL (2013) Edifying thoughts of a patent watcher: the nature of DNA. UCLA Law Rev Discourse 60:92–102

Burke S (2012) Justification for patents as applied to human-animal chimeras. Eur Intellect Prop Rev 34:237–243

Campbell K, Wilmut I (2001) Quiescent cell populations for nuclear transfer. EP0849990 B1

Campbell K, Stockman H, Wilmut I (2000) Quiescent cell populations for nuclear transfer. GB2318578 (B)

Casabona CMR (2014) Patentability, synthetic biology and human genome. In: de Miguel Beriain I, Casabona CMR (eds) Synbio and human health: a challenge to the current IP framework? Springer, Dordrecht, pp 175–186
Charter of fundamental rights of the European Union (2007/C 303/1), 2000
Cho HS, Daniel T, Dimarchi R, Hays AM, Wilson T, Sim BC, Litzinger D (2007) Modified human growth hormone polypeptides and their uses. EP1740608 A2
Cho HS, Daniel T, Dimarchi R, Hays AM, Wilson T, Sim BC, Litzinger D (2012) Modified human growth hormone polypeptides and their uses. EP1740608 B1
Common position (EC) No 4/94 adopted by the Council with a view to adopting European Parliament and Council Directive 94/.../EC of... on the legal protection of biotechnological inventions (94/C 101/01), 1994., OJ C 101, 9.4.1994, p 65–75
Convention for the protection of human rights and dignity of the human being with regard to the application of biology and medicine: Convention on human rights and biomedicine (ETS 164), 1997
Convention on the grant of European patents of 5 October 1973, 1973
Convention on the grant of European patents of 5 October 1973, as revised on 29 November 2000 (European Patent Convention - EPC), 1973
Convention on the unification of certain points of substantive law on patents for invention (ETS No. 047), 1963
Council Directive 90/220/EEC on the deliberate release into the environment of genetically modified organisms, 1990. OJ L 117, 8.5.1990, p 15–27
Covone MG (2015) Presentation: patents in the field of biotechnology. Palermo
Crespi SR (2006) The human embryo and patent law - major challenge ahead? Eur Intellect Prop Rev 28:569–575
Curley D, Sharples A (2002) Patenting biotechnology in Europe: the ethical debate moves on. Eur Intellect Prop Rev 24:565–570
de Carvalho NP (2018) The TRIPS regime of patents and test data, 5th edn. Kluwer Law International, Alphen aan den Rijn
de Miguel Beriain I (2010) Patenting human life, a question of morals? [WWW Document]. www.sybhel.org. URL http://www.sybhel.org/?p=419. Accessed 29 Feb 2020
Derclaye E (2009) Patent law's role in the protection of the environment: re-assessing patent law and its justifications in the 21st century. Int Rev Intellect Prop Compet Law 40:249–273
Deutsche Akademie der Technikwissenschaften, Deutsche Akademie der Naturforscher Leopoldina, Deutsche Forschungsgemeinschaft (2009) Synthetische Biologie: Stellungnahme. Wiley-VCH-Verlag, Weinheim
Directive 98/71/EC of the European Parliament and of the Council on the legal protection of designs, 1998
Dolder F (1984) Schranken der Patentierbarkeit biotechnologischer Erfindungen nach dem Europäischen Patentübereinkommen. Mitteilungen der deutschen Patentanwälte:1–7
Douglas T, Powell R, Savulescu J (2013) Is the creation of artificial life morally significant? Stud Hist Philos Biol Biomed Sci 44:688–696
Dworkin G (1994) HL paper 28 – UK House of Lords Select Committee on the European Community
Enlarged Board of Appeal of the EPO (EBA) (2005) G 0001/04 (Diagnostic methods)
Enlarged Board of Appeal of the EPO (EBA) (2008) G 0002/06 (Use of embryos/WARF)
Enlarged Board of Appeal of the EPO (EBA) (2010a) G 0002/08 (Dosage regime/Abbott respiratory)
Enlarged Board of Appeal of the EPO (EBA) (2010b) G 0001/07 (Treatment by surgery/Medi-Physics)
EPC Working Party (Patent Working Party) (1961) IV/2767/61-E – Proceedings of the 1st meeting of the Patent Working Party held at Brussels from 17 to 28 April 1961
EPO (2018) Guidelines for examination [WWW Document]. www.epo.org. URL https://www.epo.org/law-practice/legal-texts/guidelines.html. Accessed 3 Sept 2019

European convention for the protection of human rights and fundamental freedoms – Council of Europe (ETS 5), 1950
European Court of Human Rights (2004) Vo. v. France (Application no. 53924/00)
European Court of Justice (ECJ) (1974) (C 41/74) - Yvonne van Duyn v Home Office
European Court of Justice (ECJ) (1979) (C 34/79) - Regina v Maurice Donald Henn and John Frederick Ernest Darby
European Court of Justice (ECJ) (1998) (C 1/96) - The Queen v Minister of Agriculture, Fisheries and Food, ex parte Compassion in World Farming Ltd
European Court of Justice (ECJ) (2001) (C 377/98) Kingdom of the Netherlands v. European Parliament and Council of the European Union
European Court of Justice (ECJ) (2005) (C 456/03) Commission of the European Communities v Italian Republic
European Court of Justice (ECJ) (2011) (C 34/12) Oliver Brüstle v. Greenpeace e.V
European Court of Justice (ECJ) (2014) (C 364/13) International Stem Cell Corporation v Comptroller General of Patents, Designs and Trade Marks
European Group on Ethics in Science and New Technologies to the European Commission (2002) Opinion 16: Ethical aspects of patenting involving human stem cells
European Group on Ethics in Science and New Technologies to the European Commission (2009) Opinion 25: Ethics of synthetic biology
European Parliament (2012) European Parliament resolution of 10 May 2012 on the patenting of essential biological processes
European Parliament decision on the joint text approved by the Conciliation Committee for a European Parliament and Council Directive on the legal protection of biotechnological inventions, 1995. (1995) OJ C 68/26
Fernandez y Branas FJ (2014) Patentability of synthetic biology under the European Patent Convention (EPC). In: de Miguel Beriain I, Romeo Casabona CM (eds) Synbio and human health. Springer, Dordrecht, pp 187–199
Fletcher GL, Hew CL (2001) Gene construct for production of transgenic fish. EP0578653 B1
Ford R (1997) The morality of biotech patents: differing legal obligations in Europe? Eur Intellect Prop Rev 19:315–318
General Agreement on Trade in Services (GATS)(1869 U.N.T.S. 183), 1994
German Federal Patent Court (BPatG) (2006) Case on Patent 19756864 (Brüstle), BPatG 253
Glass JI, Young L, Lartigue C, Assad-Garcia N, Smith HO, Hutchison C, Venter JC (2008a) Installation of genomes or partial genomes into cells or cell-like systems. EP1963515 A2
Glass JI, Smith HO, Clyde A, Hutchison CA, Alperovich NY, Assad-Garcia N (2008b) Minimal bacterial genome. EP1951874 A1
Glass JI, Alperovich N, Clyde A, Hutchinson CA, Lartigue C, Merryman C, Smith HO, Vashee S, Venter JC (2010) Methods of genome installation in a recipient host cell. EP2147099 A1
Glass JI, Young L, Lartigue C, Assad-Garcia N, Smith HO, Hutchison C, Venter JC (2014) Installation of genomes or partial genomes into cells or cell-like systems. EP1963515 B1
Grosheide FW (2010) Intellectual property and human rights: a paradox. Edward Elgar, Cheltenham
Hally A (2014) Patenting natural products in the US and Europe – the divide grows [WWW Document]. www.frkelly.com. URL https://frkelly.com/patenting-natural-products-us-europe-divide-grows. Accessed 29 Feb 2020
Hansen J (2013) Method of producing isoprenoid compounds in yeast. US20130137138 A1
Harmon SH (2006) The rules re-engagement: the use of patent proceedings to influence the regulation of science ("What the salmon does when it comes back downstream"). Intellect Prop Q 4:378–403
Harmon SH, Laurie G, Courtney A (2013) Dignity, plurality and patentability: the unfinished story of Brüstle v Greenpeace. Eur Law Rev 38:92–106

Hellstadius A (2009) A comparative analysis of the national implementation of the Directive's morality clause. In: Plomer A, Torremans P (eds) Embryonic stem cell patents: European law and ethics. Oxford University Press, Oxford, pp 117–141
Hutchison CA, Montague MG, Smith HO (2012) Encoding text into nucleic acid sequences. EP2494060 A1
Hutchison CA, Montague MG, Smith HO (2016) Encoding text into nucleic acid sequences. EP2494060 B1
In the matter for an application for a patent by A. and H., 1927., R.P.C. (1927) 44 (9)
Intellectual Property Office (IPO) (2016) Examination guidelines for patent applications relating to biotechnological inventions in the Intellectual Property Office
Kaebnick GE (2009) Should moral objections to synthetic biology affect public policy? Nat Biotech 27:1106–1108
Krishnakumar R, Glass JI, Merryman CE (2010) System and method for producing synthetic microorganisms capable of translating proteins containing non-standard amino acids. EP2217700 A1
Liddell K (2012) Immorality and patents: the exclusion of inventions contrary to ordre public and morality. In: Lever A (ed) New frontiers in the philosophy of intellectual property, Cambridge intellectual property and information law. Cambridge University Press, Cambridge, pp 140–171
Lu T, Farzadfard F (2016) Genomically-encoded memory in live cells. WO2016025719 A1
Lu T, Siuti P (2015) Recombinase-based logic and memory systems. EP2932443 A1
Lykkeskov A, Jørgensen H (2005) In: Danish Council of Ethics (ed) The ethics of patenting human genes and stem cells: Conference report and summaries held in Copenhagen 28 September 2004. Danish Council of Ethics, Copenhagen
Macchia G (2012) Patentability requirements of biotech inventions at the European Patent Office: ethical issues. In: Bin R, Lorenzon S, Lucchi N (eds) Biotech innovations and fundamental rights. Springer, Milan, pp 37–43
Mayor S (2000) First UK patents for cloning issued to creators of Dolly the sheep. BMJ 320:270
McNab A, Wood L, Vogeli G, Kaytes P (n.d.) Transgenic mice for the analysis of hair growth. EP0439553
Milius D, Townend D (2008) Thoughts on the scope and operation of morality clauses in patent law [WWW Document]. URL https://www.academia.edu/2384702/Thoughts_on_the_Scope_and_Operation_of_Morality_Clauses_in_Patent_Law. Accessed 8 Nov 2016
Mills O (2010) Biotechnological inventions: moral restraints and patent law, Revised edition. Ashgate Publishing, Farnham
Min Y (2012) Morality: a equivocal area in the patent system. Eur Intellect Prop Rev 34:261–265
Moufang R (1998) The concept of "ordre public" and morality in patent law. In: Van Overwalle G (ed) Octrooirecht, ethiek en biotechnologie - Patent law, ethics, and biotechnology. Bruylant, Brussel, pp 67–77
Nuffield Council on Bioethics (ed) (1995) Human tissue, ethical and legal issues. KKS Printing, London
O'Sullivan E (2012) Is Article 53(a) EPC still of narrow interpretation? J Intellect Prop Law Pract 7:680–690
Odell-West A (2020) Invention and the human embryo. Intellect Prop Q:1–19
OECD (2014) Emerging policy issues in synthetic biology. OECD, Paris
Opposition Division of the EPO (1989) Oncomouse, OJ EPO 1989, 451
Opposition Division of the EPO (1992a) Lubrizol Genetics Inc. (EP 84302533)
Opposition Division of the EPO (1992b) Lubrizol/Plant gene expression (App. 122.791)
Opposition Division of the EPO (1995a) Relaxin (EP0112149)
Opposition Division of the EPO (1995b) T 0272/95 (Relaxin/Howard Florey Institute)
Opposition Division of the EPO (2001) Leland Stanford/Modified Animal (App. 88312222.8)
Opposition Division of the EPO (2003a Edinburgh (EP0695351B1)
Opposition Division of the EPO (2003b) Isolation, selection and propagation of animal transgenic stem cells other than embryonic stem cells (EP0695351)

Ottolia A (2011) Moral limits to biotech patents in Europe: a quest for higher harmonization. In: Arezzo E, Ghidini G (eds) Biotechnology and software patent law: a comparative review of new developments, new directions in patent law. Edward Elgar, Cheltenham, pp 309–337
Plomer A (2009) Towards systemic legal conflict: Article 6(2)(c) of the EU Directive on biotechnological inventions. In: Plomer A, Torremans P (eds) Embryonic stem cell patents: European law and ethics. Oxford University Press, Oxford, pp 174–202
Pollack A (2016) Scientists talk privately about creating a synthetic human genome [WWW Document]. www.nytimes.com. URL https://www.nytimes.com/2016/05/14/science/synthetic-human-genome.html. Accessed 29 Feb 20
Pompidou A (2006) Comments by the President of the EPO in case G2/06
Porter G (2009a) The drafting history of the European biotechnology directive. In: Plomer A, Torremans P (eds) Embryonic stem cell patents: European law and ethics. Oxford University Press, Oxford, pp 3–26
Porter G (2009b) Human embryos, patents, and global trade: assessing the scope and contents of the TRIPS morality exception. In: Plomer A, Torremans P (eds) Embryonic stem cell patents: European law and ethics. Oxford University Press, Oxford, pp 343–367
Proposal for a Council Directive on the legal protection of biotechnological inventions (COM/88/496FINAL - SYN 159), 1988
Proposal for a European Parliament and Council Directive on the legal protection of biotechnological inventions, 1995. COM/95/0661 final
Rothley W (1997) Report on the proposal for a European Parliament and Council Directive on the legal protection of biotechnological inventions
Rothschild MF, Newman S (2002) Intellectual property rights in animal breeding and genetics. CABI, Wallingford
Rowlandson M (2010) WARF/Stem cells (G2/06): the ordre public and morality exception and its impact on the patentability of human embryonic stem cells. Eur Intellect Prop Rev 32:67–76
Rutz B (2007) Synthetic biology through the prism of scenarios. Biotechnol J 2:1072–1075
Rutz B (2009) Synthetic biology and patents. A European perspective. EMBO Rep 10:S14–S17
Sarmiento RJ, Baskerville DS, Zhang X (2015) Structure based predictive modeling. US20150134315 A1
Savulescu J, Powell R (2013) Mammoth cloning: The ethics [WWW Document]. www.oxfordmartin.ox.ac.uk. URL https://theconversation.com/mammoth-cloning-the-ethics-16183. Accessed 29 Feb 2020
Schatz U (1998) Patentability of genetic engineering inventions in European patent office practice. Int Rev Intellect Prop Compet Law 29:2–16
Schatz U (2000) Patents and morality. In: Sterckx S (ed) Biotechnology, patents, and morality. Ashgate, Aldershot, pp 217–228
Schneider I (2014) Exclusions and exceptions to patent eligibility revisited: examining the political functions of the "discovery" and "ordre public" clauses in the European Patent Convention and the arenas of negotiation. In: de Miguel Beriain I, Casabona CMR (eds) Synbio and human health. Springer, Dordrecht, pp 145–173
Schuster MI (2012) ECJ ruling on the patentability of human embryonic stem-cell-related inventions. Int Rev Intellect Prop Compet Law 43:626–640
Sommer T (2008) Patenting the animal kingdom? From cross-breeding to genetic make-up and biomedical research. Int Rev Intellect Prop Compet Law 39:139–177
Sommer T (2013) Can law make life (too) simple?: from gene patents to the patenting of environmentally sound technologies, 1st edn. DJØF Publishing, Copenhagen
Spanish Bioethics Committee, Portuguese National Ethics Council for the Life Sciences (2011) Synthetic biology - a joint report by the Spanish Bioethics Committee and the Portuguese National Ethics Council for the Life Sciences. Spanish Bioethics Committee - Portuguese National Ethics Council for the Life Sciences, Lisbon - Barcelona
Stazi A (2015) Biotechnological inventions and patentability of life: the US and European experience, new directions in patent law. Edward Elgar, Cheltenham

Sterckx S (2008) The European patent convention and the (non-) patentability of human embryonic stem cells: the WARF case. Intellect Prop Q:478–495
Sterckx S, Cockbain J (2010) Assessing the morality of the commercial exploitation of inventions concerning uses of human embryos and the relevance of moral complicity. SCRIPTed J Law Technol Soc 7:83–103
Sterckx S, Cockbain J (2012) Exclusions from patentability: how far has the European Patent Office eroded boundaries? Cambridge University Press, Cambridge
Technical Board of Appeal of the EPO (TBA) (1988) T 0320/87 (Hybrid plants/Lubrizol)
Technical Board of Appeal of the EPO (TBA) (1990) T 0019/90 (Oncomouse)
Technical Board of Appeal of the EPO (TBA) (1995) T 0356/93 (Plant cells)
Technical Board of Appeal of the EPO (TBA) (1999) T 919/99 (Human stem cell/Biocyte)
Technical Board of Appeal of the EPO (TBA) (2004) T 0315/03 (Transgenic animals/Harvard)
Technical Board of Appeal of the EPO (TBA) (2005) T 0866/01 (Euthanasia Compositions/ Michigan State University)
Technical Board of Appeal of the EPO (TBA) (2006) T 1374/04 (Stem cells/WARF)
Technical Board of Appeal of the EPO (TBA) (2007) T 1213/05 (Breast and ovarian cancer/ University of Utah)
Technical Board of Appeal of the EPO (TBA) (2008a) T 0666/05 (Mutation/University of Utah)
Technical Board of Appeal of the EPO (TBA) (2008b) T 0080/05 (Method of diagnosis/University of Utah)
The Life of Dolly | Dolly the Sheep [WWW Document], n.d.. http://dolly.roslin.ed.ac.uk. URL https://dolly.roslin.ed.ac.uk/facts/the-life-of-dolly/index.html Accessed 29 Feb 2020
Thomas D, Richards GA (2004) The importance of the morality exception under the European patent convention: the oncomouse case continues.... Eur Intellect Prop Rev 26:97–104
Torremans PLC (2009a) A transnational institution confronted with a single jurisdiction model: guidance for the EPO's implementation of the Directive from a private international law perspective. In: Plomer A, Torremans P (eds) Embryonic stem cell patents: European law and ethics. Oxford University Press, Oxford, pp 271–302
Torremans PLC (2009b) The construction of the Directive's moral exclusions under the EPC. In: Plomer A, Torremans P (eds) Embryonic stem cell patents: European law and ethics. Oxford University Press, Oxford, pp 142–171
Torremans PLC (2011) Patentability of human stem cell or synthetic biology based inventions. In: Arezzo E, Ghidini G (eds) Biotechnology and software patent law: a comparative review of new developments. Edward Elgar, Cheltenham, pp 288–308
Treaty establishing the European Economic Community (EEC Treaty), 1957
Treaty on European Union (TEU) - 2008/C 115/01, 2007
Treaty on the Functioning of the European Union (TFEU) - 2008/C 115/01, 2007
Treichel P (2009) G2/06 and the verdict of immorality. Int Rev Intellect Prop Compet Law 40:459–471
U.S. Circuit Court (Massachusetts) (1817) Lowell v. Lewis, 15 Fed. Cas. 1018 (C.C.D.Mass. 1817)
Ugurlu AS (2014) Bioethics and the patent eligibility of human embryonic stem cells-related inventions in Europe. Nomos, Baden-Baden
UK Government, 1623. Statute of Monopolies 1623
UN Educational, Scientific and Cultural Organisation (UNESCO) (1997) Universal declaration on the human genome and human rights
UN General Assembly (1948) Universal declaration of human rights (217 A (III))
Van Overwalle G (2011) Policy levers tailoring patent law to biotechnology: comparing U.S. and European approaches. UC Irvine Law Rev 1:435–517
Vandergheynst D (1998) La notion d'ordre public et des bonne mœurs dans la proposition de directive européenne relative a la protection juridique des inventions biotechnologiques. In: Van Overwalle G (ed) Octrooirecht, ethiek en biotechnologie - Patent law, ethics, and biotechnology, CIR. Bruylant, Brussel, pp 82–92
Venter JC, Smith HO (2008) Synthetic genomes. EP1968994 A2

Venter JC, Smith HO, Iii CAH, Gibson DG (2013) Synthetic genomes. EP1968994 B1
Venter JC, Gibson DG, Gill JE (2015) Digital to biological converter. EP2885408 A1
Vienna convention on the law of treaties (1155 UNTS 1155 331), 1969
Viens AM (2009) Morality provisions in law concerning the commercialization of human embryos and stem cells. In: Plomer A, Torremans P (eds) Embryonic stem cell patents: European law and ethics. Oxford University Press, Oxford, pp 85–114
Visser D (2019) Visser's annotated European Patent Convention, 26th edn. Wolters Kluwer, Alphen aan den Rijn
Wade N (2010) Researchers say they created a "synthetic cell" [WWW Document]. www.nytimes.com. URL https://www.nytimes.com/2010/05/21/science/21cell.html. Accessed 29 Feb 2020
Warren-Jones A (1998) A mouse in sheep's clothing: the challenge to the patent morality criterion posed by "Dolly.". Eur Intellect Prop Rev 20:445–452
Warren-Jones A (2001) Patenting rDNA: human and animal biotechnology in the United Kingdom and Europe. Lawtext Pub, Witney
Warren-Jones A (2007) Vital parameters for patent morality—a question of form. J Intellect Prop Law Practice 2:832–846
Warren-Jones A (2008a) Morally regulating innovation: what is "commercial exploitation"? Intellect Prop Q 2:193–212
Warren-Jones A (2008b) Finding a "common morality codex" for biotech - a question of substance. Int Rev Intellect Prop Compet Law 39:638–661
WTO Appellate Body (1998) European Communities – measures concerning meat and meat products (Hormones), WT/DS26/AB/R
WTO Panel (2004) United States - measures affecting the cross-border supply of gambling and betting services (No. DS285)
Zullo S (2012) From the patentability of living matter to the ethics of biotechnological innovation: the person-body relationship. In: Bin R, Lorenzon S, Lucchi N (eds) Biotech innovations and fundamental rights. Springer, Milan, pp 87–96

# Chapter 6
# Novelty and Inventive Step

## 6.1 Introduction

To obtain a patent for a synthetic biology invention, applicants will need to fulfil further requirements in addition to the ones set in Article 52(2) and 53 EPC. The two most relevant ones are the novelty and inventiveness criteria detailed in Article 54 and 56 EPC. Although these two criteria are not within the focus of this analysis, it is important to briefly consider which issues they might raise for the patentability of synthetic biology inventions. The assessment will be limited to general considerations, as a detailed analysis of the impact of each criterion will be possible only on a case-by-case basis.

Given the limited relevance assigned to subject matter patentability, the novelty and inventiveness requirements fulfil a pivotal role in determining which applications will be rewarded with patent protection. This trend is confirmed by an analysis of the SynBio patent applications filed before the EPO. While subject matter patentability and morality issues were seldom, novelty and inventiveness were ubiquitous and often led to a revision of the scope of the patent application or even to its rejection. Such tendency is in line with the approach adopted by the EPO in other fields, where the Office focused on novelty and inventiveness and away from subject matter patentability. According to scholars, this approach should be welcomed in the field of biotechnology as "*the requirements of novelty and inventive step have a greater body of case law precedents, and are arguably more flexible and able to accommodate shifting technologies*".[1]

[1]Hawkins (2016, p. 90). *Hawkins* noted that, in the field of computer-related inventions: "*This has not led to an increase in the numbers of patents granted, but instead a change in the reasons that patents are excluded*" (Hawkins 2016, p. 90). A similar approach seems to exist in the US, as it was stated that: "*With the issue of patentability of living organisms being decided, patentability typically revolves around novelty and non-obviousness in view of the naturally occurring products on which the synthetic ones are based*" (LeGuyader 2009, p. 32).



I. de Lisa, *The Patentability of Synthetic Biology Inventions*,
https://doi.org/10.1007/978-3-030-51206-4_6

At first glance, the significant impact of those two criteria may seem surprising, given that synthetic biology has often been associated with the concepts of novelty and inventiveness. For instance, commentators argued that "*traditional patent requirements such as novelty, non-obviousness... would be satisfied by most patentable innovations in this field*".[2] This can be explained by noting that, while synthetic biology as a discipline is generally innovative and inventive in comparison to naturally existing biological systems, the patent law definition of those two concepts revolves around other aspects. Hence, specific synthetic biology inventions could be objected to on this basis. This issue and related ones will be analysed in the following sections.

## 6.2 Novelty

The novelty requirement is described in Article 54 EPC. Pursuant to patent law, an invention will be considered new if it does not form part of the state of the art. The latter should be interpreted to comprise everything that has been made available to the public, in whatever form, before the filing of the patent application.[3] The Directive contemplates novelty only indirectly. Article 3(2) and 5(2) establish that biological material, even of human origin, could be patent protected even if it previously occurred in nature or is identical to a natural element.[4] Apart from those two provisions, the Directive relies on the novelty assessment set in the EPC.

In light of this, it is foreseeable that the impact of the inapplicability of the Directive would be limited. Xenobiology and protocell inventions, which might fall outside the realm of the Directive, do not have natural equivalents and have not previously occurred in nature. Hence, the provisions set by Article 3(2) and 5(2) of the Directive would have little bearing on them, as Article 54 EPC would apply.

By contrast, these two norms might be relevant in the debate over the novelty of biological materials having a natural equivalent, such as synthetic life and de-extinction inventions. This objection is not new. Similar concerns have been raised ever since patent protection has been sought for biological materials. For instance, it has often been asked how, "*if a gene or protein is present in nature, can it be said not to be available to the public?*".[5] This critique is based on the idea that, if something previously existed in nature, it would not pass the novelty requirement.

---

[2]Bhutkar (2005, p. 24).

[3]This concept is known under the name of "absolute novelty" (Manley and Vickers 2015, p. 7).

[4]Nonetheless, it is important to remember that these provisions relate to subject matter patentability and are only indirectly relevant in assessing novelty.

[5]Nuffield Council on Bioethics (2002, p. 29).

Such objection shows a number of overlaps with the debate on the line between discovery and invention illustrated above.[6]

However, an objection to the novelty of SynBio inventions based upon natural equivalents will most likely fail. Several points lead to this conclusion.

First, the patent novelty requirement is different from the common understanding of this term. This has several reasons. On the one hand, patent novelty takes into consideration the state of the art. In the view of the EPO, this relates to information given to a person skilled in the art that would enable him to "*practise the technical teaching which is the subject of the document, taking into account also the general knowledge at that time in the field to be expected of him*".[7] This includes any subject matter deriving directly and unambiguously from that document.[8] Also, for this assessment, it is not possible to combine together different items of prior art, as the examination is done on a document-by-document basis.[9] Furthermore, this concept requires the complete conformance between the claims and one prejudicial disclosure.[10] If novelty is interpreted and assessed pursuant to the above, the prior existence of natural equivalents of synthetic life and de-extinction inventions would not *per se* hinder novelty.

Second, Article 3(2) and 5(2) of the Directive explicitly confirm the patentability of elements that are identical to natural ones or that have previously occurred in nature. While the two provisions relate to patentability rather than novelty, they could hint to a positive attitude of the legislator towards this type of inventions.

Similar conclusions have already been reached in the past. The EPO maintained that substances can be considered novel "*if the already naturally occurring identical substance was not readily available to the public*".[11] Equally, the novelty of a substance would not be hindered if that element does not exist in that exact form in nature. *Bostyn* cited as an example the difference between DNA and cDNA. While the two contain the same coding regions, the former includes also non-coding ones. Because of this difference, cDNA would be considered novel despite the presence of

[6]Yet, it is important not to confuse the two. *Pila* confirmed that: "*Since Howard Florey, the EPO Boards have also emphasized the 'essentially separate and independent' nature of the inherent patentability and novelty enquiries*" (Pila 2014, p. 536).

[7]EPO (2018, § G.VI.4).

[8]EPO (2018, § G.VI.2).

[9]EPO (2018, § G.VI.1) and Harguth and Carlson (2017, p. 74).

[10]Visser (2019, p. 85). *Visser* noted that: "*Conformance between a prior art document and an alleged invention does not require equivalent wording: for an invention to lack novelty, the state of the art must disclose the basic idea, the content of the teaching*" (Visser 2019, p. 85).

[11]Bostyn (2004, p. 17) and Technical Board of Appeal of the EPO (TBA) (1986, § 2). Similarly, in the German *Antamanid* case, it was held that: "*A substance which already existed in nature before application date, but whose existence and function were unknown, cannot destroy novelty of the later isolated substance, which can have an identical structure*" (Bostyn 2004, p. 44; German Federal Patent Court (BPatG) 1977).

DNA with identical coding regions in nature.[12] Even a difference in one nucleotide would be enough to establish novelty.[13]

This latter aspect is particularly important in the field of synthetic biology. If this idea were to extend to synthetic life inventions, it could be argued that any difference between those SynBio inventions and their natural equivalents would hinder novelty objections.

This reasoning could be applied also to de-extinction creations. First, from a technical perspective, those items are never an exact copy of their natural ancestors. Second, "*although a cloned extinct animal is not novel, intuitively speaking, inasmuch as the animal previously existed in nature, it is not unpatentable for want of novelty under the legal definition of novelty*".[14] Indeed, the concepts of state of the art and enabling disclosure suggest that the previous existence of the now extinct animal would not impede its patentability on patent novelty grounds.[15]

Even though the novelty of a SynBio invention could not be objected to by referring to the presence of natural equivalents or ancestors, SynBio inventions could still face novelty hurdles. An increasing number of scientific publications and patents have been submitted on topics relating to synthetic biology, thus widening the information available in the state of the art. As a consequence, SynBio patent applications might lack novelty due to prior enabling disclosures done by other patent applicants or scientists. As it will be examined in more detail below, this constitutes one of the main grounds of rejection of SynBio patent applications and shows that the innovative nature of SynBio inventions is not a guarantee of their novelty pursuant to Article 54 EPC.

This conclusion contrasts with the opinion expressed by other scholars, who believe that the novelty assessment of SynBio inventions will be straightforward. *Torremans* considered that:

> Synthetic live forms, such as an empty host bacterium with a genome created in the laboratory in it, are by definition new in the sense that they did not exist in nature before. The novelty requirement applies therefore in a straightforward way in the same sense it would apply to any new product (not involving living material).[16]

Yet, this view is reductive. While SynBio inventions are often without a natural counterpart and thus new compared to what exists in the environment, patent novelty is assessed on different criteria and therefore the mere absence of natural equivalents

---

[12]McDonell et al. (2016, p. 622) and Bostyn (2004, p. 45). *Bostyn* noted that: "*Even if one would accept the reasoning… that a product can never be novel if it already existed in nature before, there is still the fact that the process for producing the product which is as such known in nature, can be patented. And according to a general principle of patent law, protection for a patented process extends to the product immediately produced by the patented process (Art. 64(2) EPC). That means that protection can extend to products which are as such not new*" (Bostyn 2004, p. 45).

[13]Stolzenburg et al. (2003, p. 635).

[14]Hagglund (2007, p. 411).

[15]For an overview of the issues concerning prior art and enablement for de-extinction inventions, Rohrbaugh (1997, pp. 391–396).

[16]Torremans (2011, p. 289).

may not be enough to overcome the Article 54 EPC threshold. Xenobiology inventions perfectly exemplify this situation. Those inventions are often intuitively associated with novelty, as they go beyond the paradigm of what is available in nature. However, the legal definition of novelty is not tied to natural availability, but rather to the concept of enabling disclosure. For this reason, even innovative research endeavours such as this could lack novelty due to prior patents or publications. In the latter case, even the applicant himself could destroy the novelty of his inventions, if he discloses their content ahead of time. This often occurs when scientists seek credit for their achievements via publications or conferences. Once the information is released, it becomes prior art and could be invoked to demonstrate the lack of novelty of the invention.[17] Hence, SynBio inventions are not immune from novelty issues, as demonstrated by the novelty objections raised against some of the SynBio sample patents examined in previous chapters.[18]

A further layer of complexity in the evaluation of the novelty of SynBio inventions is introduced by their synthetic nature. It has been argued that, if synthetically obtained products are identical to their natural equivalent, they could not be considered new and inventive.[19] This approach would hinder the formation of a new wave of patents on natural products that have been produced synthetically (e.g. synthetic human insulin).[20] So far, it is unclear whether this objection will acquire practical relevance in the field of synthetic biology, especially considering that minimal changes between substances are capable of establishing novelty.[21]

In light of the above, it would be incorrect to dismiss the novelty assessment of SynBio inventions as straightforward. Both the points presented here and the analysis of the sample applications presented before the EPO show that novelty is pivotal in the patent assessment of those inventions and in shaping the form and scope of the granted patents. This is unsurprising considering that this requirement has often played an important role before the EPO, especially due to the limited relevance of the eligibility criteria.

[17]Harguth and Carlson (2017, pp. 68–69). Publications were seen as a challenge to the novelty of a number of patent applications presented by Venter and his team (Palombi 2009a, p. 381).

[18]Most patent applications raised some novelty issues during their examination procedure. For an example of the issues raised, Cho et al. (2006, Form 1507 of January 2009 pp. 7–9) and Venter and Smith (2008, Written Opinion of the International Searching Authority of 23 July 2008, Box n° V).

[19]Palombi (2009b, p. 222). By contrast, their manufacturing process might fulfil those two requirements.

[20]Palombi (2009a, p. 385).

[21]For product-by-process claims, the current approach of the EPO, which denies the novelty of a known substance that has been obtained via a new process, would seem to apply (Intellectual Property Office (IPO) 2016, p. 10).

## 6.3 Inventive Step

In order to receive patent protection, inventions need to display an inventive step. Pursuant to Article 56 EPC, an inventive step exists if, "*having regard to the state of the art, it is not obvious to a person skilled in the art*". Inventiveness is governed by the norms of the EPC, as the Directive does not set additional provisions in this field. For this reason, the same normative framework will apply to both SynBio inventions covered by the Directive and those excluded from it. This requirement, whose analysis will begin once novelty is established, is different from the one in Article 54 EPC given that, "*while the novelty examination ensures that there is a quantitative difference between the invention and the state of the existing knowledge, non-obviousness ensures that this difference is of a quality deserving of patent protection*".[22] Therefore, the inventive step requirement constitutes an effective hurdle on the way to patentability.

While the inventiveness of each patent application will need to be considered on its own merit, the analysis of the SynBio applications presented before the EPO shows that this requirement is often problematic. Most of the examined applications were objected to on Article 56 EPC grounds due to either prior patent applications or scientific publications. Just as for the novelty requirement, this shows that the generally innovative nature of SynBio inventions does not guarantee their patentability on the merit. Indeed, while those inventions may be inventive in comparison to naturally existing items, Article 56 EPC requires a legal assessment that is based on the prior art and not on innovation vis-à-vis nature.

This clarification is important considering that misunderstandings continue to exist on this point. Authors noted that the use of known research techniques and tools and the similarities between inventions and their natural equivalents could indicate a lack of inventiveness. By extending this argument to synthetic biology, it was held that:

> Similar issues can arise in relation to basic research tools and extremely simple synthetic biology inventions, but once both areas of science really take off and develop further the major inventions should not be affected.[23]

As noted for novelty, this approach is reductive considering that the legal criterion of inventiveness is judged on the basis of the prior art and not on similarity with nature.

The number and type of prior art as well as the status of a discipline have an effect on its inventiveness assessment. As a field develops, techniques and results that would have been considered inventive at the beginning may no longer fulfil this criterion.[24] Equally, unexplored disciplines provide less prior art and less

---

[22] Bently et al. (2018, p. 578).

[23] Torremans (2011, p. 290).

[24] *Storz* noted that: "*Probably due to the rapid technological progress in the biotechnology industry, arguments that were accepted in support of sufficient inventiveness in the past now may be rejected*

expectations of success in carrying out a project. As higher expectations of success are a hurdle for inventiveness, new developments in synthetic biology might benefit from the uncertainty and youth of this field. By contrast, the establishment of routine and standardised processes might increase the inventiveness threshold. Equally, this may also occur due to an increased use of bioinformatics tools.[25]

Specific SynBio inventions have also been objected to on inventiveness grounds. In discussing the synthetic life work of Venter, commentators argued that:

> Although it is indeed a stupendous technological achievement, it could also be argued that conceptually it was somewhat... routine. Furthermore, the three steps... have been individually successfully demonstrated by Venter... hence... the combination... would be considered... non-inventive since the whole... has not been greater than the sum of its parts... One should also keep in mind that the synthetic genome used in his work was virtually identical in its sequence to that of a natural bacterium.[26]

Others considered the work of Venter as reverse engineering of the *Mycoplasma genitalium* and, as such, "*perhaps not straightforward, but then, hardly inventive*".[27]

In conclusion, synthetic biology inventions do not seem to raise general inventiveness issues. However, specific inventions might be objected to on the basis of their inventiveness vis-à-vis the prior art, as shown by the patent analysis. In light of this, and just as for novelty, the impact of this requirement in practice must not be underestimated. It is necessary to move away from the idea that this criterion will be

---

*by the patent authorities as falling under the routine of a skilled artisan... In other words: The biotech industry is... a victim of its own success*" (Storz 2014, pp. 11–12).

[25]The use of laborious and costly procedures is not sufficient to establish inventiveness (Intellectual Property Office (IPO) 2016, p. 15).

[26]Gowrishankar (2010). The same author observed that: "*Our accumulated knowledge... in the last 50 years would have predicted or foreseen the present results that were obtained by Venter, once the technological hurdles were overcome... Hence, there is certainly no 'Eureka' moment here. One should also keep in mind that the synthetic genome used in this work was virtually identical in its sequences to that of a natural bacterium... with very few 'cosmetic' modifications*" (Gowrishankar 2010). Similarly, other scholars pointed out that such claims were very broad, which might subject them to attacks on their validity. Authors have mentioned the risk that such patents would inhibit research in this field (Varela 2010, p. 259). Yet, it needn't be forgotten that claims can be narrowed significantly in the course of their examination and prosecution. Moreover, while Venter argued that the only unknown thing is when and not if this technology can be used for commercial purposes, one could also argue that this state of development may not be sufficient to pass the threshold of industrial applicability in Europe. This test requires the application to show an immediate and concrete way to exploit the invention (England 2011, p. 49). Lastly, it needs to be considered that: "*Whether e.g. a sequence comprises an inventive step is determine in a similar fashion to that which applies to chemical compounds, i.e. whilst identity of structure will be enough to prove lack of novelty, similarity of structure will not be enough to prove lack of inventive step unless the activity is identical in at least qualitative terms*" (Intellectual Property Office (IPO) 2016, p. 13). This is attributable to the fact that minimal variations can have a major impact on the functionality and activity of a biological system. Hence, even quantitatively minimal changes could have a relevant qualitative effect. For this reason, the inventiveness analysis should concentrate on the actual results and technical solutions achieved by the invention rather than focus on the amount of modification (D'Antonio 2004, pp. 160–161).

[27]Palombi (2009a, p. 374).

fulfilled simply because of the innovative nature of this discipline and to recognise the strong impact it has on SynBio patent applications and their scope.

## References


Bently L, Sherman B, Gangjee D, Johnson P (2018) Intellectual property law, 5th edn. Oxford University Press, Oxford

Bhutkar A (2005) Synthetic biology: navigating the challenges ahead. J Biolaw Bus 8:19–29

Bostyn SJR (2004) Patenting DNA sequences (polynucleotides) and scope of protection in the European Union: an evaluation. Office for Official Publications of the European Communities, Luxembourg

Cho H, Daniel T, Hays AM, Wilson T, Litzinger D, Mariani R, Kimmel B, Keefe W (2006) Biosynthetic polypeptides utilizing non-naturally encoded amino acids. US20060019347 A1

D'Antonio V (2004) Invenzioni biotecnologiche e modelli giuridici: Europa e Stati Uniti. Jovene, Napoli

England P (2011) Patents and biofuels – foundations of the business. Intellect Property Mag 2011:47–50

EPO (2018) Guidelines for examination [WWW Document]. www.epo.org. URL https://www.epo.org/law-practice/legal-texts/guidelines.html. Accessed 9 Mar 19

German Federal Patent Court (BPatG) (1977) Naturstoffe, 16 W (Pat) 64/75

Gowrishankar J (2010) Craig Venter, and the claim for "synthetic life.". Curr Sci 99:152

Hagglund R (2007) Patentability of cloned extinct animals. Geo Mason Law Rev 15:381–446

Harguth A, Carlson S (2017) Patents in Germany and Europe: procurement, enforcement and defense: an international handbook, 2nd edn. Kluwer Law International B.V, Alphen aan den Rijn

Hawkins N (2016) A red herring – invalidity of human gene sequence patents. Eur Intellect Property Rev 38:83–91

Intellectual Property Office (IPO) (2016) Examination guidelines for patent applications relating to biotechnological inventions in the Intellectual Property Office

LeGuyader J (2009) Presentation: patents and synthetic biology. Presented at the patenting synthetic biology – a transatlantic perspective workshop, Washington DC

Manley MI, Vickers M (eds) (2015) Navigating European pharmaceutical law. Oxford University Press, Oxford

McDonell LA, Haley JF, Hosoda Y, Jaenichen H-R, Meier J (2016) From clones to claims: an encyclopedia of the European Patent Office's case law on the patentability of biotechnology inventions with a comparison to the United States and Japanese practice, 6th edn. Carl Heymanns Verlag, Cologne

Nuffield Council on Bioethics (2002) The ethics of patenting DNA: a discussion paper. Nuffield Council on Bioethics, London

Palombi L (2009a) Beyond recombinant technology: synthetic biology and patentable subject matter. J World Intellect Property 12:371–401

Palombi L (2009b) Gene cartels: biotech patents in the age of free trade. Edward Elgar, Cheltenham

Pila J (2014) Isolated human genes: the patent equivalent of a non-copyrightable sound recording. Law Q Rev:180–185

Rohrbaugh ML (1997) The patenting of extinct organisms: revival of lost arts. AIPLA Q J 25:371–417

Stolzenburg F, Weinzierl G, Jaenichen H-R (2003) Patenting of genome research results. Pharmacogenomics 4:633–642

Storz U (2014) Patentability requirements of biotech patents. In: Storz U, Quodbach M, Marty SD, Constantine DE, Parker M (eds) Biopatent law: European vs. US patent law. Springer, Berlin, pp 1–21

Technical Board of Appeal of the EPO (TBA) (1986) T 0206/83 (Herbicides)

Torremans PLC (2011) Patentability of human stem cell or synthetic biology based inventions. In: Arezzo E, Ghidini G (eds) Biotechnology and software patent law: a comparative review of new developments. Edward Elgar, Cheltenham, pp 288–308

Varela JC (2010) Synthetic cells and their regulatory challenges. Eur J Risk Regul 1:259–263

Venter JC, Smith HO (2008) Synthetic genomes. EP1968994 A2

Visser D (2019) Visser's annotated European Patent convention, 26th edn. Wolters Kluwer, Alphen aan den Rijn

# Chapter 7
# Conclusions

Synthetic biology is a discipline full of contradictions. It promises to remediate the damages done to the environment, but it could also be accused of doing the exact opposite by destroying biodiversity. It investigates the origin of life on the planet, while it is accused of subverting natural evolution. It tries to reduce biological complexity and it is blamed for simplifying life. It attracts major investments, but little public attention.

All of these paradoxes translate in the patent analysis of this discipline. Instead of numerous patents, studies show that the synthetic biology patent panorama is still limited and in a developing phase. In spite of this, a number of conclusions could be drawn from the above analysis.

The patentability of SynBio inventions seems established under the EPC, as those items generally overcome the exclusions from patentability set in Article 52(2) EPC. All the examined patent applications presented before the EPO confirm this finding. Yet, the patentability of SynBio inventions in Europe presents a hidden level of complexity, as the application of the Directive to a number of lines of research is unclear. The repercussions of this situation could only be assessed in theory, given that Boards and Courts are yet to address the matter. Nevertheless, it could already be noted that this situation may have repercussions not only on the patentability of a single invention, but also on the structure and coherence of the European patent system for biotechnologies as a whole. From a European perspective, the unity of the biotech patent system may already have been undermined by the opposing approaches adopted by the EPO and the USPTO on the eligibility of some SynBio inventions.

Conversely, the morality assessment presents fewer uncertainties, at least for the moment. Despite the morality and ethical issues raised against synthetic biology, none of the examined patent applications was rejected on morality grounds. Still, this might change in the future, as this discipline leaves the realm of microorganisms and transitions towards higher life forms. At that point, moral concerns are to be expected.



I. de Lisa, *The Patentability of Synthetic Biology Inventions*,
https://doi.org/10.1007/978-3-030-51206-4_7

The patent analysis showed instead that novelty and inventiveness issues are already being moved against SynBio patent applications. In spite of suggestions that synthetic biology inventions would not raise these types of issues, the patent assessment clearly revealed that Article 54 and 56 EPC play a pivotal role in this discipline. This central role is bound to become even more prominent in the future, considering both the growing state of the art in this field and the focus placed by the EPO on those criteria.

In conclusion, it is evident that synthetic biology has an immense potential, which has not fully been realised until now. Whether those promises will be fulfilled is not only of scientific importance, but also of a legal one. Once synthetic biology transitions to systems that challenge the contemporary notion of biological or concern higher life forms, the patent system will be faced with difficult questions. Those issues would be rendered more complex by the policy and politically charged nature of these matters as well as by public opinion. In light of this, the patent system (in particular patent offices, as they will be the first to be confronted with this situation) should thoroughly consider the patent issues raised by synthetic biology in order to be prepared for the scientific developments that this discipline promises to realise and their related legal questions.

# Index





I. de Lisa, *The Patentability of Synthetic Biology Inventions*,
https://doi.org/10.1007/978-3-030-51206-4

The manufacturer's authorised representative in the EU is Springer Nature Customer Service Centre GmbH, Europaplatz 3, 69115 Heidelberg, Germany. If you have any concerns regarding our products, please contact ProductSafety@springernature.com

Printed and bound by CPI Group (UK) Ltd, Croydon, CR0 4YY
30/09/2026
02225696-0015